VOLUME ONE HUNDRED AND TWO

ADVANCES IN PROTEIN CHEMISTRY AND STRUCTURAL BIOLOGY

Personalized Medicine

Edited by

ROSSEN DONEV

Biomed Consult Ltd
Swansea, United Kingdom

ELSEVIER

AMSTERDAM • BOSTON • HEIDELBERG • LONDON
NEW YORK • OXFORD • PARIS • SAN DIEGO
SAN FRANCISCO • SINGAPORE • SYDNEY • TOKYO
Academic Press is an imprint of Elsevier

Academic Press is an imprint of Elsevier
50 Hampshire Street, 5th Floor, Cambridge, MA 02139, USA
525 B Street, Suite 1800, San Diego, CA 92101-4495, USA
The Boulevard, Langford Lane, Kidlington, Oxford OX5 1GB, UK
125 London Wall, London, EC2Y 5AS, UK

First edition 2016

Notices

Knowledge and best practice in this field are constantly changing. As new research and experience broaden our understanding, changes in research methods, professional practices, or medical treatment may become necessary.

Practitioners and researchers must always rely on their own experience and knowledge in evaluating and using any information, methods, compounds, or experiments described herein. In using such information or methods they should be mindful of their own safety and the safety of others, including parties for whom they have a professional responsibility.

To the fullest extent of the law, neither the Publisher nor the authors, contributors, or editors, assume any liability for any injury and/or damage to persons or property as a matter of products liability, negligence or otherwise, or from any use or operation of any methods, products, instructions, or ideas contained in the material herein.

ISBN: 978-0-12-804795-8
ISSN: 1876-1623

For information on all Academic Press publications visit our website at http://store.elsevier.com

CONTENTS

CONTRIBUTORS

Leonidas Alexopoulos*
School of Mechanical Engineering, National Technical University of Athens, Athens, Greece. ProtAtonce Ltd., Athens, Greece

Shabana Kouser Ali
Medical Biotechnology Division, School of Biosciences and Technology, VIT University, Vellore, Tamil Nadu, India

Jane P.F. Bai
Office of Clinical Pharmacology, Office of Translational Science, Center for Drug Evaluation and Research, U.S. Food and Drug Administration, Silver Spring, Maryland, USA

Sandya R. Beeram
Department of Chemistry, University of Nebraska-Lincoln, Lincoln, Nebraska, USA

Cong Bi
Department of Chemistry, University of Nebraska-Lincoln, Lincoln, Nebraska, USA

Mohammad Bohlooly
Astrazeneca, Discovery Sciences, Reagents and Assay Development, Transgenics Group, Molndal, Sweden

Ilse Du Preez
School for Physical and Chemical Sciences, Human Metabolomics, North-West University, Potchefstroom, South Africa

C. George Priya Doss
Medical Biotechnology Division, School of Biosciences and Technology, VIT University, Vellore, Tamil Nadu, India

Ellen Y. Guo
Molecular and Cellular Biology, Liberal Arts and Sciences, University of Illinois at Urbana–Champaign, Urbana–Champaign, Illinois, USA

David S. Hage
Department of Chemistry, University of Nebraska-Lincoln, Lincoln, Nebraska, USA

Junguk Hur
Department of Biomedical Sciences, School of Medicine & Health Sciences, University of North Dakota, Grand Forks, North Dakota, USA

Kewal K. Jain
Jain PharmaBiotech, Basel, Switzerland

*Present address: School of Mechanical Engineering, National Technical University of Athens, Athens, Greece.

Nadia Koen
School for Physical and Chemical Sciences, Human Metabolomics, North-West University, Potchefstroom, South Africa

Zhao Li
Department of Chemistry, University of Nebraska-Lincoln, Lincoln, Nebraska, USA

Du Toit Loots
School for Physical and Chemical Sciences, Human Metabolomics, North-West University, Potchefstroom, South Africa

Ioannis N. Melas
Astrazeneca, Discovery Sciences, Group of Quantitative Biology, Molndal, Sweden

V.S. Priyadharshini
Instituto Nacional de Enfemedades Respiratorias, Delegación Tlalpan, Mexico

Theodore Sakellaropoulos[†]
Office of Clinical Pharmacology, Office of Translational Science, Center for Drug Evaluation and Research, U.S. Food and Drug Administration, Silver Spring, Maryland, USA

Donald R.J. Singer
Fellowship of Postgraduate Medicine, London, United Kingdom

P. Sneha
Medical Biotechnology Division, School of Biosciences and Technology, VIT University, Vellore, Tamil Nadu, India

D. Suresh[‡]
Department of Chemistry, University of Nebraska-Lincoln, Lincoln, Nebraska, USA

S.A. Syed Haneef
Medical Biotechnology Division, School of Biosciences and Technology, VIT University, Vellore, Tamil Nadu, India

Luis M. Teran
Instituto Nacional de Enfemedades Respiratorias, Delegación Tlalpan, Mexico

D. Thirumal Kumar
Medical Biotechnology Division, School of Bioscience and Technology, VIT University, Vellore, Tamil Nadu, India

Zoulikha M. Zaïr
Warwick Medical School, University of Warwick, Coventry, United Kingdom

Xiwei Zheng
Department of Chemistry, University of Nebraska-Lincoln, Lincoln, Nebraska, USA

[†]Present address: School of Mechanical Engineering, National Technical University of Athens, Athens, Greece.

[‡]Home department: Department of Chemistry, Tumkur University, Tumakuru, Karnataka, India.

PREFACE

Personalized medicine encompasses the use of biological information for each patient in order to provide customized health care tailored to the individual patient. Such an individualized approach for medical decisions, practices, and treatment became clearly necessary due to a number of observations. For instance, the course of disease differs from one person to another. The effects of drugs used for treatment of a disease also vary largely between different patients. It is now well established that individual differences between patients, such as genetic polymorphisms, have significant effect on the onset of diseases as well as on absorption of drugs and their metabolism in patient's body. Personalized medicine takes such variations into account and tries to take advantage of them. It is believed that such customized therapies will improve patients' response rates to the treatment while reducing significantly any adverse effects the drugs might have.

Personalized medicine is based on the dynamics of systems biology and uses predictive tools to evaluate health risks and to design personalized health plans to help patients to minimize risks, prevent disease, and to treat it with precision when it occurs. Some of the most contemporary and very promising tools employed in personalized treatment of patients are discussed in the first three chapters of this volume. First chapter discusses the general principles of high-performance affinity chromatography and the various approaches that have been used in this technique to examine drug–protein binding and in work related to personalized medicine. This technique is a great asset to personalized medicine because the binding of drugs with proteins and other agents in serum can affect the dosage and action of drugs. The extent of this binding may also vary with a given disease state. Second article in this thematic volume focuses on the advances in proteomic technologies that have made important contribution to the development of personalized medicine by facilitating detection of protein biomarkers, proteomics-based molecular diagnostics as well as protein biochips and pharmacoproteomics. Application of nanobiotechnology in proteomics, nanoproteomics, has further enhanced applications in personalized medicine. Proteomics has already proved to be a good bridge between diagnostics and design of therapeutics. The integration of last two processes shows to be of a great importance for advancing personalized medicine. The third chapter in this volume reviews in detail metabolomics as a tool used in personalized medicine.

Metabolomics is the newest addition to the "omics" domain and the closest to the observed phenotype. It reflects changes occurring at all molecular levels, as well as influences resulting from other internal and external factors. By comparing the metabolite profiles of two or more disease phenotypes, metabolomics can be applied to identify biomarkers related to the perturbation being investigated. These biomarkers can, in turn, be used to develop personalized prognostic, diagnostic, and treatment approaches, and can also be applied to the monitoring of disease progression, treatment efficacy, predisposition to drug-related side effects, and potential relapse.

The fourth chapter discusses the importance of a range of new approaches to developing new and reprofiled medicines to treat common and serious diseases, and rare diseases: new network pharmacology approaches, adaptive trial designs with enriched populations more likely to respond safely to treatment, as assessed by companion diagnostics for response and toxicity risk and use of "real-world" data. Case studies are described of single and multiple protein–drug targets in several important therapeutic areas. These case studies also illustrate the value and complexity in use of selective biomarkers of clinical response and risk of adverse drug effects, either singly or in combination.

Chapters 5 and 6 in this volume give in-depth analyses of applying the tools of personalized medicine in some of the most common diseases. The fifth article in this volume highlights the contributions of proteomics toward the understanding of personalized medicine in respiratory disease and its potential applications in the clinic. The sixth chapter is focused on the challenges of treating different cancer types, which behave like moving targets due to mutation and evolution, and the current state-of-the-art research in this area. A special emphasis is made on the computational approaches to accelerating novel medicine and better personalized patient care from bedside to benchtop.

Chapter 7 in this volume focuses on high-end computational methods, such as molecular dynamics (MD) simulation that has proved to be a constitutive approach for detecting the minor changes associated with single nucleotide polymorphisms (SNPs) in nucleic acids for better understanding of their role in protein structural and functional alterations. MD along with docking analysis can reveal the synergetic effect of an SNP in protein–ligand interaction and provides a foundation for designing a particular drug molecule for an individual. This compelling application of computational power and the advent of other technologies have paved a promising way toward personalized medicine. In the Eighth article of this thematic volume, authors discuss the available clinical strategies and different methods how pharmacological chaperones can be personalized and used as a next-generation approach to address different lysosomal storage disorders.

The final two chapters in this volume exemplify the applicability of molecular modeling and simulation approaches in personalized medicine by exploring the inhibitory activity of Wortmannin toward oncogenic mutations in phosphatidylinositol-4,5-bisphosphate 3-kinase, catalytic subunit alpha (PIK3CA) (Chapter 9), and the importance of mutations in A1-domain of von Willebrand factor (VWD) gene for the structural and functional alterations related to thrombosis, compared to the native VWD protein (Chapter 10).

The aim of this volume is to promote further research and development in the design of new personalized therapeutics and treatments using biological information for each patient via translation of recent genomic, genetic, proteomics, and metabolomics advances into clinical context.

Dr. Rossen Donev
Biomed Consult Ltd
United Kingdom

CHAPTER ONE

High-Performance Affinity Chromatography: Applications in Drug–Protein Binding Studies and Personalized Medicine

Zhao Li, Sandya R. Beeram, Cong Bi, D. Suresh[1], Xiwei Zheng, David S. Hage[2]
Department of Chemistry, University of Nebraska-Lincoln, Lincoln, Nebraska, USA
[2]Corresponding author: e-mail address: dhage1@unl.edu

Contents

Abstract

The binding of drugs with proteins and other agents in serum is of interest in personalized medicine because this process can affect the dosage and action of drugs. The extent of this binding may also vary with a given disease state. These interactions may involve serum proteins, such as human serum albumin or α_1-acid glycoprotein, or other agents, such as lipoproteins. High-performance affinity chromatography

[1]Home department: Department of Chemistry, Tumkur University, Tumakuru, Karnataka, India.

Advances in Protein Chemistry and Structural Biology, Volume 102
ISSN 1876-1623
http://dx.doi.org/10.1016/bs.apcsb.2015.09.007

(HPAC) is a tool that has received increasing interest as a means for studying these interactions. This review discusses the general principles of HPAC and the various approaches that have been used in this technique to examine drug–protein binding and in work related to personalized medicine. These approaches include frontal analysis and zonal elution, as well as peak decay analysis, ultrafast affinity extraction, and chromatographic immunoassays. The operation of each method is described and examples of applications for these techniques are provided. The type of information that can be obtained by these methods is also discussed, as related to the analysis of drug–protein binding and the study of clinical or pharmaceutical samples.

1. INTRODUCTION

Personalized medicine is a field of healthcare in which the treatment of an individual is based on information related to their clinical status, genetics, or environment (Gunsburg & Willard, 2009; Ziegler, Koch, Krockenberger, & Grobhennig, 2012). The overall goal of personalized medicine is to optimize the medical care and outcome for an individual (Gunsburg & Willard, 2009). Work in this area has been carried out to promote the health and wellness of a patient by aiding in disease prevention, assist in the detection of disease, and optimize the selection of treatment for a patient. As a result of these possible applications, this field is believed to hold great promise in the future of medicine and healthcare (Buchner, Blonski, & Lichtenstein, 2011; Cottone, Orlando, & Renna, 2010; Gunsburg & Willard, 2009; Trusheim, Berndt, & Douglas, 2007).

One area that has been of interest in personalized medicine is in how the dosage and action of drugs may vary from one person to the next or between healthy individuals and those with a given disease (Gunsburg & Willard, 2009; Ziegler et al., 2012). A process that can affect the apparent dosage of many pharmaceutical agents is the binding of drugs with proteins and other carrier agents in blood. This binding may involve serum proteins, such as human serum albumin (HSA) or α_1-acid glycoprotein (AGP), or other agents, such as lipoproteins. This process is significant for many drugs, with 43% of the 1500 most common drugs being at least 90% bound in serum (Kratochwil, Huber, Muller, Kansy, & Gerber, 2002). This binding can influence the eventual activity and fate of drugs once they have entered the circulation and plays a central role in determining the absorption, distribution, and rate of excretion or metabolism of many drugs in the bloodstream (Hage, 2002; Kwong, 1985; Lindup, 1987; Peters, 1996).

High-performance affinity chromatography (HPAC) is a tool that has received increasing interest in recent years as a means of studying and characterizing drug interactions with serum proteins (Hage et al., 2009; Kim & Wainer, 2008; Schiel & Hage, 2009). This method is a type of high-performance liquid chromatography (HPLC) that uses an immobilized binding agent (e.g., a serum protein) as the stationary phase (Hage et al., 2009). The basic components of HPAC are shown in Fig. 1. In this technique, the binding agent is immobilized onto a chromatographic support, which is suitable for use in HPLC, and is placed into a column. This column and support can be used to bind and retain drugs that interact with the binding agent. The resulting elution profile can then be used to provide information on the strength, rate, and nature of the drug's interaction with the binding agent (Hage et al., 2009; Kim & Wainer, 2008; Schiel & Hage, 2009).

There are several features that have made HPAC of interest for drug–protein binding studies. For instance, this method has good precision, is relatively fast, and is easy to automate (Hage, 2001). It can also be used with many types of detectors and has been shown in numerous studies to provide good correlation with reference methods when it is used to study drug interactions with serum proteins. Another useful feature of HPAC is its ability to often reuse the same protein or binding agent for a large number of experiments, with even hundreds of studies being possible in some cases (Hage et al., 2009; Kim & Wainer, 2008; Schiel & Hage, 2009).

This review will discuss the principles of HPAC and of the methods that have been used in this technique to examine drug interactions with serum binding agents. Various applications of these methods will be presented as

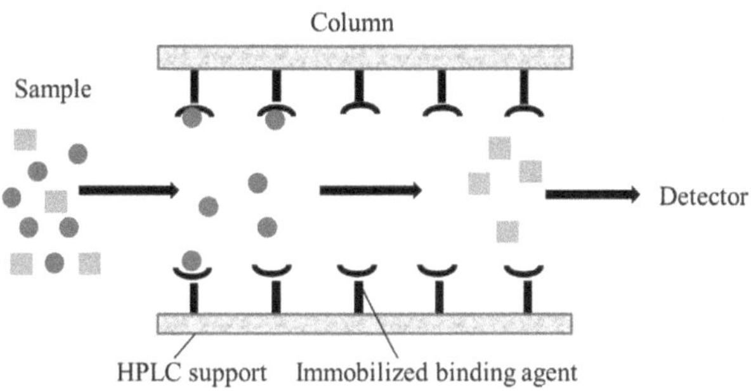

Figure 1 General components of a system for carrying out high-performance affinity chromatography. The squares represent nonretained or weakly retained components of the original sample, and the circles represent more strongly retained sample components.

related to personalized medicine and drug–protein binding studies. These examples will include the use of HPAC to examine simple or complex interactions between drugs and serum proteins or other binding agents. The type of information that can be provided by each technique will be discussed, and the use of each method with clinical or pharmaceutical samples will be considered.

1.1 Drug–Protein Interactions in Blood

Drugs can bind to various proteins and agents in the bloodstream. One protein that acts as a binding agent for many drugs and small solutes in blood is HSA (Lindup, 1987; Mohamadi-Nejad, Moosavi-Movahedi, Hakimelahi, & Sheibani, 2002; Peters, 1996). HSA, whose structure is shown in Fig. 2A, is the most abundant protein in serum and has a normal concentration that ranges from 30 to 50 g/L (Peters, 1996; Tietz, 1986). This protein has a molar mass of 66.5 kDa and consists of a single chain of 585 amino acids with 17 internal sulfide bonds (Peters, 1996). This protein has various roles, including its action to help regulate osmotic pressure and control pH in the bloodstream (Fanali et al., 2012; Peters, 1996; Wang et al., 2008). HSA also acts as a transport protein for a large number of low mass substances within blood, including some hormones, fatty acids, and many drugs (Barzegar et al., 2007; Ghuman et al., 2005; Herve, Urien, Albengres, Duche, & Tillement, 1994; Lindup, 1987; Mohamadi-Nejad et al., 2002; Peters, 1996; Seedher & Kanojia, 2008; Vidal, Nielsen, & Welinder, 1992).

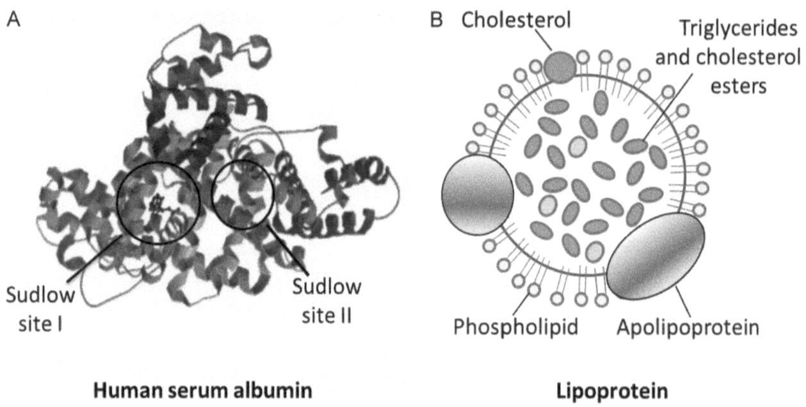

Figure 2 (A) Structure of human serum albumin (HSA), including the locations of the two major drug-binding sites on this protein (i.e., Sudlow sites I and II), and (B) the general structure of a lipoprotein. The structure in (A) was generated by using PDB file 2BXD and includes a molecule of warfarin that is bound at Sudlow site I (Ghuman et al., 2005).

HSA has a number of relatively well-defined binding regions for drugs. The two major binding regions for drugs on HSA are Sudlow sites I and II (Otagiri, 2005; Peters, 1996; Sudlow, Birkett, & Wade, 1975). Sudlow site I is located in subdomain IIA of HSA and tends to bind bulky heterocyclic compounds such as warfarin, phenylbutazone and azapropazone (Loun & Hage, 1994; Otagiri, 2005; Peters, 1996; Sudlow et al., 1975). Sudlow site II is located in subdomain IIA and can bind to aromatic compounds such as L-tryptophan, ibuprofen, and benzodiazepines (Loun & Hage, 1994; Otagiri, 2005; Peters, 1996; Sudlow et al., 1975). Examples of other, minor binding sites for drugs on HSA include the digitoxin and tamoxifen sites, which bind to digitoxin- and tamoxifen-related compounds, respectively (Hage & Sengupta, 1998; Peters, 1996; Sengupta & Hage, 1999a, 1999b; Sudlow et al., 1975).

AGP is another serum protein that is involved in the transportation and distribution of many drugs (Board, Jones, & Bentley, 1986; Founier, Medjoubi, & Porquet, 2000; Israili & Dayton, 2001; Schmid, Nimberg, Kimura, Yamaguchi, & Binette, 1977; Schonfeld, Ravelli, Mueller, & Skerra, 2008; Van Dijk, Havenaar, & Brinkman-van der Linden, 1995). AGP has a carbohydrate content of 45% (w/w) and an approximate molar mass of 41 kDa (Schmid et al., 1977). This protein has a single polypeptide chain and can contain five carbohydrate groups. It has been estimated that there may be 12–20 different forms of AGP that can occur in human serum due to variations in the amino acid sequence of this protein and in the types and numbers of carbohydrate groups that are attached to its polypeptide chain (Founier et al., 2000; Van Dijk et al., 1995). AGP tends to bind best to basic and neutral drugs in blood (Founier et al., 2000; Israili & Dayton, 2001).

Lipoproteins such as high-density lipoprotein (HDL) and low-density lipoprotein (LDL) are other binding agents that can interact with some drugs in blood (Barklay, 1972; Durrington, 1989; Harmony, Aleson, & McCarthy, 1986; Havel & Kane, 1995; Jonas, 2002; Kwong, 1985; Mbewu & Durrington, 1990; Skipski, 1972; Wasan & Cassidy, 1998). The general structure of a lipoprotein is shown in Fig. 2B. Lipoproteins are soluble macromolecular complexes of various lipids (e.g., triglycerides and cholesterol esters) that are coated with a layer of phospholipids and that contain some proteins (i.e., apolipoproteins). An important function of lipoproteins is to transport cholesterol and other lipids in the bloodstream; however, lipoproteins can also bind and transport many hydrophobic drugs (Barklay, 1972; Jonas, 2002; Mahley, Innerarity, Rall, & Weisgraber,

1984; Wasan & Cassidy, 1998). The different compositions of lipoproteins such as HDL, LDL, and very low-density lipoprotein (VLDL) (Mahley et al., 1984; Morrisett, Jackson, & Gotto, 1975; Skipski, 1972) have been proposed to give them some differences in their binding strength and types of interactions that occur for drugs (Chen, Sobansky, & Hage, 2010; Sobansky & Hage, 2012a, 2012b, 2014).

There are various ways in which a disease can affect drug–protein interactions in blood. One possible way this binding can be altered is when there is a large change in the concentration of the protein or binding agent. For instance, it is known that the levels of AGP can increase from two- to five-fold during the acute phase of some diseases (Hochepied et al., 2002; Kremer, Wilting, & Janssen, 1988). This increase in the level of AGP will cause a shift in the extent of its binding to drugs such as disopyramide, phenobarbital, and carbamazepine (Ceciliani & Pocacqua, 2007; Kuroda, Matsumoto, Shibukawa, & Nakagawa, 2003). Changes in the concentrations of other binding agents in blood (e.g., HSA) or in the composition of these agents (e.g., lipoproteins) can also occur in disorders such as kidney disease, liver disease, and heart disease (Kwan, Kronenberg, Beddhy, & Cheung, 2007; Natarajan, Ray, Christopher, & Cannon, 2010; Sakai et al., 1999; Tietz, 1986).

Another way drug–protein binding can be altered is if the structure of a protein is modified as a result of a disease process. For example, diabetes is known to result in an increase in the nonenzymatic glycation of serum proteins such as HSA (Anguizola et al., 2013; Iberg & Fluckiger, 1986; Koyama, Sugioka, Uno, Mori, & Nakajima, 1997; Mendez, Jensen, McElroy, Pena, & Esquerra, 2005; Nakajou, Watanabe, Kragh-Hansen, Maruyama, & Otagiri, 2003). Glycation is a reaction in which glucose combines with free amine groups on a protein to form a reversible Schiff base, as illustrated in Fig. 3. Some of this product will then rearrange to form a more stable Amadori product, or fructosamine (Lapolla et al., 2005; Lapolla, Fedele, Seraglia, & Traldi,

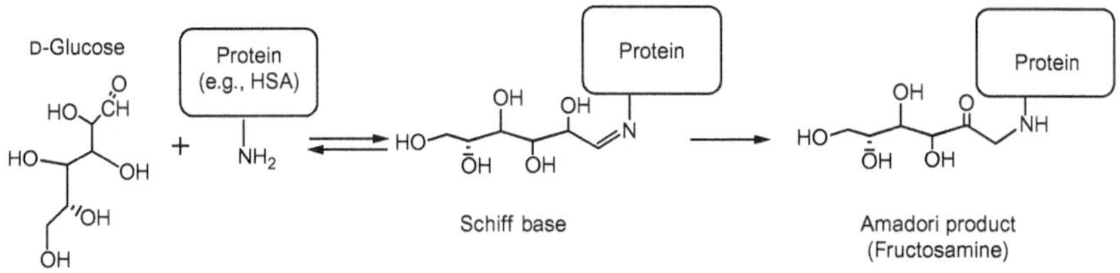

Figure 3 The general process of protein glycation, as can occur with human serum albumin (HSA).

2006; Zhang, Ames, Smith, Baynes, & Metz, 2009). Advanced glycation end-products can also be formed on proteins during diabetes by further oxidation, dehydration, and cross-linking processes (Anguizola, Matsuda, et al., 2013). In the case of HSA, some of these modifications can occur at or near Sudlow sites I and II and have been found to affect the binding of various drugs with this protein (Iberg & Fluckiger, 1986; Joseph, Anguizola, & Hage, 2011; Joseph & Hage, 2010a; Koyama et al., 1997; Matsuda, Anguizola, Joseph, & Hage, 2011, 2012; Mendez et al., 2005; Nakajou et al., 2003).

The presence of endogenous agents that can interact with drugs at their binding sites is another way disease can affect drug–protein binding in serum. For instance, a variety of fatty acids can bind to HSA; these fatty acids can, in turn, lead to allosteric effects or direct competition during the binding of some drugs to HSA (Anguizola, Matsuda, et al., 2013; Noctor, Wainer, & Hage, 1992; Peters, 1996; Simard, Zunszain, Hamilton, & Curry, 2006; Spector, 1976). Bilirubin is another endogenous agent that can bind to HSA (Gary & Stroupe, 1978). It is also known that a number of drugs, such as sulfisoxazole and other sulfonamides, can be displaced from serum proteins by bilirubin (Bratlid, 1972; Johnson, Sarmiento, Blanc, & Day, 1959; Kantor et al., 1961; Odell, 1959; Seedher & Agarwal, 2013; Seedher & Kanojia, 2013; Zamet & Chunga, 1971).

1.2 Preparation of HPAC Columns for Drug–Protein Binding Studies

Several factors must be considered when preparing columns for HPAC-based drug–protein binding studies. The support material is one important item to select. In many studies of drug–protein binding by HPAC, HPLC-grade porous silica is used as the support material (Gustavsson & Larsson, 2006; Hage, Anguizola, Barnaby, et al., 2011). This support can provide good mass transfer properties and is stable at the pressures and flow rates that are used in HPAC. It is also important to use a pore size for the silica, or for any HPAC support, that is sufficiently large to allow the binding agent to be immobilized and for the target to be able to access this immobilized binding agent (Gustavsson & Larsson, 2006). This is often accomplished in work with immobilized proteins by using silica with a nominal pore size of 300 or 500 Å (Clarke et al., 2000; Gustavsson & Larsson, 2006; Hage, 2002; Sobansky & Hage, 2012b). Even wider pore sizes (e.g., 1000 Å) may be needed for larger agents (e.g., the lipoprotein VLDL) (Sobansky & Hage, 2014). One disadvantage of silica is that this support in its original form can have strong nonspecific binding for proteins and related biomolecules.

However, this effect can be minimized or eliminated by first converting the silica into a diol-bonded form, which has low nonspecific binding for many biological compounds (Gustavsson & Larsson, 2006; Hage, Anguizola, Barnaby, et al., 2011).

Monolithic supports have also been used in HPAC for drug–protein binding studies. Attractive features of these supports include their good mass transfer properties, ability to often provide low backpressure, and capability of being made in a variety of shapes and sizes. Supports that have been used in HPAC have included organic-based monoliths based on copolymers of glycidyl methacrylate and ethylene glycol dimethacrylate (Mallik & Hage, 2006). These organic monoliths contain an epoxy group that can either be converted into a diol group or used directly or activated for the immobilization of proteins and other binding agents (Jiang, Mallik, & Hage, 2005; Kalashnikova, Ivanova, & Tennikova, 2007; Mallik & Hage, 2006; Plantonova, Pankova, Ill'ina, Vlasosv, & Tennikova, 1999). Silica monoliths have also been utilized in HPAC for drug–protein binding studies (Cabrera, 2004; Mallik & Hage, 2006; Mallik, Xuan, & Hage, 2007; Yoo & Hage, 2010). Other possible options for supports include monoliths based on agarose and cryogels, which offer good chemical and physical stabilities for use in affinity methods (Dogan, Ozkara, Sari, Uzun, & Denizli, 2012; Gustavsson & Larsson, 2006; Hermanson, Mallia, & Smith, 1992; Mallik & Hage, 2006).

Various methods can be employed to immobilize proteins or other binding agents to support for HPAC. One common technique that is used with amine-containing agents is the Schiff base method, as is shown in Fig. 4A. The Schiff base method is carried out by oxidizing a support that contains diol groups to form aldehydes. This activated support is then allowed to react with free amine groups on a protein or binding agent to form a reversible Schiff base, which can be reduced by using sodium cyanoborohydride to form a stable secondary amine linkage (Kim & Hage, 2006). A stronger reducing agent, such as sodium borohydride, can later be added to remove any remaining aldehyde groups on the support. This method works well for the immobilization of HSA because it tends to couple this protein through the N-terminus or at residues that are not located near Sudlow sites I and II or at major sites of glycation (Anguizola et al., 2013; Barnaby, Wa, Cerny, Clarke, & Hage, 2010; Joseph & Hage, 2010a, 2010b; Wa, Cerny, Clarke, & Hage, 2007; Wa, Cerny, & Hage, 2006). This approach has also been used to immobilize various types of lipoproteins (Chen et al., 2010; Sobansky & Hage, 2012a, 2014) and agents such as antibodies for HPAC methods (Ohnmacht, Schiel, & Hage, 2006; Pfaunmiller, Moser, & Hage, 2012).

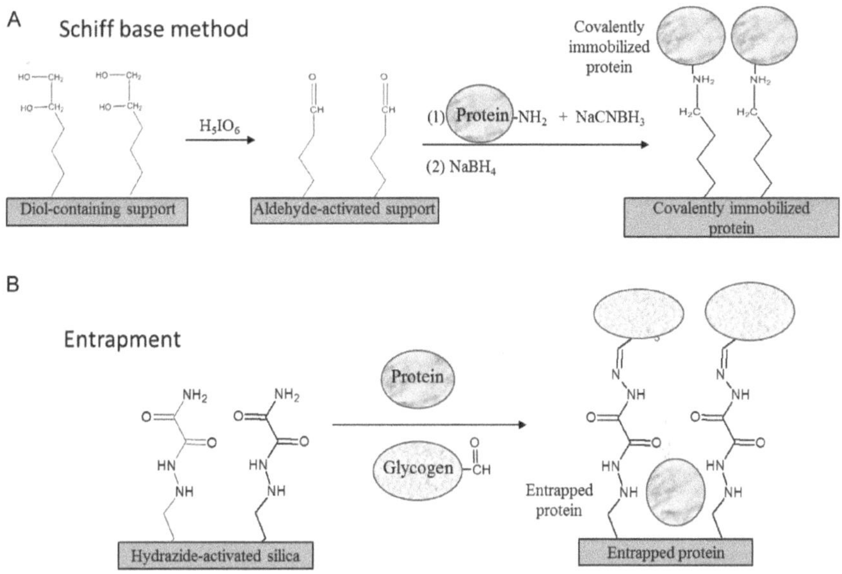

Figure 4 Reactions involved in the immobilization of proteins (A) onto a diol-bonded or diol-containing support by the Schiff base method or (B) within a hydrazide-activated support through the use of entrapment and mildly oxidized glycogen as a capping agent.

Several other covalent immobilization techniques have been used to make HPAC supports for drug–protein binding studies. Examples of other amine-based coupling methods include the *N*-hydroxysuccinimide (NHS), carbonyldiimidazole (CDI), and epoxy methods (An et al., 2014; Gustavsson & Larsson, 2006; Mallik & Hage, 2006). An alternative method that has been employed with HSA is to couple this protein through its lone free sulfhydryl group to silica (Mallik, Wa, & Hage, 2007). In addition, AGP and antibodies have been immobilized by oxidizing these glycoproteins under mild conditions and then attaching their oxidized forms to hydrazide-activated supports (Ruhn, Garver, & Hage, 1994; Xuan & Hage, 2005).

A possible issue with covalent immobilization methods is that they can lead to an apparent or actual loss of activity for the binding agent due to effects such as multisite attachment, improper orientation, and steric effects (Kim & Hage, 2006). Many of these problems can be avoided by instead using entrapment to immobilize the binding agent. One way of accomplishing this is shown in Fig. 4B (Jackson, Anguizola, Pfaunmiller, & Hage, 2013; Jackson, Xuan, & Hage, 2010). In this method, mildly oxidized glycogen is allowed to combine with a hydrazide-activated support in the presence of the soluble protein or binding agent. This method results in the protein or binding agent being entrapped at the surface or within the pores of the support in a form that retained its initial activity. This method has been used to entrap HSA and AGP in silica supports, and to place samples of glycated HSA within similar materials for use in drug–protein binding studies (Jackson et al., 2010, 2013).

Column size is another factor to consider in preparing HPAC columns for the study of drug–protein interactions. Work in this area has been carried out by using traditional HPLC columns, affinity microcolumns, and affinity disks (Hage, 2002; Pfaunmiller et al., 2012; Yoo & Hage, 2010; Yoo, Schiel, & Hage, 2010; Zacharis, Kalaitzantonakis, Podgornik, & Theodoridis, 2006; Zheng, Li, Beeram, et al., 2014). One advantage of affinity microcolumns and affinity disks is their short residence times, which makes them suitable for applications such as rapid drug–protein binding studies and the high-throughput screening of drugs (Pfaunmiller et al., 2012; Yoo et al., 2010; Zheng, Li, Beeram, et al., 2014). Open tubular capillary columns have also been employed for drug–protein binding studies (Moaddel, Bullock, & Wainer, 2004). Affinity microcolumns, affinity disks, and affinity capillary columns all require the use of only small amounts of binding agents and can often use this agent for hundreds of experiments. These features make such columns attractive for work with proteins or agents that are expensive, difficult to obtain, or that are isolated from individual patients (Moaddel et al., 2004, 2013).

2. FRONTAL ANALYSIS STUDIES OF DRUG–PROTEIN INTERACTIONS

2.1 General Principles of Frontal Analysis

Frontal analysis is one approach that has been employed in HPAC for drug–protein interaction studies and work related to personalized medicine (Anguizola, Joseph, et al., 2013; Calleri, Temporini, & Massolini, 2011; Hage, Anguizola, Barnaby, et al., 2011; Hage, Anguizola, Jackson, et al., 2011; Matsuda, Kye, Anguizola, & Hage, 2014; Matsuda, Li, Zheng, & Hage, 2015a, 2015b; Schriemer, 2004; Vuignier, Schappler, Veuthey, Carrupt, & Martel, 2010; Zheng, Li, Beeram, et al., 2014). In this method, solutions with known concentrations of a given target compound or drug (D) are continuously applied to a column that contains the immobilized protein or binding agent. As the solution of the target passes through the column, some or most of the binding sites for this target will become occupied. This process will lead to an increase in the amount of target that elutes from the column over time. The result is a breakthrough curve, as is illustrated in Fig. 5 for the application of several solutions containing the drug carbamazepine to a column containing AGP (Xuan et al., 2010).

If the target and binding agent have relatively fast association and dissociation kinetics on the timescale of the experiment, the position of the breakthrough curve can be used to obtain the equilibrium constants and moles of binding sites for the interaction of the target with the immobilized agent. This can be done by first determining the apparent moles of analyte ($m_{L,app}$) that are required to reach the mean position of the breakthrough curve, and by measuring this value over a relatively wide range of concentrations for the applied target or drug ([D]) (Chen et al., 2010; Hage, Anguizola, Barnaby, et al., 2011; Hage, Anguizola, Jackson, et al., 2011; Matsuda et al., 2015a, 2015b; Sobansky & Hage, 2012a; Zheng, Li, Beeram, et al., 2014). The change in the value of $m_{L,app}$ as a function of [D] can then be fit to various binding models, such as those shown in Table 1. The fit of the data to these models can be used to determine the type of interaction that is taking place, to estimate the moles of binding sites for this interaction, and to estimate the equilibrium constant(s) for this

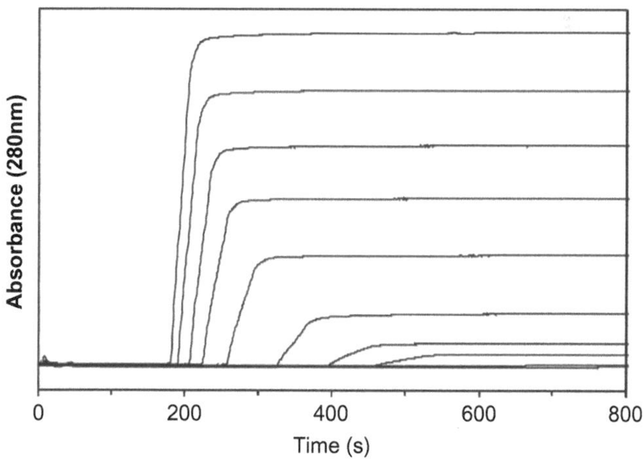

Figure 5 Frontal analysis studies for the binding of carbamazepine to immobilized AGP at pH 7.4 and 37 °C. The concentrations of carbamazepine in these experiments were (left-to-right) 21.8, 18.2, 14.6, 10.9, 7.3, 3.6, 1.46, 0.73, 0.073, and 0.036 μM. The flow rate was 1.0 mL/min and a 5.0 cm × 4.6 mm I.D. column was used. *Reproduced with permission (Xuan, Joseph, Wa, & Hage, 2010).*

Table 1 Examples of Binding Models Used in Frontal Analysis to Examine Drug Interactions with Binding Agents in Serum

Type of Binding Model	Predicted Response	
Single type of interaction		
One set of saturable sites (one-site model)	$m_{L,app} = \frac{m_{L1} K_{a1}[D]}{1 + K_{a1}[D]}$	(1)
	$\frac{1}{m_{L,app}} = \frac{1}{K_{a1} m_{L1}[D]} + \frac{1}{m_{L1}}$	(2)
One set of nonsaturable sites (nonsaturable model)	$m_{L,app} = m_{L1} K_a[D]$	(3)
Two types of interactions		
Two groups of saturable sites (two-site model)	$m_{L,app} = \frac{m_{L1} K_{a1}[D]}{1 + K_{a1}[D]} + \frac{m_{L2} K_{a2}[D]}{1 + K_{a2}[D]}$	(4)
A set of saturable sites and nonsaturable sites (mixed-mode model)	$m_{L,app} = \frac{m_{L1} K_{a1}[D]}{1 + K_{a1}[D]} + m_{L2} K_a[D]$	(5)

Terms: $m_{L,app}$, moles of applied analyte required to reach the mean position of the breakthrough curve at a given drug concentration, [D]; m_{L1} and m_{L2}, moles of binding sites 1 and 2; K_{a1} and K_{a2}, association equilibrium constants for site 1 and 2.

process (Chen et al., 2010; Hage, Anguizola, Barnaby, et al., 2011; Sobansky & Hage, 2012a, 2012b).

Frontal analysis has been employed in studying interactions with association equilibrium constants that have ranged from 10^2 to 10^{12} M^{-1} (Chen, Ohnmacht, & Hage, 2004; Chen et al., 2010; Hage, Anguizola, Barnaby, et al., 2011; Matsuda et al., 2015a, 2015b; Vuignier et al., 2010). One advantage of this method is that the equilibrium constant for the target with the binding agent can be determined at the same time as an estimate is obtained for the moles of binding sites that are present (Hage, Anguizola, Barnaby, et al., 2011; Kim & Wainer, 2008; Zheng, Li, Beeram et al., 2014). This makes this approach useful in examining the stoichiometry of a reaction and in comparing binding constants that are obtained from columns that may contain different amounts of the binding agent. However, this method can require large amounts of the target and, because of the number of experiments involved, does tend to take longer to carry out than some of the alternative methods that are described later in this review (Kim & Wainer, 2008; Yoo et al., 2010; Zheng, Li, Beeram, et al., 2014).

2.2 Characterization of Simple Interactions by Frontal Analysis

The simplest binding models that can be used to examine frontal analysis data are those that involve only a single type of interaction. As an example, Eq. (1) describes the behavior that would be expected for the interaction of an applied target with a single type of saturable binding site in a column. In this case, a nonlinear fit to a plot of $m_{L,app}$ versus [D] can be made according to Eq. (1), or a linear fit can be made to a plot of $1/m_{L,app}$ versus 1/[D] according to Eq. (2). For instance, a system with a single set of saturable sites would be expected to give a straight line when fit to Eq. (2); the inverse of the intercept for this fit would then provide the moles of binding sites in the column (m_{L1}), and the ratio of the intercept over the slope would give the association equilibrium constant (K_{a1}) for this system (An et al., 2014; Hage, Anguizola, Barnaby, et al., 2011; Zheng, Li, Beeram, et al., 2014).

A single-site binding model has been used to study many types of processes using HPAC. For instance, this method has been used to characterize the overall binding of targets with various types of affinity columns (An et al., 2014; Joseph & Hage, 2010b; Kim & Wainer, 2008; Matsuda et al., 2015a, 2015b; Sobansky & Hage, 2012a; Soman, Yoo, Jang, & Hage, 2010; Xuan et al., 2010; Yoo, Smith, & Hage, 2009) and to study the interactions of targets with specific regions on immobilized binding agents (Chattopadhyay, Tian, Kortum, & Hage, 1998; Joseph & Hage, 2010a; Joseph, Moser, Basiaga, Schiel, & Hage, 2009; Loun & Hage, 1992, 1995; Tweed, Loun, & Hage, 1997). In the case of personalized medicine, this type of model has been found to be useful in examining the effects of glycation on the binding of HSA with warfarin and L-tryptophan, which are often used as probes for Sudlow sites I and II of this protein (Joseph & Hage, 2010a; Matsuda et al., 2014, 2015a). Figure 6 shows some examples of plots that were prepared according to Eq. (2) for L-tryptophan on HPAC columns that contained several samples of *in vitro* glycated HSA. From these results, it was determined that L-tryptophan had a 4.7- to 5.8-fold increase in its affinity at Sudlow site II for glycated HSA when compared with normal HSA. Similar experiments with warfarin gave no significant changes in the binding of this probe at Sudlow site I for the same samples (Joseph & Hage, 2010a).

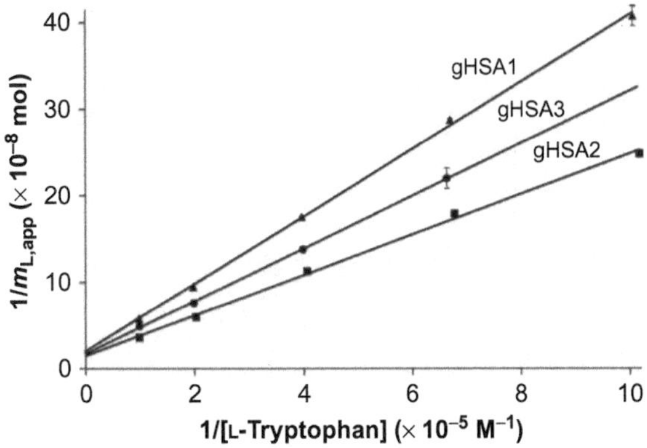

Figure 6 Frontal analysis studies and use of Eq. (2) to examine the binding of L-tryptophan with columns containing three samples of *in vitro* glycated HSA (i.e., gHSA1, gHSA2, and gHSA3) with various levels of modification. *Reproduced with permission (Joseph & Hage, 2010a).*

A single-site model has also been used with frontal analysis to correct for the nonspecific binding between drugs and supports in HPAC columns (Kim, Mallik, & Hage, 2006; Xuan et al., 2010). One approach for making such a correction is to subtract the breakthrough time, or value of $m_{L,app}$, for a drug on a control column from the results that are obtained under the same conditions for the drug on a column containing the immobilized binding agent. This method can be used when the degree of nonspecific binding does not change due to the presence of the binding agent. For instance, such an approach has been used to compare two immobilization methods (i.e., the NHS and Schiff base methods) when they were both used in binding studies where carbamazepine was the applied drug and HSA was the binding agent (Kim et al., 2006).

A more general approach for correcting for nonspecific binding by the support is to carry out detailed frontal analysis studies with the drug on a control column. The best-fit parameters for the binding of the drug to this column can then be used when analyzing data for the same drug on a column that contains an immobilized binding agent on the same support. This method has been employed when examining the interaction of carbamazepine with immobilized AGP (Xuan et al., 2010) and can be modified for use in situations where the extent of nonspecific binding is affected by the presence of the immobilized agent.

Another type of single interaction that has been examined by frontal analysis is one that involves a nonsaturable process. This model is represented by Eq. (3) in Table 1 and has been used in studying the binding of *S*-propranolol with immobilized LDL (Chen et al., 2010; Sobansky & Hage, 2012a). It was found in this particular study that only a simple nonsaturable interaction was present between *S*-propranolol and LDL, while the *R*-enantiomer of propranolol underwent a more complex mixed-mode interaction with the same type of lipoprotein (Sobansky & Hage, 2012a).

2.3 Characterization of Complex Interactions by Frontal Analysis

Other binding models have been used with frontal analysis to examine combinations of interactions. These combinations have included the presence of two types of saturable binding sites or a combination of saturable sites and nonsaturable sites, as represented by Eqs. (4) and (5) in Table 1. These combined models have been employed in work that has investigated the binding of drugs with serum agents such as HSA, AGP, and lipoproteins (Anguizola, Matsuda, et al., 2013; Chen et al., 2010; Joseph, Anguizola, Jackson, & Hage, 2010; Kim & Hage, 2005; Mallik, Xuan, Guiochon, & Hage, 2008; Matsuda, Anguizola, Hoy, & Hage, 2015; Matsuda et al., 2015a, 2015b; Sobansky & Hage, 2012a, 2012b; Xuan et al., 2010; Yoo et al., 2009).

There are many examples in which interactions that involve two independent and saturable sites have been noted for drugs with serum proteins. This situation has been seen for various sulfonylurea drugs during their binding to normal HSA and *in vitro* glycated HSA (Joseph et al., 2010, 2011; Joseph & Hage, 2010b; Matsuda, Anguizola, et al., 2015; Matsuda et al., 2011, 2012, 2015a, 2015b). The same model has also been used to compare the overall affinities and number of binding sites that are present for these drugs with normal HSA and glycated forms of HSA (Matsuda et al., 2015a, 2015b).

Frontal analysis has further been used with HPAC affinity microcolumns to examine the changes in drug interactions that occur with *in vivo* glycated HSA that has been isolated from diabetic patients (Anguizola, Joseph, et al., 2013; Anguizola, Matsuda, et al., 2013). These columns were used to characterize the binding of glycated HSA with the drugs acetohexamide, tolbutamide, and gliclazide. A two-site binding model was found to give a good fit to the frontal analysis data for each type of drug and glycated HSA. The results indicated that these drugs had a group of high-affinity sites on both glycated HSA and normal HSA, as well as a set of lower affinity regions. These results were in good agreement with those found when using *in vitro* glycated HSA that had similar levels of modification (Anguizola, Joseph, et al., 2013; Anguizola, Matsuda, et al., 2013).

Some systems have been observed to include a combination of saturable binding sites and nonsaturable interactions (see Fig. 7). This type of behavior has been seen during the binding of *R*- and *S*-propranolol with immobilized

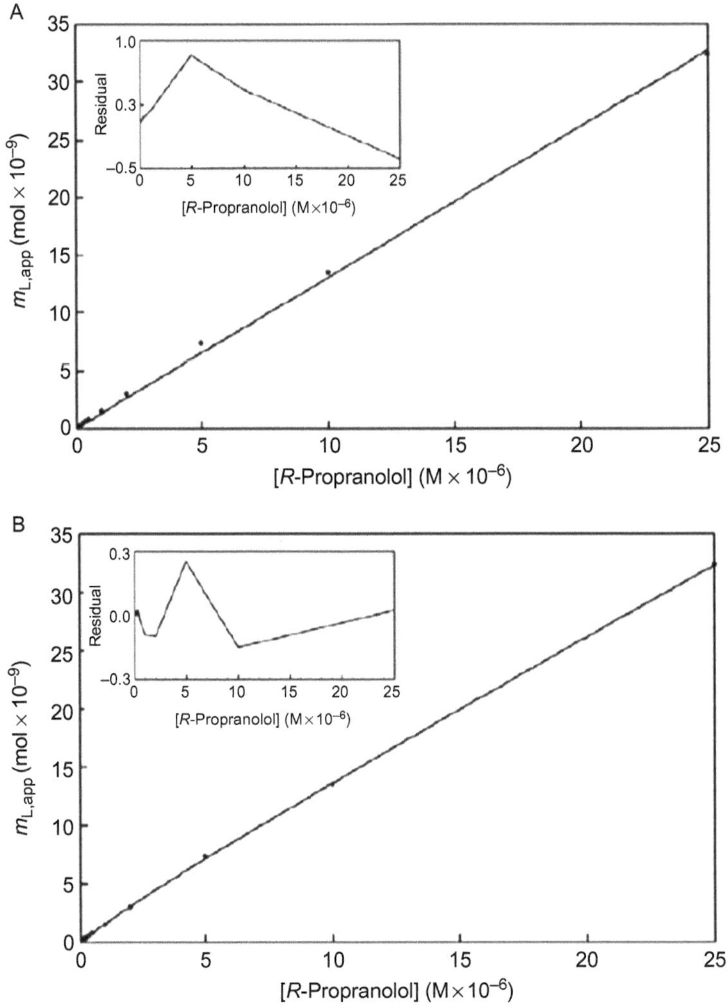

Figure 7 Fit according to (A) a nonsaturable model, as described by Eq. (3), or (B) a mixed-mode model, as described by Eq. (5), for frontal analysis data that were obtained at pH 7.4 and 37 °C for the binding of *R*-propranolol with an HDL column. The insets show the corresponding residual plots for each fit. *Adapted with permission (Chen et al., 2010).*

AGP (Mallik et al., 2008). In addition, the same model has been utilized to describe the interactions of HDL, LDL, or VLDL with drugs such *R*- or *S*-propranolol and verapamil (Hage, Anguizola, Barnaby, et al., 2011; Sobansky & Hage, 2012a, 2014). For these lipoproteins, the saturable drug interaction has been proposed to be due to binding by apolipoproteins. This type of interaction has been found through frontal analysis experiments to be stereoselective for *R*- and *S*-propranolol on LDL, but not when HDL or VLDL is the binding agent (Sobansky & Hage, 2012a, 2014). The non-saturable interactions of these lipoproteins are believed to involve partitioning of a drug into the nonpolar core and have been noted to vary with the size and nonpolar content of the lipoprotein (Sobansky & Hage, 2012a, 2012b, 2014).

3. ZONAL ELUTION STUDIES OF DRUG–PROTEIN INTERACTIONS

3.1 General Principles of Zonal Elution

Zonal elution is another approach that can be used in HPAC to obtain information on a drug–protein interaction and to carry out studies that are related to personalized medicine. In this method, a small amount of a test compound is injected onto a column that contains an immobilized binding agent, and the retention time (or retention volume) for the injected compound is determined. The retention factor (k) for this compound can then be calculated from the compound's retention time (t_R) and the void time of the system (t_M), where $k = (t_R - t_M)/t_M$. This value for k can also be related to the various interactions that have created the compound's retention on the column, as is demonstrated by the following expression.

$$k = \frac{(n_1 K_{a1} + \ldots n_n K_{an}) m_L}{V_M}. \tag{6}$$

In this equation, K_{a1} through K_{an} are the association equilibrium constants for the injected compound at a series of independent sites 1 through n in the column, and the values n_1 through n_n represent the relative moles for each type of site. The term m_L is the total moles of all binding sites that are present for the compound in the column, and V_M is the void volume of the column. This expression indicates that the value of k will be a direct measure of the overall affinity for the injected compound at all of its binding sites in the column, as well as the amount of such sites that are present (Hage, 2002).

One way zonal elution can be used in biological interaction studies is to characterize the binding of a drug at a specific site on a protein (Hage, 2002; Hage, Anguizola, Barnaby, et al., 2011; Joseph & Hage, 2010b; Matsuda et al., 2015a; Patel, Wainer, & Lough, 2006; Schiel & Hage, 2009). This can be accomplished by placing the drug (or solute of interest) into the mobile phase, passing this mobile phase through the column, and then injecting a small plug of a site-specific probe onto the column (Hage, 2002; Patel et al., 2006). If the probe and mobile phase additive have interactions at common or related sites within the column, the retention time of the probe will shift as the concentration of the additive is varied. The degree of this shift can give information on the type of interaction that is occurring between these agents, and the binding constants for the agents at the sites that are involved in this interaction (Chen & Hage, 2006a; Patel et al., 2006; Schiel & Hage, 2009).

Various binding models can be used to describe how the retention factor for an injected target or analyte (A) may change in the presence of a mobile phase additive or competing agent (I). Equation (7) is one example of such an equation. This expression represents a system in which A and I have direct competition at a single type of binding site, and A has no other significant interactions with the column (Chen & Hage, 2006a; Hage, Anguizola, Barnaby, et al., 2011; Patel et al., 2006).

$$\frac{1}{k}=\frac{K_{\mathrm{aI}}\,V_{\mathrm{M}}\,[\mathrm{I}]}{K_{\mathrm{aA}}\,m_{\mathrm{L}}}+\frac{V_{\mathrm{M}}}{K_{\mathrm{aA}}\,m_{\mathrm{L}}}. \tag{7}$$

In this equation, K_{aA} and K_{aI} are the association equilibrium constants for the analyte and competition agent at their site of competition, V_{M} is the column void volume, and m_{L} is the total moles of common binding sites that A and I have in the column. If a plot is made according to Eq. (7) and gives a linear response with a positive slope, the system can be said to follow a model in which A and I have direct competition at a single type of site. In this case, the association equilibrium constant for the competing agent at its site of direct competition with the injected analyte can be found by using the ratio of the slope versus the intercept (Chen & Hage, 2006a; Patel et al., 2006; Schiel & Hage, 2009).

Deviations from a linear response can occur for a plot that is made according to Eq. (7) if the injected compound can bind to multiple sites or has allosteric interactions with the mobile phase additive (Chen & Hage, 2006a; Hage, 2002). The types of deviations that are present (e.g., curvature with a positive or negative slope) can give important clues as to how A and I may be interacting. It is possible in some of these cases to use an alternative expression to Eq. (7) to analyze the data. An example is shown in Eq. (8), which can be used to describe either direct competition at a single-site or allosteric effects between a competing agent and injected compound at two separate but allosterically linked binding regions (Chen & Hage, 2004).

$$\frac{k_0}{k-k_0}=\frac{1}{\beta_{\mathrm{I}\rightarrow\mathrm{A}}-1}\cdot\left(\frac{1}{K_{\mathrm{aIL}}\,[\mathrm{I}]}+1\right). \tag{8}$$

The term K_{aIL} in this equation is the association equilibrium constant for the competing agent with the immobilized binding agent (or ligand, L), while k_0 and k are the retention factors measured for the injected analyte in the absence and presence of a given concentration of the competing agent, respectively. The term $\beta_{\mathrm{I}\rightarrow\mathrm{A}}$ is the coupling constant for the allosteric effect of I on the interaction of A with the immobilized binding agent. Equation (8) predicts that a plot of $k_0/(k-k_0)$ versus $1/[\mathrm{I}]$ should result in a linear relationship for a simple allosteric interaction or a system with direct competition. In these two cases, the slope and intercept of the plot can be used to determine the coupling constant and the association equilibrium constant for the competing agent (see Fig. 8). If a value for $\beta_{\mathrm{I}\rightarrow\mathrm{A}}$ is obtained that is greater than 1, this indicates that a positive allosteric effect is occurring between I and A, while a value that ranges from $0<\beta_{\mathrm{I}\rightarrow\mathrm{A}}<1$ would represent a negative allosteric effect. A value of 0 for $\beta_{\mathrm{I}\rightarrow\mathrm{A}}$ will result if direct competition is taking place between A and I, and a system with no competition should provide a value for $\beta_{\mathrm{I}\rightarrow\mathrm{A}}$ that is equal to 1 (Chen, Fitos, & Hage, 2006; Chen & Hage, 2004). Similar expressions can be written for more complex systems with simple allosteric effects (Chen & Hage, 2006b).

To study a system that involves multiple binding sites, competition studies can be performed by injecting an analyte onto a column and in the presence of a mobile phase that has a known concentration of a site-specific probe (Hage, 2002). In this type of experiment, the overall retention factor for the analyte (k_{A}) should be equal to the sum of the individual retention factors for the same compound at all independent binding sites that are present in the column. This relationship is shown in Eq. (9), where k_1 through k_n are the individual retention factors due to sites 1 through n.

$$k_{\mathrm{A}}=k_1+\ldots k_n. \tag{9}$$

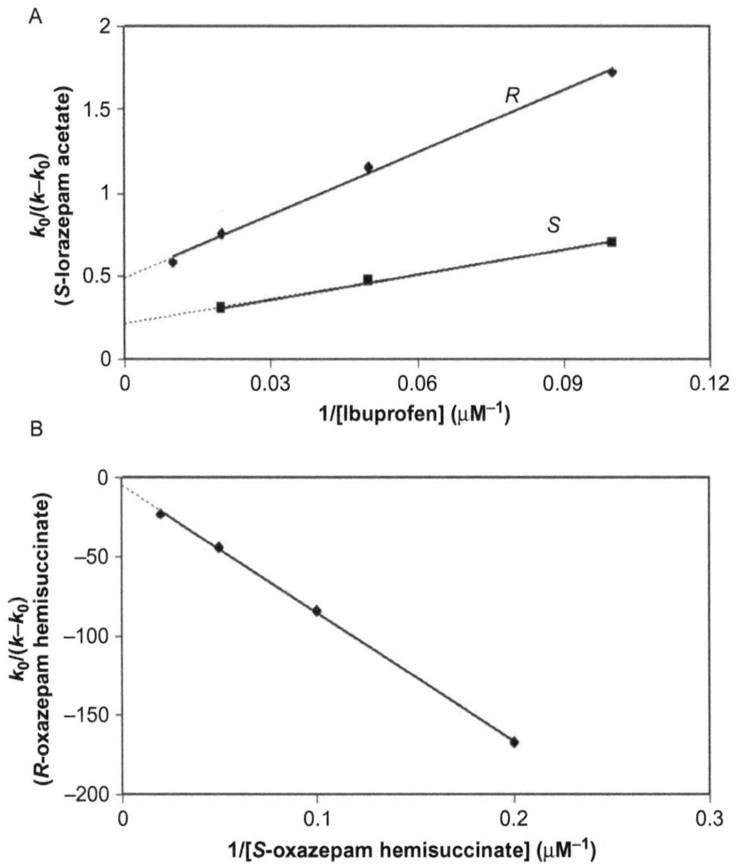

Figure 8 Analysis of zonal elution data according to Eq. (8) for (A) the effects of *R*- or *S*-ibuprofen on the retention of *S*-lorazepam acetate on an HSA column and (B) the effects of *S*-oxazepam hemisuccinate on the binding *R*-oxazepam hemisuccinate on an HSA column. *Adapted with permission (Chen & Hage, 2004).*

This equation can be expanded into the form given in Eq. (10) if a mobile phase additive I can compete at one or more of these sites.

$$k_A = \frac{K_{aA1} m_{L1}}{V_M(1 + K_{aI1}[I])} \ldots + \frac{K_{aAn} m_{Ln}}{V_M(1 + K_{aIn}[I])}. \quad (10)$$

In this expanded expression, K_{aA1} through K_{aAn} are the association equilibrium constants for the analyte at sites 1 through n, and K_{aI1} through K_{aIn} are the association equilibrium constants for the competing agent at the same sites (Hage, 2002).

3.2 Characterization of Drug–Protein Interactions by Zonal Elution

One way zonal elution has been employed in drug-binding studies and in work related to personalized medicine is to use the retention of a drug or solute to follow the stability of HPAC columns or to compare immobilization methods. For instance, this approach has been used during the creation of columns that contained the lipoproteins HDL, LDL, or VLDL, in which the retention for *R*- or *S*-propranolol was monitored over time to determine the usable lifetimes of such supports (Chen et al., 2010; Sobansky & Hage, 2012a, 2014).

Zonal elution and retention factor measurements have also been used to directly examine drug interactions with columns that contained entrapped proteins (Jackson et al., 2010, 2013). In these studies, the retention factor was used with the known protein content of a column to estimate the binding constants for various drugs with entrapped samples of normal HSA or glycated HSA. The results were comparable to those obtained in solution-phase methods or when using covalently immobilized proteins (Jackson et al., 2010, 2013). A related technique, based on a plot of log(k) versus log(K_a) for compounds with known binding constants, has also been tested for use with an immobilized AGP column to quickly estimate the affinities of various drugs for this protein (Xuan & Hage, 2005).

Another application of zonal elution has been in studying the individual or combined effects of several factors on drug–protein binding. For instance, this method has been used to investigate the effects of both glycation and the presence of fatty acids on the interactions between drugs and HSA, as is illustrated in Fig. 9 (Basiaga & Hage, 2010). Simple retention measurements were used to compare the changes in binding that could occur due to either of these factors, and to see how the size of these effects changed with the type of drug or fatty acid that was being examined (Anguizola, Basiaga, & Hage, 2013; Basiaga & Hage, 2010).

Zonal elution has been used in many reports to examine drug–drug competition for binding sites on proteins (Chen & Hage, 2006a, 2006b; Hage, 2002; Patel et al., 2006). This method has been utilized in experiments with AGP columns to examine the competition between carbamazepine, lidocaine, and verapamil with propranolol, where the latter was used as a binding site probe for this protein (Chen et al., 2010; Soman et al., 2010; Xuan et al., 2010). A number of recent studies have used *R*-warfarin and L-tryptophan as probes for Sudlow sites I and II of HSA to study the effects of glycation on the binding of various drugs at these regions (Anguizola, Basiaga, & Hage,

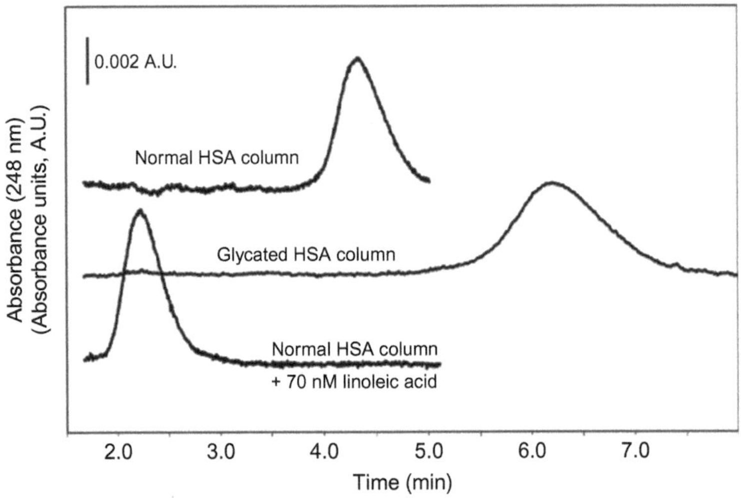

Figure 9 Zonal elution studies carried out at 37 °C by HPAC and examining the retention of acetohexamide on columns containing normal HSA or glycated HSA and in the presence of only pH 7.4, 0.067 M phosphate buffer or the same buffer that also contained 70 nM linoleic acid. The flow rate was 0.5 mL/min, and a 2.0 cm × 2.1 mm I.D. column was used. *Reproduced with permission (Basiaga & Hage, 2010).*

2013; Anguizola, Joseph, et al., 2013; Joseph et al., 2010, 2011; Joseph & Hage, 2010b; Matsuda et al., 2011, 2012). Table 2 shows some results that have been obtained in such experiments for several sulfonylurea drugs and when using *in vivo* glycated HSA (Anguizola, Joseph, et al., 2013); these results demonstrated that changing the glycation level of HSA can affect the binding of drugs with this protein and can lead to either an increase or decrease in affinity at a given binding region. Other probes such as digitoxin or tamoxifen have also been used in zonal elution experiments to examine the local effects of glycation on the binding of drugs with HSA (Matsuda et al., 2012, 2015a, 2015b).

4. OTHER METHODS FOR EXAMINING DRUG–PROTEIN INTERACTIONS

4.1 Peak Decay Method

The peak decay method is an HPAC method that can be used to determine the dissociation rate constant for the release of a drug or target from an immobilized binding agent (Moore & Walters, 1987; Walters, 1987). In

Table 2 Association Equilibrium Constants Measured at Sudlow Sites I and II for Various Sulfonylurea Drugs with Normal HSA and *In Vivo* Glycated HSA

	Association Equilibrium Constant ($\times 10^4\ M^{-1}$)[a]		
Type of HSA	**Tolbutamide**	**Gliclazide**	**Acetohexamide**
Results for Sudlow site I			
Normal HSA	5.5 (±0.2)	1.9 (±0.04)	4.2 (±0.4)
Glycated HSA, clinical sample 1	6.8 (±0.5)	2.3 (±0.2)	4.7 (±0.3)
Glycated HSA, clinical sample 2	7.9 (±0.4)	1.4 (±0.1)	4.0 (±0.4)
Results for Sudlow site II			
Normal HSA	5.3 (±0.2)	6.0 (±0.5)	13 (±1.0)
Glycated HSA, clinical sample 1	7.3 (±0.5)	11 (±1.0)	13 (±0.2)
Glycated HSA clinical sample 2	7.9 (±0.3)	9.2 (±0.7)	17 (±0.4)

[a]These binding parameters were calculated from data obtained at pH 7.4 and 37 °C, with the values in parentheses representing a range of ±1S.D.
Adapted with permission (Anguizola, Joseph, et al., 2013).

the original form of this technique, a small pulse of the target is first applied to a column that contains the binding agent of interest. This step is followed by the continuous application of a high concentration of a competing agent to promote dissociation of the target and to prevent the released target from rebinding to the column. This creates a situation in which the target is washed from the column as soon as it dissociates from the binding agent. The resulting elution profile for the target is a decay curve that can, under appropriate experimental conditions, be used to estimate the dissociation rate constant for the target from the column (Moore & Walters, 1987; Schiel & Hage, 2009; Walters, 1987).

A variation on this approach is a noncompetitive peak decay method, in which a high concentration of the target is initially applied to the column (see Fig. 10). This condition also minimizes rebinding of the retained target as this compound later dissociates from the immobilized binding agent (Chen, Schiel, & Hage, 2009; Yoo & Hage, 2011a). An advantage of the non-competitive method is it does not require any change in the mobile phase between the application of the target and its elution; in addition, no competing agent is needed in this approach (Yoo & Hage, 2011a). The non-competitive method tends to use higher flow rates and smaller columns than the traditional peak decay method, which makes this newer approach

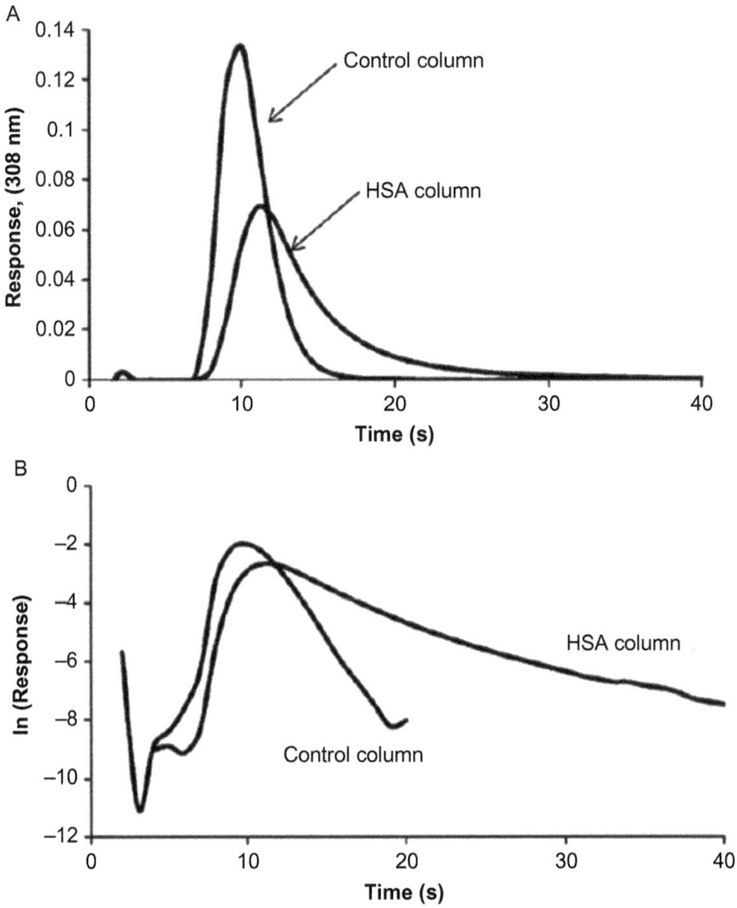

Figure 10 (A) Chromatograms and (B) their natural logarithmic forms, as obtained during peak decay analysis for the injection of racemic warfarin at 4 mL/min onto 1.0 cm × 4.6 mm I.D. columns that contained immobilized HSA or a control support. These studies were carried out at pH 7.4 and 37 °C. *Reproduced with permission (Yoo & Hage, 2011a).*

of potential interest for the high-throughput screening drug–protein interactions (Chen et al., 2009; Schiel & Hage, 2009; Yoo & Hage, 2011a).

$$\ln\left(\frac{\mathrm{d}m_{\mathrm{Ee}}}{\mathrm{d}t}\right) = \ln(m_{\mathrm{Eo}}k_{\mathrm{d}}) - k_{\mathrm{d}}t. \tag{11}$$

In either the traditional or noncompetitive peak decay methods, the elution profile for the dissociated target can be represented in a logarithmic form by using Eq. (11). The term m_{Eo} in this equation is the initial moles of target that are retained by the immobilized binding agent, m_{Ee} is the moles of target that elutes from the column at time t, and k_d is the dissociation rate constant for the target from the binding agent. According to Eq. (11), a plot of the natural logarithm of the target's elution profile versus time should give a linear relationship. In addition, the slope can be used to provide the value of k_d under the correct experimental conditions. These conditions include the need for negligible rebinding of the target with the immobilized agent, and the presence of a situation in which dissociation of the target from this agent is the rate-limiting step in the release of target from the column (Moore & Walters, 1987; Schiel & Hage, 2009).

The dissociation rate constants that have been measured by the peak decay method have ranged from 10^{-2} to 10^{1} s^{-1} (Moore & Walters, 1987; Nelson, Moser, & Hage, 2010; Schiel & Hage, 2009; Yoo & Hage, 2011a, 2011b; Zheng, Li, Beeram, et al., 2014). This method can be used to examine drug–protein interactions and other processes that have either weak or moderate binding (Chen et al., 2009; Moore & Walters, 1987; Nelson et al., 2010; Yoo & Hage, 2011a, 2011b). This technique has also been used to determine the elution conditions that are needed for systems with stronger binding, such as the interactions of antibodies or aptamers with their targets (Hage, Xuan, & Nelson, 2006; Pfaunmiller et al., 2012).

A number of recent studies have used the noncompetitive peak decay method to estimate the dissociation rate constants for drugs with serum proteins. An example of such work is shown in Fig. 10. In this example, noncompetitive peak decay experiments were carried out with racemic warfarin on silica monolith columns that contained immobilized HSA or an inert control support (Yoo & Hage, 2011a). The same method has been used to estimate the dissociation rate constants for imipramine, diazepam, acetohexamide, and tolbutamide with HSA. This general approach was also employed with silica monoliths to measure dissociation rate constants for amitriptyline, nortriptyline, and lidocaine with AGP (Yoo & Hage, 2011b). In each case, the results gave good agreement with previous values that have been reported for the same drugs or other compounds with similar structures and/or affinities for HSA and AGP (Yoo & Hage, 2011a, 2011b).

4.2 Ultrafast Affinity Extraction

Ultrafast affinity extraction is an HPAC technique that can be used to directly examine the interactions between drugs or hormones with serum proteins in clinical or pharmaceutical samples (Clarke & Hage, 2001; Clarke, Schiel, Moser, & Hage, 2005; Mallik, Yoo, Briscoe, & Hage, 2010; Ohnmacht et al., 2006; Schiel, Tong, Sakulthaew, & Hage, 2011; Zheng, Matsuda, & Hage, 2014). This method measures the free (or nonprotein bound) forms of such solutes in solutions or samples that also contain binding agents for these solutes (Clarke & Hage, 2001). The general scheme for ultrafast affinity extraction is shown in Fig. 11A.

In this type of experiment, the solute of interest (in the presence of a soluble protein or binding agent) is injected onto an affinity microcolumn that can quickly bind to this solute (i.e., on a time scale of a few seconds to a few hundred milliseconds) (Clarke & Hage, 2001; Clarke et al., 2005; Schiel et al., 2011). The binding agent/protein in the sample and any solute that has formed a complex with this binding agent is eluted as a nonretained peak,

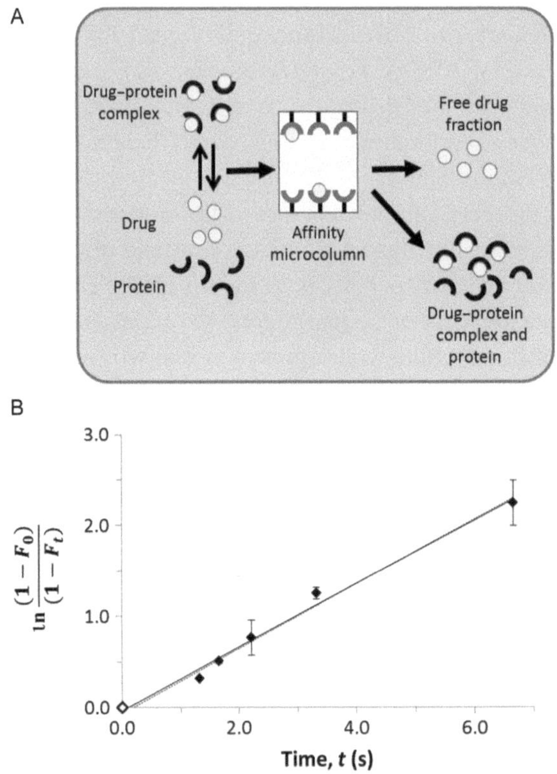

Figure 11 (A) General scheme for ultrafast affinity extraction and (B) use of this method to examine the dissociation of verapamil from soluble HSA according to Eq. (12). *Adapted with permission (Zheng, Li, Beeram, et al., 2014; Zheng, Li, Podariu, & Hage, 2014).*

while at least part of the free solute fraction is retained by the column. The free solute fraction in the original sample is determined by comparing the retained peak for this solute with a peak that is obtained for the same compound when no soluble binding agent is present. The columns that are used to carry out ultrafast affinity extraction may contain either specific binding agents for the target, such as antibodies, or more general binding agents, such as HSA (Clarke & Hage, 2001; Clarke et al., 2005; Mallik et al., 2010; Ohnmacht et al., 2006; Schiel et al., 2011).

One application of ultrafast affinity extraction has been its use in characterizing the kinetics of drug–protein interactions (Zheng, Li, Podariu, et al., 2014). In this type of study, the effect of changing the flow rate and column residence time is examined on the measured free drug fraction. The use of a low-to-moderate flow rate will allow part of the protein-bound portion of the drug in the sample to dissociate as it passes through the column and will result in an increase in the apparent free fraction. The dissociation rate constant (k_{d}) for the drug–protein complex can then be determined from this data by using Eqs. (12) or (13),

$$\ln \frac{(1-f_0)}{(1-f_t)} = k_{\mathrm{d}} t \tag{12}$$

$$\ln \frac{1}{(1-f_t)} = k_{\mathrm{d}} t - \ln(1-f_0), \tag{13}$$

where f_t represents the apparent free drug fraction that is measured at column residence time t, and f_0 is the original free drug fraction in the sample. According to Eqs. (12) and (13), a linear relationship should be obtained for a system that follows first-order dissociation kinetics when a plot is made of either $\ln[(1-f_0)/(1-f_t)]$ or $\ln[1/(1-f_t)]$ versus t. The value of the dissociation rate constant k_{d} can then be estimated from the slope of either plot. An example of such a plot is shown in Fig. 11B. This method has been used to obtain the k_{d} values for HSA with drugs such as verapamil, warfarin, tolbutamide, acetohexamide, gliclazide, and chlorpromazine (Zheng, Li, Podariu, et al., 2014).

Ultrafast affinity extraction can also be used to determine the association equilibrium constant (K_{a}, for a single-site interaction) or the global affinity constant (nK_a, for a multisite interaction) for a drug–protein interaction (Mallik et al., 2010; Zheng, Li, Beeram, et al., 2014; Zheng, Li, Podariu, et al., 2014). In this type of experiment, the free drug fraction in the original sample (f_0) is determined at high flow rates, where a short column residence time is used to avoid dissociation of the drug from its complexes with proteins in the sample. This value for f_0 can then be used with Eq. (14) and the known concentrations of the drug ([D]) and protein ([P]) in the sample to estimate K_a or nK_a for the drug and protein.

$$nK_a' \,\text{or}\, K_a = \frac{1 - f_0}{f_0([P] - [D] + [D]\,f_0)}. \tag{14}$$

A multidimensional HPAC system has been used with ultrafast affinity extraction to make it easier for this method to measure free drug fractions and estimate binding constants for drugs and proteins at physiological or therapeutic levels (Zheng, Li, Podariu, et al., 2014). This approach has made it possible to conduct measurements under conditions in which a protein may be present in a large excess versus the drug (i.e., a situation which can make it difficult to obtain adequate resolution between the peaks for the drug–protein complex and drug on a single-column system for ultrafast affinity extraction). This problem can be avoided by using a multidimensional system, where a second affinity column is placed online after the free drug fraction has begun to elute from the first column. Equation (14) can then be used to estimate the association equilibrium constant or global affinity constant for the drug–protein interaction. This type of system has been used to measure the free fractions for several sulfonylurea drugs in samples that contained normal HSA or glycated HSA (Zheng, Li, Podariu, et al., 2014). In addition, a multidimensional system that combined ultrafast affinity extraction with a chiral stationary phase was used to simultaneously measure the free fractions of warfarin enantiomers in human serum and to determine the binding constant for each enantiomer with HSA (Zheng, Yoo, & Hage, 2013).

4.3 Chromatographic Immunoassays

Antibodies have been used as immobilized agents in ultrafast affinity extraction, resulting in a method known as "ultrafast immunoextraction." This method, in turn, has been combined with chromatographic-based immunoassays to measure the free fractions of drugs and hormones in clinical samples (Clarke, Choudhuri, & Hage, 2001; Clarke et al., 2005; Ohnmacht et al., 2006).

A chromatographic-based displacement immunoassay, as is illustrated in Fig. 12, is one format that has been employed in such work. In this method, an affinity microcolumn containing immobilized antibodies is loaded with a

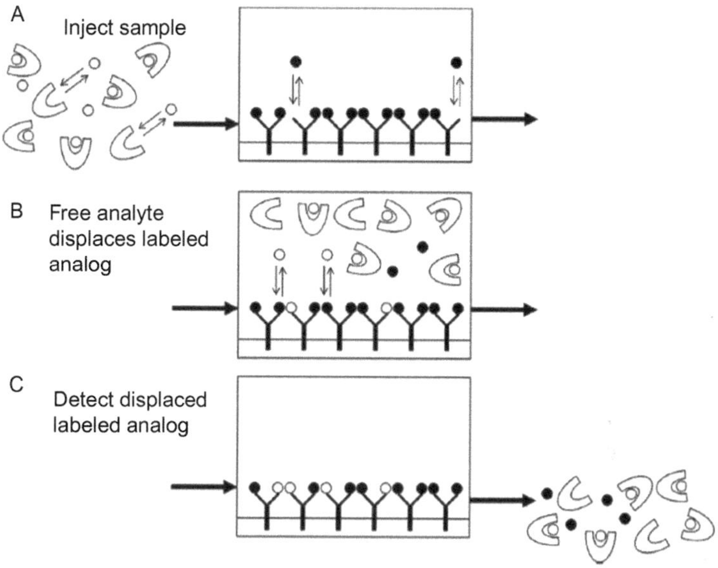

Figure 12 General scheme for a displacement immunoassay that is combined with ultrafast immunoextraction for the detection of a free drug or hormone. The dark circles represent the labeled analog of the drug/hormone, and the open circles represent the drug or hormone from the original sample. The column contains immobilized or adsorbed antibodies that can bind to both the labeled analog and the drug/hormone. Serum proteins in the sample are represented by the open "C"-shaped structures. *Reproduced with permission (Clarke et al., 2005).*

labeled analog of the target. A sample that contains the target is then injected onto this column. The free fraction of the target in the sample will be able to bind to antibodies as the labeled analog is momentarily dissociated from these binding agents. This results in a displacement peak for the labeled analog, which will have an area that is proportional to the target's free concentration. The labeled analog and retained analyte can later be eluted from the column and the system regenerated before the next sample is measured. This approach has been combined with chemiluminescent-labeled analogs for the measurement of free thyroxine in serum (Clarke et al., 2005). A similar technique has been used with near-infrared fluorescent-labeled analogs to measure the free fraction of phenytoin in serum (Ohnmacht et al., 2006).

An alternative strategy is to use ultrafast immunoextraction with a reverse displacement immunoassay (Schiel et al., 2011). This approach employs an affinity microcolumn that contains an immobilized analog of the target. A labeled binding agent (e.g., labeled antibodies or labeled F_{ab} fragments) is then applied to the affinity microcolumn. The sample is next injected and the free drug is allowed to displace some of the labeled binding agent from the column. This method has been utilized with near-infrared fluorescent labels to measure the free fraction of phenytoin in serum (Schiel et al., 2011). An advantage of this approach versus the normal displacement immunoassay format is that can potentially be used without the need for an elution and regeneration step for the affinity microcolumn between each sample injection.

5. CONCLUSION

This review examined the use of HPAC for the study of drug interactions with serum binding agents, within an emphasis on applications that were related to personalized medicine. The importance of drug–protein interactions in blood were first considered, along with ways in which such interactions may be affected by disease. The preparation of HPAC columns for the characterization of drug–protein interactions was then discussed. This included the general types of supports and immobilization methods that can be used to make such columns. The various types of columns that have been employed in such work were also described.

Several approaches were next considered that can be used in HPAC for drug–protein binding studies and personalized medicine. One of these methods was frontal analysis, which has been used in a number of reports to determine and compare the overall binding constants and number of sites for drugs with various binding agents in serum. These agents have included proteins such as HSA and AGP, modified forms of these proteins (e.g., *in vitro* or *in vivo* glycated HSA), and carrier agents such as lipoproteins. It was further shown how both simple and complex biological systems have been studied through the use of frontal analysis.

Other methods have also been used in HPAC for work that is related to drug–protein binding and personalized medicine. For instance, zonal elution has been used to evaluate HPAC columns, to estimate the affinity of drugs with entrapped or immobilized proteins, and to determine the binding constants for drugs and other solutes at specific sites on proteins or modified forms of these proteins. In addition, zonal elution has been employed as a tool to examine the individual or combined effects of several factors on drug–protein binding, such as the effects of both glycation and fatty acids on the interactions of drugs with HSA. The peak decay method and ultrafast affinity extraction have been utilized to examine the rates of drug–protein interactions. Ultrafast affinity extraction and chromatographic immunoassays have also been used to measure the free fractions of drugs and hormones in clinical samples and to examine the interactions of soluble proteins with these solutes.

There are several advantages to HPAC that have made it a promising technique in these reports for the study of drug–protein interactions and as a tool for personalized medicine. These advantages have included the speed and precision of this technique, its ability to be used with a variety of formats and detection modes, its ease of automation, and its ability to work with small amounts of samples and binding agents (Hage et al., 2009; Kim & Wainer, 2008; Schiel & Hage, 2009). This review has demonstrated how these features have led to the use of HPAC with a relatively wide variety of drugs and serum agents. These features have also made it possible for HPAC to be used directly with clinical samples and with protein samples that have been obtained from individual patients. As HPAC continues to be developed for such work, it is expected that even more applications and formats will continue to appear for this method in the field of personalized medicine.

ACKNOWLEDGMENTS

Portions of this work were supported by the NIH under grants R01 GM044931 and R01 DK069629, and the NSF/EPSCoR program under grant EPS-1004094. D.S. was supported by the University Grants Commission in India through a Raman Postdoctoral Fellowship.

REFERENCES

An, Y., Li, Q., Chen, J., Gao, X., Chen, H., Xiao, C., et al. (2014). Binding of caffeic acid to human serum albumin by the retention data and frontal analysis. *Biomedical Chromatography, 28*, 1881–1886.

Anguizola, J., Basiaga, S. B. G., & Hage, D. S. (2013). Effects of fatty acids and glycation on drug interactions with human serum albumin. *Current Metabolomics, 1*, 239–250.

Anguizola, J., Joseph, K. S., Barnaby, O. S., Matsuda, R., Alvarado, G., Clarke, W., et al. (2013). Development of affinity microcolumns for drug-protein binding studies in personalized medicine: Interactions of sulfonylurea drugs with *in vivo* glycated human serum albumin. *Analytical Chemistry, 85*, 4453–4460.

Anguizola, J., Matsuda, R., Barnaby, O. S., Hoy, K. S., Wa, C., Debolt, E., et al. (2013). Review: Glycation of human serum albumin. *Clinica Chimica Acta, 425*, 64–76.

Barklay, M. (1972). Lipoprotein class distribution in normal and disease states. In G. J. Nelson (Ed.), *Blood lipids and lipoproteins: Quantitation, composition, and metabolism* (pp. 587–603). New York: Wiley.

Barnaby, O. S., Wa, C., Cerny, R. L., Clarke, W., & Hage, D. S. (2010). Quantitative analysis of glycation sites on human serum albumin using $^{16}O/^{18}O$-labeling and matrix-assisted laser desorption/ionization time-of-flight mass spectrometry. *Clinica Chimica Acta, 411*, 1102–1110.

Barzegar, A., Moosavi-Movahedi, A. A., Sattarahmady, N., Hosseinpour-Fazi, M. A., Aminbakhsh, M., Ahmad, F., et al. (2007). Spectroscopic studies of the effects of glycation of human serum albumin on L-Trp binding. *Protein and Peptide Letters, 14*, 13–18.

Basiaga, S. B. G., & Hage, D. S. (2010). Chromatographic studies of changes in binding of sulfonylurea drugs to human serum albumin due to glycation and fatty acids. *Journal of Chromatography B, 878*, 3193–3197.

Board, P. G., Jones, I. M., & Bentley, A. K. (1986). Molecular cloning and nucleotide sequence of human alpha1 acid glycoprotein cDNA. *Gene, 44*, 127–131.

Bratlid, D. (1972). The effect of antimicrobial agents on bilirubin binding by human erythrocytes. *Scandinavian Journal of Clinical and Laboratory Investigation, 30*, 331–337.

Buchner, A. M., Blonski, W., & Lichtenstein, G. R. (2011). Update on the management of Crohn's disease. *Current Gastroenterology Reports, 13*, 465–474.

Cabrera, K. (2004). Applications of silica-based monolithic HPLC columns. *Journal of Separation Science, 27*, 843–852.

Calleri, E., Temporini, C., & Massolini, G. (2011). Frontal affinity chromatography in characterizing immobilized receptors. *Journal of Pharmaceutical and Biomedical Analysis, 54*, 911–925.

Ceciliani, F., & Pocacqua, F. (2007). The acute phase protein α_1-acid glycoprotein: A model for altered glycosylation during diseases. *Current Protein and Peptide Science, 8*, 91–108.

Chattopadhyay, A., Tian, T., Kortum, L., & Hage, D. S. (1998). Development of tryptophan-modified human serum albumin columns for site-specific studies of drug-protein interactions by high-performance affinity chromatography. *Journal of Chromatography B, 715*, 183–190.

Chen, J., Fitos, I., & Hage, D. S. (2006). Chromatographic analysis of allosteric effects between ibuprofen and benzodiazepines on human serum albumin. *Chirality, 18*, 24–36.

Chen, J., & Hage, D. S. (2004). Quantitative analysis of allosteric drug-protein binding by biointeraction chromatography. *Nature Biotechnology, 22*, 1445–1448.

Chen, J., & Hage, D. S. (2006a). Quantitative affinity chromatography: practical aspects. In D. S. Hage (Ed.), *Handbook of affinity chromatography* (2nd ed.). Boca Raton: CRC Press, 595–628.

Chen, J., & Hage, D. S. (2006b). Quantitative studies of allosteric effects by biointeraction chromatography: Analysis of protein binding to low solubility drugs. *Analytical Chemistry, 78*, 2672–2683.

Chen, J., Ohnmacht, C., & Hage, D. S. (2004). Studies of phenytoin binding to human serum albumin by high-performance affinity chromatography. *Journal of Chromatography B, 809*, 137–145.

Chen, J., Schiel, J. E., & Hage, D. S. (2009). Noncompetitive peak decay analysis of drug-protein dissociation by high-performance affinity chromatography. *Journal of Separation Science, 32*, 1632–1641.

Chen, S., Sobansky, M. R., & Hage, D. S. (2010). Analysis of drug interactions with high-density lipoprotein by high-performance affinity chromatography. *Analytical Biochemistry, 397*, 107–114.

Clarke, W., Beckwith, J. D., Jackson, A., Reynolds, B., Karle, E. M., & Hage, D. S. (2000). Antibody immobilization to high-performance liquid chromatographic supports: Characterization of maximum loading capacity for intact immunoglobulin G and Fab fragments. *Journal of Chromatography A, 888*, 13–22.

Clarke, W., Choudhuri, A. R., & Hage, D. S. (2001). Analysis of free drug fractions by ultrafast immunoaffinity chromatography. *Analytical Chemistry, 73*, 2157–2164.

Clarke, W., & Hage, D. S. (2001). Development of sandwich HPLC microcolumns for analyte adsorption on the millisecond time scale. *Analytical Chemistry, 73*, 1366–1373.

Clarke, W., Schiel, J. E., Moser, A., & Hage, D. S. (2005). Analysis of free hormone fractions by an ultrafast immunoextraction/displacement immunoassay: Studies using free thyroxine as a model system. *Analytical Chemistry*, *77*, 1859–1866.

Cottone, M., Orlando, A., & Renna, S. (2010). Investigational agents for Crohn's disease. *Expert Opinion on Investigational Drugs*, *19*, 1147–1159.

Dogan, A., Ozkara, S., Sari, M. M., Uzun, L., & Denizli, A. (2012). Evaluation of human interferon adsorption performance of Cibacron Blue F3GA attached cryogels and interferon purification by using FPLC system. *Journal of Chromatography B*, *893–894*, 69–76.

Durrington, P. N. (1989). Hyperlipidemia: diagnosis and management. In P. N. Durrington (Ed.), *Lipoproteins and lipids* (pp. 255–277). London: Wright.

Fanali, G., di Masi, A., Trezza, V., Marina, M., Fasano, M., & Ascenzi, P. (2012). Human serum albumin: From bench to bedside. *Molecular Aspects of Medicine*, *33*, 209–290.

Founier, T., Medjoubi, N. N., & Porquet, D. (2000). Alpha-1-acid glycoprotein. *Biochimica et Biophysica Acta*, *1482*, 157–171.

Gary, D. R., & Stroupe, D. S. (1978). Kinetics and mechanism of bilirubin binding to human serum albumin. *The Journal of Biological Chemistry*, *253*, 4370–4377.

Ghuman, J., Zunszain, P. A., Petitpas, I., Bhattacharya, A. A., Otagiri, M., & Curry, S. (2005). Structural basis of the drug-binding specificity of human serum albumin. *Journal of Molecular Biology*, *353*, 38–52.

Gunsburg, G. S., & Willard, H. F. (2009). Genomic and personalized medicine: Foundations and applications. *Translational Research*, *154*, 277–287.

Gustavsson, P. E., & Larsson, P. O. (2006). Support materials for affinity chromatography. In D. S. Hage (Ed.), *Handbook of affinity chromatography* (2nd ed.). Boca Raton: CRC Press. 15–33.

Hage, D. S. (2001). Chromatographic and electrophoretic studies of protein binding to chiral solutes. *Journal of Chromatography A*, *906*, 459–481.

Hage, D. S. (2002). High-performance affinity chromatography: A powerful tool for studying serum protein binding. *Journal of Chromatography B*, *768*, 3–30.

Hage, D. S., Anguizola, J., Barnaby, O., Jackson, A., Yoo, M. J., Papastavros, E., et al. (2011a). Characterization of drug interactions with serum proteins by using high-performance affinity chromatography. *Current Drug Metabolism*, *12*, 313–328.

Hage, D. S., Anguizola, J. A., Jackson, A. J., Matsuda, R., Papastavros, E., Pfaunmiller, E., et al. (2011b). Chromatographic analysis of drug interactions in the serum proteome. *Analytical Methods*, *3*, 1–24.

Hage, D. S., Jackson, A., Sobansky, M., Schiel, J. E., Yoo, M. J., & Joseph, K. S. (2009). Characterization of drug-protein interactions in blood using high-performance affinity chromatography. *Journal of Separation Science*, *32*, 835–853.

Hage, D. S., & Sengupta, A. (1998). Studies of protein binding to non-polar solutes by using zonal elution and high-performance affinity chromatography: Interactions of cis- and trans-clomiphene with human serum albumin in the presence of β-cyclodextrin. *Analytical Chemistry*, *70*, 4602–4609.

Hage, D. S., Xuan, H., & Nelson, M. A. (2006). Application and elution in affinity chromatography. In D. S. Hage (Ed.), *Handbook of affinity chromatography* (2nd ed.). Boca Raton: CRC Press, 79–97.

Harmony, J. A. K., Aleson, A. K., & McCarthy, B. M. (1986). Lipoprotein structure. In A. M. Scanu & A. A. Spector (Eds.), *Biochemistry and biology of plasma lipoproteins* (pp. 403–430). New York: Marcel Dekker.

Havel, R. J., & Kane, J. P. (1995). Structure and metabolism of plasma proteins. In C. R. Scriver, A. L. Beaudet, W. S. Sly, D. Valle, B. Childs, K. W. Kinzler, & B. Vogelstein (Eds.), *The metabolic and molecular basis of inherited disease* (pp. 1129–1138). New York: McGraw-Hill.

Hermanson, G. T., Mallia, A. K., & Smith, P. K. (1992). *Immobilized affinity ligand techniques* (pp. 137–279). New York: Academic Press.

Herve, F., Urien, S., Albengres, E., Duche, J. C., & Tillement, J. P. (1994). Drug binding in plasma: A summary of recent trends in the study of drug and hormone binding. *Clinical Pharmacokinetics, 26*, 44–58.

Hochepied, T., Wullaert, A., Berger, F. G., Baumann, H., Brouckaert, P., Steidler, L., et al. (2002). Overexpression of α_1-acid glycoprotein in transgenic mice leads to sensitization to acute colitis. *Gut, 51*, 398–404.

Iberg, N., & Fluckiger, R. (1986). Nonenzymatic glycosylation of albumin in vivo. Identification of multiple glycosylated sites. *Journal of Biological Chemistry, 261*, 13542–13545.

Israili, Z. H., & Dayton, P. G. (2001). Human alpha-1-glycoprotein and its interactions with drugs. *Drug Metabolism Reviews, 33*, 161–235.

Jackson, A. J., Anguizola, J., Pfaunmiller, E. L., & Hage, D. S. (2013). Use of entrapment and high-performance affinity chromatography to compare the binding of drugs and site-specific probes with normal and glycated human serum albumin. *Analytical and Bioanalytical Chemistry, 405*, 5833–5841.

Jackson, A. J., Xuan, H., & Hage, D. S. (2010). Entrapment of proteins in glycogen-capped and hydrazide-activated supports. *Analytical Biochemistry, 404*, 106–108.

Jiang, T., Mallik, R., & Hage, D. S. (2005). Affinity monoliths for ultrafast immunoextraction. *Analytical Chemistry, 77*, 2362–2372.

Johnson, L., Sarmiento, F., Blanc, W. A., & Day, R. (1959). Kernicterus in rats with an inherited deficiency of glucuronyl transferase. *American Journal of Diseases of Children, 97*, 591–608.

Jonas, A. (2002). Biochemistry of lipids, lipoproteins, and membranes. In D. E. Vance (Ed.), (pp. 483–504). Amsterdam: Elsevier.

Joseph, K. S., Anguizola, J., & Hage, D. S. (2011). Binding of tolbutamide to glycated human serum albumin. *Journal of Pharmaceutical and Biomedical Analysis, 54*, 426–432.

Joseph, K. S., Anguizola, J., Jackson, A. J., & Hage, D. S. (2010). Chromatographic analysis of acetohexamide binding to glycated human serum albumin. *Journal of Chromatography B, 878*, 2775–2781.

Joseph, K. S., & Hage, D. S. (2010a). The effects of glycation on the binding of human serum albumin to warfarin and L-tryptophan. *Journal of Pharmaceutical and Biomedical Analysis, 53*, 811–818.

Joseph, K. S., & Hage, D. S. (2010b). Characterization of the binding of sulfonylurea drugs to HSA by high-performance affinity chromatography. *Journal of Chromatography B, 878*, 1590–1598.

Joseph, K. S., Moser, A. C., Basiaga, S. B., Schiel, J. E., & Hage, D. S. (2009). Evaluation of alternatives to warfarin as probes for Sudlow site I of human serum albumin: Characterization by high-performance affinity chromatography. *Journal of Chromatography A, 1216*, 3492–3500.

Kalashnikova, I. L., Ivanova, N., & Tennikova, T. B. (2007). Macroporous monolithic layers as efficient 3-D microarrays for quantitative detection of virus-like particles. *Analytical Chemistry, 79*, 5173–5180.

Kantor, H. I., Sutherland, D. A., Leonard, J. T., Kamholz, J. H., Fry, N. D., & White, W. L. (1961). Effect on bilirubin metabolism in the newborn of sulfisoxazole administered to the mother. *Obstetrics & Gynecology, 17*, 494–500.

Kim, H. S., & Hage, D. S. (2005). Chromatographic analysis of carbamazepine binding to human serum albumin. *Journal of Chromatography B, 816*, 57–66.

Kim, H. S., & Hage, D. S. (2006). Immobilization methods for affinity chromatography. In D. S. Hage (Ed.), *Handbook of affinity chromatography* (2nd ed.). Boca Raton: CRC Press, 35–78.

Kim, H. S., Mallik, R., & Hage, D. S. (2006). Chromatographic analysis of carbamazepine binding to human serum albumin. II. Comparison of the Schiff base and N-hydroxysuccinimide immobilization methods. *Journal of Chromatography B, 837*, 138–146.

Kim, H. S., & Wainer, I. W. (2008). Rapid analysis of the interactions between drugs and human serum albumin (HSA) using high-performance affinity chromatography (HPAC). *Journal of Chromatography B, 870*, 22–26.

Koyama, H., Sugioka, N., Uno, A., Mori, S., & Nakajima, K. (1997). Effects of glycosylation of hypoglycemic drug binding to serum albumin. *Biopharmaceuticals and Drug Disposition, 18*, 791–801.

Kratochwil, N. A., Huber, W., Muller, F., Kansy, M., & Gerber, P. R. (2002). Predicting plasma protein binding of drugs: A new approach. *Biochemical Pharmacology, 64*, 1355–1374.

Kremer, J. M. H., Wilting, J., & Janssen, L. H. M. (1988). Drug binding to human $alpha_1$-acid glycoprotein in health and disease. *Pharmacological Reviews, 40*, 1–47.

Kuroda, Y., Matsumoto, S., Shibukawa, A., & Nakagawa, T. (2003). Capillary electrophoretic study on pH dependence of enantioselective disopyramide binding to genetic variants of human α_1-acid glycoprotein. *Analyst, 128*, 1023–1027.

Kwan, B. C. H., Kronenberg, F., Beddhy, S., & Cheung, A. K. (2007). Lipoprotein metabolism and lipid management in chronic kidney disease. *Journal of the American Society of Nephrology, 18*, 1246–1261.

Kwong, T. C. (1985). Free drug measurements: Methodology and clinical significance. *Clinical Chimica Acta, 151*, 193–216.

Lapolla, A., Fedele, D., Reitano, R., Bonfante, L., Guizzo, M., Seraglia, R., et al. (2005). Mass spectrometric study of in vivo production of advanced glycation end-products/peptides. *Journal of Mass Spectrometry, 40*, 969–972.

Lapolla, A., Fedele, D., Seraglia, R., & Traldi, P. (2006). The role of mass spectrometry in the study of non-enzymatic protein glycation in diabetes: An update. *Mass Spectrometry Reviews, 25*, 775–797.

Lindup, W. E. (1987). Plasma protein binding of drug—Some basic and clinical aspects. *Progress in Drug Metabolism, 10*, 141–185.

Loun, B., & Hage, D. S. (1992). Characterization of thyroxine-albumin binding using high-performance affinity chromatography. I. Interactions at the warfarin and indole sites of albumin. *Journal of Chromatography, 579*, 225–235.

Loun, B., & Hage, D. S. (1994). Chiral separation mechanisms in protein-based HPLC columns. I. Thermodynamic studies of (R)- and (S)- warfarin binding to immobilized human serum albumin. *Analytical Chemistry, 66*, 3814–3822.

Loun, B., & Hage, D. S. (1995). Characterization of thyroxine-albumin binding using high-performance affinity chromatography. II. Comparison of the binding of thyroxine, triiodothyronines and related compounds at the warfarin and indole sites of human serum albumin. *Journal of Chromatography B, 665*, 303–314.

Mahley, W. R., Innerarity, L. T., Rall, S. C., & Weisgraber, K. H. (1984). Plasma lipoproteins: Apolipoprotein structure and function. *Journal of Lipid Research, 25*, 1277–1294.

Mallik, R., & Hage, D. S. (2006). Affinity monolith chromatography. *Journal of Separation Science, 12*, 1686–1704.

Mallik, R., Wa, C., & Hage, D. S. (2007). Development of sulfhydryl-reactive silica for protein immobilization in high-performance affinity chromatography. *Analytical Chemistry, 79*, 1411–1424.

Mallik, R., Xuan, H., Guiochon, G., & Hage, D. S. (2008). Immobilization of $alpha_1$-acid glycoprotein for chromatographic studies of drug-protein binding II. Correction for errors in association constant measurements. *Analytical Biochemistry, 376*, 154–156.

Mallik, R., Xuan, H., & Hage, D. S. (2007). Development of an affinity silica monolith containing α_1-acid glycoprotein for chiral separations. *Journal of Chromatography A, 1149*, 294–304.

Mallik, R., Yoo, M. J., Briscoe, C. J., & Hage, D. S. (2010). Analysis of drug–protein binding by ultrafast affinity chromatography using immobilized human serum albumin. *Journal of Chromatography A, 1217*, 2796–2803.

Matsuda, R., Anguizola, J., Hoy, K. S., & Hage, D. S. (2015). Analysis of drug-protein interactions by high-performance affinity chromatography: Interactions of sulfonylurea drugs with normal and glycated human serum albumin. *Methods in Molecular Biology, 1286*, 255–277.

Matsuda, R., Anguizola, J., Joseph, K. S., & Hage, D. S. (2011). High-performance affinity chromatography and that analysis of drug interactions with modified proteins: Binding of gliclazide with glycated human serum albumin. *Analytical and Bioanalytical Chemistry, 401*, 2811–2819.

Matsuda, R., Anguizola, J., Joseph, K. S., & Hage, D. S. (2012). Analysis of drug interactions with modified proteins by high-performance affinity chromatography: Binding of glibenclamide to normal and glycated human serum albumin. *Journal of Chromatography A, 1265*, 114–122.

Matsuda, R., Kye, S., Anguizola, J., & Hage, D. S. (2014). Studies of drug interactions with glycated human serum albumin by high-performance affinity chromatography. *Reviews in Analytical Chemistry, 33*, 79–94.

Matsuda, R., Li, Z., Zheng, X., & Hage, D. S. (2015a). Analysis of glipizide binding to normal and glycated human serum albumin by high-performance affinity chromatography. *Analytical and Bioanalytical Chemistry, 407*, 5309–5321.

Matsuda, R., Li, Z., Zheng, X., & Hage, D. S. (2015b). Analysis of multi-site drug-protein interactions by high-performance affinity chromatography: Binding by glimepiride to normal or glycated human serum albumin. *Journal of Chromatography A, 1408*, 133–144.

Mbewu, A. D., & Durrington, P. N. (1990). Lipoprotein(a): Structure, properties, and possible involvement in thrombogenesis and atherogenesis. *Atherosclerosis, 85*, 1–14.

Mendez, D. L., Jensen, R. A., McElroy, L. A., Pena, J. M., & Esquerra, R. M. (2005). The effect of non-enzymatic glycation on the unfolding of human serum albumin. *Archives of Biochemistry and Biophysics, 444*, 92–99.

Moaddel, R., Bullock, P. L., & Wainer, I. W. (2004). Development and characterization of an open tubular column containing immobilized P-glycoprotein for rapid on-line screening for P-glycoprotein substrates. *Journal of Chromatography B, 799*, 255–263.

Mohamadi-Nejad, A., Moosavi-Movahedi, A. A., Hakimelahi, G. H., & Sheibani, N. (2002). Thermodynamic analysis of human serum albumin interactions with glucose: Insights into the diabetic range of glucose concentration. *The International Journal of Biochemistry & Cell Biology, 34*, 1115–1124.

Moore, R. M., & Walters, R. R. (1987). Peak-decay method for the measurement of dissociation rate constants by high-performance affinity chromatography. *Journal of Chromatography, 384*, 91–103.

Morrisett, D. J., Jackson, L. R., & Gotto, M. A. (1975). Lipoproteins: Structure and function. *Annual Review of Biochemistry, 44*, 183–207.

Nakajou, K., Watanabe, H., Kragh-Hansen, U., Maruyama, T., & Otagiri, M. (2003). The effect of glycation on the structure, function and biological fate of human serum albumin as revealed by recombinant mutants. *Biochimica et Biophysica Acta, 1623*, 88–97.

Natarajan, P., Ray, K. K., Christopher, P., & Cannon, C. P. (2010). High-density lipoprotein and coronary heart disease: Current and future therapies. *Journal of the American College of Cardiology, 55*, 1283–1299.

Nelson, M. A., Moser, A., & Hage, D. S. (2010). Biointeraction analysis by high-performance affinity chromatography: Kinetic studies of immobilized antibodies. *Journal of Chromatography B, 878*, 165–171.

Noctor, T. A. G., Wainer, I. W., & Hage, D. S. (1992). Allosteric and competitive displacement of drugs from human serum albumin by octanoic acid, as revealed by high-performance liquid affinity chromatography on a human serum albumin-based stationary phase. *Journal of Chromatography*, *577*, 305–315.

Odell, G. B. (1959). Studies in kernicterus. I. The protein binding of bilirubin. *Journal of Clinical Investigation*, *38*, 823–833.

Ohnmacht, C. M., Schiel, J. E., & Hage, D. S. (2006). Analysis of free drug fractions using near-infrared fluorescent labels and an ultrafast immunoextraction/displacement assay. *Analytical Chemistry*, *78*, 7547–7556.

Otagiri, M. (2005). A molecular functional study on the interactions of drugs with plasma proteins. *Drug Metabolism and Pharmacokinetics*, *20*, 309–323.

Patel, S., Wainer, I. W., & Lough, W. J. (2006). Chromatographic studies of molecular recognition and solute binding to enzymes and plasma proteins. In D. S. Hage (Ed.), *Handbook of affinity chromatography* (2nd ed.). Boca Raton: CRC Press, 663–683.

Peters, T. (1996). *All about albumin: Biochemistry, genetics, and medical applications*. San Diego: Academic Press.

Pfaunmiller, E., Moser, A. C., & Hage, D. S. (2012). Biointeraction analysis of immobilized antibodies and related agents by high-performance immunoaffinity chromatography. *Methods*, *56*, 130–135.

Plantonova, G. A., Pankova, G. A., Ill'ina, I. Y., Vlasosv, G. P., & Tennikova, T. B. (1999). Quantitative fast fractionation of a pool of polyclonal antibodies by immunoaffinity memebrane chromatography. *Journal of Chromatography A*, *852*, 129–140.

Ruhn, P. F., Garver, S., & Hage, D. S. (1994). Development of dihydrazide-activated silica supports for high-performance affinity chromatography. *Journal of Chromatography A*, *669*, 9–19.

Sakai, T., Maruyama, T., Sako, T., Ahmed, S., Zuidema, X., Fujiyama, S., et al. (1999). Stereoselective serum protein binding of ketoprofen in liver disease. *Enantiomer*, *4*, 477–482.

Schiel, J. E., & Hage, D. S. (2009). Kinetic studies of biological interactions by affinity chromatography. *Journal of Separation Science*, *32*, 1507–1522.

Schiel, J. E., Tong, Z., Sakulthaew, C., & Hage, D. S. (2011). Development of a flow-based ultrafast immunoextraction and reverse displacement immunoassay: Analysis of free drug fractions. *Analytical Chemistry*, *83*, 9384–9390.

Schmid, K., Nimberg, R. B., Kimura, A., Yamaguchi, H., & Binette, J. P. (1977). The carbohydrate unites of human plasma α_1-acid glycoprotein. *Biochimica et Biophysica Acta*, *492*, 291–302.

Schonfeld, D. L., Ravelli, R. B. G., Mueller, U., & Skerra, A. (2008). The 1.8-Å crystal structure α_1-acid glycoprotein (orosomucoid) solved by UV RIP reveals the broad drug-binding activity of this human plasma lipocalin. *Journal of Molecular Biology*, *384*, 393–405.

Schriemer, D. C. (2004). Biosensor alternative: Frontal affinity chromatography. *Analytical Chemistry*, *76*, 440A–448A.

Seedher, N., & Agarwal, P. (2013). Competitive binding of fluoroquinolone antibiotics and some other drugs to human serum albumin: A luminescence spectroscopic study. *Luminescence*, *28*, 562–568.

Seedher, N., & Kanojia, M. (2008). Reversible binding of antidiabetic drugs, repaglinide and gliclazide, with human serum albumin. *Chemical Biology & Drug Design*, *72*, 290–296.

Seedher, N., & Kanojia, M. (2013). Fluorescence spectroscopic study for competitive binding of antidiabetic drugs and endogenous substances on serum albumin. *Drug Metabolism and Personalized Therapy*, *28*, 107–114.

Sengupta, A., & Hage, D. S. (1999a). Characterization of the binding of digitoxin and acetyldigitoxin to human serum albumin by high-performance affinity chromatography. *Journal of Chromatography B*, *725*, 91–100.

Sengupta, A., & Hage, D. S. (1999b). Characterization of minor site probes for human serum albumin by high-performance affinity chromatography. *Analytical Chemistry, 71*, 3821–3827.

Simard, J. R., Zunszain, P. A., Hamilton, J. A., & Curry, S. (2006). Location of high and low affinity fatty acid binding sites on human serum albumin revealed by NMR drug-competition analysis. *Journal of Molecular Biology, 361*, 336–351.

Skipski, V. R. (1972). Plasma lipoproteins: composition, structure and biochemistry. In G. J. Nelson (Ed.), *Blood lipids and lipoproteins: Quantitation, composition, and metabolism* (pp. 471–583). New York: Wiley.

Sobansky, M. R., & Hage, D. S. (2012a). Identification and analysis of stereoselective drug interactions with low-density lipoprotein by high-performance affinity chromatography. *Analytical and Bioanalytical Chemistry, 403*, 563–571.

Sobansky, M. R., & Hage, D. S. (2012b). Analysis of drug interactions with lipoproteins by high-performance affinity chromatography. In L. V. Berhardt (Ed.), *Advances in medicine and biology*. Vol. 53. Hauppauge: Nova Science, 199–216.

Sobansky, M. R., & Hage, D. S. (2014). Analysis of drug interactions with very low density lipoprotein by high-performance affinity chromatography. *Analytical and Bioanalytical Chemistry, 406*, 6203–6211.

Soman, S., Yoo, M. J., Jang, Y. J., & Hage, D. S. (2010). Analysis of lidocaine interactions with serum proteins using high-performance affinity chromatography. *Journal of Chromatography B, 878*, 705–708.

Spector, A. A. (1976). Fatty acid binding to plasma albumin. *Journal of Lipid Research, 16*, 165–179.

Sudlow, G., Birkett, D. J., & Wade, D. N. (1975). The characterization of two specific drug binding sites on human serum albumin. *Molecular Pharmacology, 11*, 824–832.

Tietz, N. W. (1986). *Textbook of clinical chemistry*. Philadelphia: Saunders.

Trusheim, M. R., Berndt, E. R., & Douglas, F. L. (2007). Stratified medicine strategic and economic implications of combining drugs and clinical biomarkers. *Nature Reviews Drug Discovery, 6*, 287–293.

Tweed, S. A., Loun, B., & Hage, D. S. (1997). Effects of ligand heterogeneity in the characterization of affinity columns by frontal analysis. *Analytical Chemistry, 69*, 4790–4798.

Van Dijk, W., Havenaar, E. C., & Brinkman-van der Linden, E. C. (1995). Alpha$_1$-acid glycoprotein (orosomucoid): Pathophysiological changes in glycosylation in relation to its function. *Glycoconjugate Journal, 12*, 227–233.

Vidal, P., Nielsen, E., & Welinder, B. S. (1992). Effect of glycation on the heterogeneity of human serum albumin analysed by reversed-phase high-performance liquid chromatography in a solvent containing formic acid. *Journal of Chromatography, 573*, 201–206.

Vuignier, K., Schappler, J., Veuthey, J. L., Carrupt, P. A., & Martel, S. (2010). Drug–protein binding: A critical review of analytical tools. *Analytical and Bioanalytical Chemistry, 398*, 53–66.

Wa, C., Cerny, R. L., Clarke, W. A., & Hage, D. S. (2007). Characterization of glycation adducts on human serum albumin by matrix-assisted laser desorption/ionization time-of-flight. *Clinica Chimica Acta, 385*, 48–60.

Wa, C., Cerny, R. L., & Hage, D. S. (2006). Obtaining high sequence coverage in matrix-assisted laser desorption time-of-flight mass spectrometry for studies of protein modification: Analysis of human serum albumin as a model. *Analytical Biochemistry, 349*, 229–241.

Walters, R. R. (1987). Practical approaches for the measurement of rate constants by affinity chromatography. In I. M. Chaiken (Ed.), *Analytical affinity chromatography*. Boca Raton: CRC Press, 117–156.

Wang, T., Xiang, B., Wang, Y., Chen, C., Dong, Y., Fang, H., et al. (2008). Spectroscopic investigation on the binding of bioactive pyridazinone derivative to human serum albumin and molecular modeling. *Colloids and Surfaces B: Biointerfaces, 65*, 113–119.

Wasan, K. M., & Cassidy, S. M. (1998). Role of plasma lipoproteins in modifying the biological activity of hydrophobic drugs. *Journal of Pharmaceutical Sciences*, *87*, 411–424.

Xuan, H., & Hage, D. S. (2005). Immobilization of alpha$_1$-acid glycoprotein for chromatographic studies of drug-protein binding. *Analytical Biochemistry*, *346*, 300–310.

Xuan, H., Joseph, K. S., Wa, C., & Hage, D. S. (2010). Biointeraction analysis of carbamazepine binding to alpha$_1$-acid glycoprotein by high-performance affinity chromatography. *Journal of Separation Science*, *33*, 2294–2301.

Yoo, M. J., & Hage, D. S. (2010). Affinity monolith chromatography: principles and recent developments. In P. Wang (Ed.), *Monolithic chromatography and its modern applications*. St. Albans: ILM, 3–25.

Yoo, M. J., & Hage, D. S. (2011a). Use of peak decay analysis and affinity microcolumns containing silica monoliths for rapid determination of drug-protein dissociation rates. *Journal of Chromatography A*, *1218*, 2072–2078.

Yoo, M. J., & Hage, D. S. (2011b). High-throughput analysis of drug dissociation from serum proteins using affinity silica monoliths. *Journal of Separation Science*, *34*, 2255–2263.

Yoo, M. J., Schiel, J. E., & Hage, D. S. (2010). Evaluation of affinity microcolumns containing human serum albumin for rapid analysis of drug-protein binding. *Journal of Chromatography B*, *878*, 1707–1713.

Yoo, M. J., Smith, Q. R., & Hage, D. S. (2009). Studies of imipramine binding to human serum albumin by high-performance affinity chromatography. *Journal of Chromatography B*, *877*, 1149–1154.

Zacharis, C. K., Kalaitzantonakis, E. A., Podgornik, A., & Theodoridis, G. (2006). Sequential injection affinity chromatography utilizing an albumin immobilized monolithic column to study drug-protein interactions. *Journal of Chromatography A*, *1144*, 126–134.

Zamet, P., & Chunga, F. (1971). Separation by gel filtration and microdetermination of unbound bilirubin. II. Study of sera in icteric newborn infants. *Acta Paediatrica*, *60*, 33–38.

Zhang, Q., Ames, J. M., Smith, R. D., Baynes, J. W., & Metz, T. O. (2009). A perspective on the Maillard reaction and the analysis of protein glycation by mass spectrometry: Probing the pathogenesis of chronic disease. *Journal of Proteome Research*, *8*, 754–769.

Zheng, X., Li, Z., Beeram, S., Podariu, M., Matsuda, R., Pfaunmiller, E. L., et al. (2014). Analysis of biomolecular interactions using affinity microcolumns: A review. *Journal of Chromatography B*, *968*, 49–63.

Zheng, X., Li, Z., Podariu, M. I., & Hage, D. S. (2014). Determination of rate constants and equilibrium constants for solution-phase drug-protein interactions by ultrafast affinity extraction. *Analytical Chemistry*, *86*, 6454–6460.

Zheng, X., Matsuda, R., & Hage, D. S. (2014). Analysis of free drug fractions by ultrafast affinity extraction: Interactions of sulfonylurea drugs with normal or glycated human serum albumin. *Journal of Chromatography A*, *1371*, 82–89.

Zheng, X., Yoo, M. J., & Hage, D. S. (2013). Analysis of free fractions for chiral drugs using ultrafast extraction and multi-dimensional high-performance affinity chromatography. *Analyst*, *138*, 6262–6265.

Ziegler, A., Koch, A., Krockenberger, K., & Grobhennig, A. (2012). Personalized medicine using DNA biomarkers: A review. *Human Genetics*, *131*, 1627–1638.

CHAPTER TWO

Role of Proteomics in the Development of Personalized Medicine☆

Kewal K. Jain[1]
Jain PharmaBiotech, Basel, Switzerland
[1]Corresponding author: e-mail address: jain@pharmabiotech.ch

Contents

Abstract

Advances in proteomic technologies have made import contribution to the development of personalized medicine by facilitating detection of protein biomarkers, proteomics-based molecular diagnostics, as well as protein biochips and pharmacoproteomics. Application of nanobiotechnology in proteomics, nanoproteomics, has further enhanced applications in personalized medicine. Proteomics-based molecular diagnostics will have an important role in the diagnosis of certain conditions and understanding the pathomechanism of disease. Proteomics will be a good bridge between diagnostics and therapeutics; the integration of these will be important for advancing personalized medicine. Use of proteomic biomarkers and combination of pharmacoproteomics with pharmacogenomics will enable stratification of clinical trials and improve monitoring of patients for development of personalized therapies. Proteomics is an important component of several interacting technologies used for development of personalized

☆Submitted on 1 September 2015 for publication in the thematic volume on "Personalized Medicine" of the "Advances in Protein Chemistry and Structural Biology" published by Elsevier.

Advances in Protein Chemistry and Structural Biology, Volume 102
ISSN 1876-1623
http://dx.doi.org/10.1016/bs.apcsb.2015.09.002

medicine, which is depicted graphically. Finally, cancer is a good example of applications of proteomic technologies for personalized management of cancer.

1. INTRODUCTION

Personalized medicine simply means the prescription of specific therapeutics best suited for an individual. It is usually based on pharmacogenetic, pharmacogenomic, transcriptomic, pharmacoproteomic, and pharmacometabolomic information, but other individual variations in patients and environmental factors are also taken into consideration (Jain, 2015). Two terms sometimes used to describe personalized medicine, genomic medicine and precision medicine, do not adequately cover the contribution of other omics to personalized medicine. This chapter will deal with the role of proteomics in personalized medicine starting with the definition of proteomics.

The term "proteomics" indicates PROTEins expressed by a genOME and is the systematic analysis of protein profiles of tissues (Wilkins, Sanchez, Gooley, et al., 1995). The term "proteome" refers to all proteins produced by a species, much as the genome is the entire set of genes. Unlike the genome, the proteome varies with time and is defined as "the proteins present in one sample (tissue, organism, cell culture) at a certain point in time." Proteomics parallels the related field of genomics. Now that the human genome has been sequenced, we face the greater challenge of making use of this information for improving healthcare and discovering new drugs. There is an increasing interest in proteomics technologies now because DNA sequence information provides only a static snapshot of the various ways in which the cell might use its proteins, whereas the life of the cell is a dynamic process. With this background, DNA/RNA sequences, *per se*, are not enough for the clear identification of a therapeutic target because proteins and not DNA/RNA are the basis of mode of action of drugs. Figuring out which proteins are involved in which disorders and in which individuals with any given disorder holds the promise of highly individualized therapeutics. Advances in proteomic technologies have made a very import contribution of development of personalized medicine by facilitating detection of protein biomarkers, development of proteomics-based molecular diagnostics, and pharmacoproteomics, which will be discussed in the following sections.

2. PROTEIN BIOMARKERS

There are several definitions of biomarkers. A biomarker is a characteristic that can be objectively measured and evaluated as an indicator of a physiological as well as a pathological process or pharmacological response to a therapeutic intervention. Classical biomarkers are measurable alterations in blood pressure, blood lactate levels following exercise, and blood glucose in diabetes mellitus. Any specific molecular alteration of a cell on DNA, RNA, metabolite, or protein level can be referred to as a molecular biomarker. In the era of molecular biology, biomarkers usually mean molecular biomarkers and can be divided into three broad categories (Jain, 2010):

1. Those that track disease progression over time and correlate with known clinical measures.
2. Those that detect the effect of a drug.
3. Those that serve as surrogate endpoints in clinical trials.

A biomarker can be as simple as a laboratory test or as complex as a pattern of proteins. From a practical point of view, the biomarker would specifically and sensitively reflect a disease state and could be used for diagnosis as well as for disease monitoring during and following therapy. For a biomarker to be valid, it should be measurable in a test system with established performance characteristics, and there should be reliable evidence for its clinical significance. Biomarkers are closely linked to development of personalized medicine, as they may serve as the common basis for diagnosis and drug discovery. Protein biomarkers are useful for early detection of disease to implement preventive strategies. They can also be used for monitoring of therapy.

2.1 Proteomic Technologies for Discovery of Biomarkers

Although analysis of proteins has been an integral part of the field of clinical chemistry for decades, recent advances in technology and complete identification of the human genome sequence have opened up new opportunities for analysis of proteins for clinical diagnostic purposes. New analytical methods allow the simultaneous analysis of a large number of proteins in biological fluids such as serum and plasma, offering partial views of the complete set of proteins or proteome. Besides two-dimensional gel electrophoresis and mass spectrometry (MS), several proteomics approaches have been used to identify novel biomarkers and these are described in detail in a special report on this topic (Jain, 2015a). Selected technologies relevant to biomarkers are shown in Table 1.

Table 1 Proteomic Technologies Used for Biomarker Detection

Technology	**Application Relevant to Biomarkers**
Antibody-based affinity chromatography	Removal of high-abundant proteins from blood for characterization of protein biomarkers (Haudenschild, Eldridge, Lein, & Chromy, 2014)
CyTOF (Fluidigm) is mass cytometry TOF using transition element isotopes to label antibodies	To quantify protein expression levels on a single-cell level and deciphering a drug's pharmacokinetics and pharmacodynamics relationship (Atkuri, Stevens, & Neubert, 2015)
Inductively coupled plasma (ICP)-MS as supplement to stable isotope tagging methods	Absolute protein quantification using suitable elemental-based tags (Chahrour, Cobice, & Malone, 2015)
Isotope-coded affinity tag (ICAT) peptide labeling	Combines accurate quantification and concurrent sequence identification of individual proteins in complex mixtures
Linear ion trap-Fourier transform (LTQ-FT) + linear ion trap (LTQ)-orbitrap MS	For in-depth characterization of biomarkers in body fluids and analysis of the human urinary proteome
Liquid chromatography–mass spectrometry (MS)/MS	Improves the throughput and reliability of peptide mapping
Lucid Proteomics System™ (Bio-Rad) combines surface-enhanced laser desorption/ionization (SELDI) and time-of-flight (TOF) technologies	Combines top-down and bottom-up protein profiling for protein biomarker discovery (Jourdain, Bulman, & Dalmasso, 2010)
MASStermind™ (MyCartis) is a biomarker discovery engine-based proteomics technologies	Fast, automated and flexible, and sensitive technique without using gels or affinity tags
Matrix-assisted laser desorption/ ionization MS (MALDI-MS)	For biomarker discovery in complex mixtures while maintaining the spatial localization of the proteins detected
MS with arrays containing nanopores with specific affinities for single biomarkers	Interaction between a nanometer-scale pore and analytes for real-time characterization of protein biomarkers
Multiple reaction monitoring MS (MRM-MS) assays	Used clinically for quantification of protein biomarkers, including biomarker panels (Hunter & Paramithiotis, 2010)

Table 1 Proteomic Technologies Used for Biomarker Detection—cont'd

Technology	Application Relevant to Biomarkers
ProMag™ (Bangs Laboratories), a magnetic 3 μ bead with hydrophilic surface to reduce nonspecific binding, is coated with a ligand of interest	To identify protein biomarkers faster, more efficiently, and at a lower cost than other methods
Protein Tomography™ for imaging of molecular events	Ideal method for study of proteins as biomarkers by enabling visualization of the different states of the same molecule
ProteoCarta® (Caprion Proteomics') integrated proteomics discovery platform	Disease target identification, particularly cancer, and predictive medicine
Selected reaction monitoring (SRM) MS	To determine the relative quantitation of several plasma proteins commonly observed in biomarker studies over various time courses (Randall, McKay, Pascovici, et al., 2012)
Serum protein pattern profiling by protein biochip SELDI-TOF MS with bioinformatics	Biomarkers of diffuse large B-cell lymphoma to predict response to therapy (Xu, Hu, He, et al., 2015)
Stable isotope standards and capture by antipeptide antibodies (SISCAPA) combines immuno-enrichment with SRM	SISCAPA-SRM is a sensitive assay for detecting low-abundance proteins in plasma and is aimed at biomarker quantification (Katafuchi, Esterházy, Lemoff, et al., 2015)
Stable isotope tagging methods: several samples tagged with different numbers of stable isotopes are mixed to eliminate sample-to-sample recovery differences	For determining the relative expression level of individual proteins between samples in a mass spectrometer with high precision
Tandem mass spectrometers	Accurate measurements in complex biological mixtures in femtomolar without chromatography

3. PROTEIN BIOCHIPS

Most of the biochips use nucleic acids as information molecules but protein chips are also proving to be useful. Miniaturized and highly parallel immunoassays greatly improve efficiency by increasing the amount of information acquired with single examination and reduce cost by decreasing reagent consumption.

Two classes of protein microarrays are currently available: analytical and functional protein microarrays. In the case of analytical protein microarrays, well-characterized molecules with specific activity, such as antibodies, peptide-MHC complexes, or lectins, are used as immobilized probes. These arrays have become one of the most powerful multiplexed detection platforms. Functional protein microarrays are being increasingly applied to many areas of biological discovery, including drug target identification/validation and studies of protein interaction, biochemical activity, and immune responses. Great progress has been achieved in both classes of protein microarrays in terms of sensitivity and specificity, and new protein microarray technologies are continuing to emerge. Microfluidic "laboratory-on-a-chip" is another practical microarray technology for low cost POC use. The advent of miniaturized immunoassays and gene microarray technology has led to the development of reverse phase protein array technology, which provides low-throughput as well as high-throughput methods for quantifying proteins and their posttranslationally modified forms. Finally, protein microarrays have found novel applications in both scientific research and clinical diagnostics. The most important applications are in protein biomarker discovery. Protein microarray technology is a powerful tool for biomarker discovery and validation.

Protein array technology can also provide versatile tools for functional genomics, enabling the translation of gene expression patterns of normal and diseased tissues into proteins. Protein function, such as enzyme activity, antibody specificity, or other ligand–receptor interactions and binding of nucleic acids or small molecules can be analyzed on a whole-genome level. As the array technology develops, an ever-increasing variety of formats become available (e.g., nanoplates, patterned arrays, 3D pads, flat-surface spot arrays, or microfluidic chips), and proteins can be arrayed onto different surfaces (e.g., membrane filters, polystyrene film, glass, silane, or gold). Various techniques are being developed for producing arrays, and robot-controlled, pin-based, or inkjet printing heads are the preferred tools for manufacturing protein arrays. CCD cameras or laser scanners are used for signal detection; atomic force microscopy and mass spectrometry are upcoming options. The emerging future array systems will be used for high-throughput functional annotation of gene products, their involvements in molecular pathways, and their response to medical treatment and become the physician's indispensable diagnostics tools.

3.1 Role of Protein Biochips in Personalized Medicine

Protein biochips/microarrays are well-established tools for research and some products for *in vitro* diagnostics are available commercially. Profiling proteins on biochips will be useful for distinguishing the proteins of normal cells from early-stage cancer cells and from malignant metastatic cancer cells. In comparison with the DNA microarrays, the protein microarrays/chips offer the possibility of developing a rapid global analysis of the entire proteome leading to protein-based diagnostics and therapeutics. Of all the applications of protein microarrays, molecular diagnostics is most clinically relevant and would fit in with the coming trend in personalized medicine. These technologies have an advantage in diagnosis of some conditions. For example, different proteins such as antibodies, antigens, and enzymes can be immobilized within protein biochips. Protein microarrays are reliable tools for detection of multiple biomarkers with only a minimal quantity of sample and have enormous potential in applications for personalized medicine. Reverse phase protein arrays are being used in cancer clinical trials for profiling and comparing the functional state of protein signaling pathways within tumors, between patients, or within the same patients before/after treatment, thus facilitating development of personalized medicine (Gallagher & Espina, 2014).

3.2 Nanoproteomics

Nanoproteomics—application of nanobiotechnology to proteomics—improves on most current protocols including protein purification and automated identification schemes that yield unacceptably low recoveries with reduced sensitivity and speed while requiring more starting material (Jain, 2012). Low-abundant proteins and proteins that can only be isolated from limited source material (e.g., biopsies) can be subjected to nanoscale protein analysis—nanocapture of specific proteins and complexes, and optimization of all subsequent sample handling steps leading to analysis of peptide fragments. This is a focused approach, also termed targeted proteomics, and involves examination of subsets of the proteome, e.g., those proteins that are either specifically modified, or bind to a particular DNA sequence, or exist as members of higher order complexes, or any combination thereof. Nanoproteomics has refined single molecule proteomics. Nanoproteomic technologies will further enhance applications of proteomics to the development of personalized medicine. Nanoproteomics can be combined with nanobiochips. Applications of nanoproteomics relevant to personalized medicine include amplified immunoassays for detection of biomarkers and targeted therapy with automatically adjusted drug delivery (Kobeissy, Gulbakan, Alawieh, et al., 2014).

Nucleic acid programmable protein arrays with a cell-free expression system for production of proteins *in situ* can be used to quantify regulatory protein networks in humans. Label-free alternative to fluorescent-labeled approaches appears to be capable of analyzing protein function and protein–protein interaction in studies that are relevant to personalized medicine (Nicolini, Bragazzi, & Pechkova, 2012).

4. ROLE OF PROTEOMICS-BASED MOLECULAR DIAGNOSTICS IN PERSONALIZED MEDICINE

Proteomic technologies play an important role in molecular diagnostics, which is an important basis of personalized medicine. Proteodiagnostics have certain advantages over nucleic acid diagnostics and complement the latter. Refinement of molecular diagnosis enables a better selection of appropriate therapy. POC and companion diagnostics are important components of personalized medicine.

Proteomics had a great impact on diagnosis during the first decade of the twenty-first century. By the end of the second decade, knowledge gained from genomics and proteomics will be combined to provide optimal detection of disease at an early stage for prevention or early intervention. Proteomics-based molecular diagnostics will have an important role in the diagnosis of certain conditions and proteomics-based medicines would be integrated in the total healthcare of a patient.

4.1 Integration of Diagnostics with Therapeutics

Integration of diagnosis with therapeutics is an important feature of personalized medicine. A misconstructed word "theranostics" has been introduced as a key word in literature. Originally meant to describe therapy guided by diagnosis, it is not widely understood and is often misused. Several combinations of diagnosis with therapy include: (1) use of a diagnostic to select suitable therapeutic; (2) use of a biomarker for diagnosis as well as a drug target; (3) use of a diagnostic (usually based on a biomarker) to monitor effects of therapy; and (4) use of a diagnostic such as a nanoparticle–drug–antibody complex to localize a lesion and for its use as a therapeutic after it has accumulated in the lesion. This is of particular value in the treatment of deep-seated tumors. Proteomics will be a good bridge between diagnostics and therapeutics.

5. PHARMACOPROTEOMICS

Pharmacoproteomics is parallel to pharmacogenomics and is used for subtyping patients on the basis of protein analysis. Proteomics-based characterization of multifactorial diseases may help to match a particular target-based therapy to a particular biomarker in a subgroup of patients. By classifying patients as responders and nonresponders, this approach may accelerate the drug development process. Proteomic technologies are already being integrated into the drug discovery process as complimentary to genomic approaches. This offers the possibility of utilizing information from the genome, expressed mRNAs and their respective proteins as well as subcellular localization. By focusing on protein activity levels or expression levels, it is possible to learn more about the role proteins play in causing and treating disease. By combining pharmacoproteomics and pharmacogenomics, it will be possible to define smaller, more rapidly responsive clinical trial populations and to monitor drug responses more effectively. This will lead to reduction in trial time and cost of drug development. As a result of these developments, pharmacological therapy is likely to become increasingly tailored to the genetic and molecular profiles of specific patient groups.

5.1 Pharmacoproteomics and Personalized Medicine

Pharmacoproteomics is a more functional representation of patient-to-patient variation than that provided by genotyping. Because it includes the effects of posttranslational modification, pharmacoproteomics connects the genotype with the phenotype—a connection that is not always predicted by genotyping alone. For example, a silent SNP can give rise to two or more variants forms of mRNAs that do not produce an altered amino acid sequence in the proteins that are encoded, but it can alter a phenotype by inducing change in the mRNA folding. An understanding of mRNA conformational changes could lead to new drug targets such as an allele-specific target.

Proteomics-based characterization of multifactorial diseases may help to match a particular target-based therapy to a particular biomarker in a subgroup of patients. The commercial pharmaceutical sector is taking a lead in developing this area. Individualized therapy may be based on differential protein expression rather than a genetic polymorphism.

Advantages of use of pharmacoproteomics in personalized medicine are:

- Pharmacoproteomics is a more functional representation of patient-to-patient variation than that provided by genotyping.
- Because it includes the effects of posttranslational modification, pharmacoproteomics connects the genotype with the phenotype.
- By classifying patients as responders and nonresponders, this approach may accelerate the drug development process.
- Protein biomarkers facilitate integration of diagnostics with therapeutics from development stages to translation into clinical applications.

In the future, this clinical proteomics approach may help to improve patient selection, ensure more precise clinical phenotyping for clinical trials, and individualize patient treatment.

An example of clinical application of pharmacoproteomics is cerebrovascular disease, which often involves damage to multiple organs and systems. For example, an ischemic stroke related to patent foramen ovale involves not only the brain but also the heart, the lung, and the peripheral vascular circulation. Due to lack of tools for patient selection or for monitoring therapeutic efficacy of treatments such as therapeutic hypothermia and intravenous thrombolysis treatments, the use of these treatments is underutilized (Ning, Lopez, Sarracino, et al., 2013). Clinical proteomic approaches may be promising for monitor systemic changes in complex multiorgan diseases—particularly those where the disease process can be assessed in clinically accessible fluid samples such as blood, urine, and cerebrospinal fluid.

6. CONCLUDING REMARKS ABOUT APPLICATION OF PROTEOMICS FOR PERSONALIZED MEDICINE

Proteomics is an important component of several interacting technologies used for development of personalized medicine as shown in Fig. 1.

Proteomics plays an important role in systems biology because most biological systems involve proteins. Proteins that are disturbed by disease and gene regulatory networks differ from their normal counterparts, and these differences may be detected by multiparameter measurements of the blood. This will have a major role in creating a predictive, personalized, preventive, and participatory approach to medicine.

A good example of application of proteomics for personalized medicine is cancer as shown in Table 2.

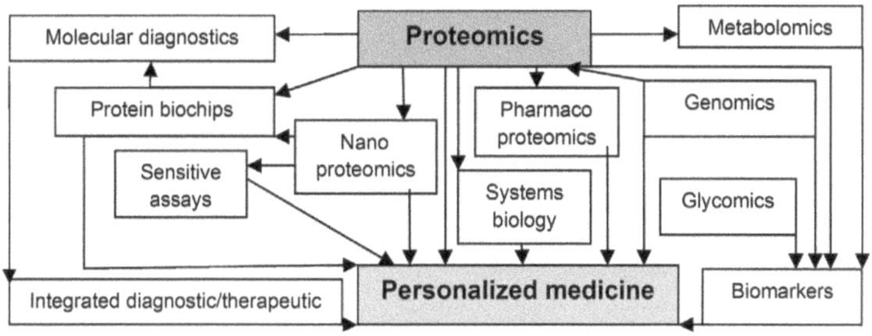

Figure 1 Interrelationship of proteomics with other technologies for development of personalized medicine. ©*JainPharmaBiotech*.

Table 2 Role of Proteomics in Personalizing Strategies for Cancer Therapy

Strategy	**Role of Proteomics**
Targeted cancer therapy for selective action against molecular targets expressed in tumors	The biochemical signature that distinguishes cancer cells from normal cells is often carried on the outside of the cell membrane in the form of glycoproteins
Antibody-based therapies for cancer	Both biochemical and proteomic approaches are used for the identification of the target antigen
Biomarker-based approaches for diagnosis and therapy of cancer	Use of proteomic technologies for identification of biomarkers
Screening for personalized anticancer drugs	Determination of structure of protein kinases, which are important drug targets in human cancers

© JainPharmaBiotech.

REFERENCES

Atkuri, K. R., Stevens, J. C., & Neubert, H. (2015). Mass cytometry: A highly multiplexed single-cell technology for advancing drug development. *Drug Metabolism and Disposition, 43*, 227–233.

Chahrour, O., Cobice, D., & Malone, J. (2015). Stable isotope labelling methods in mass spectrometry-based quantitative proteomics. *Journal of Pharmaceutical and Biomedical Analysis, 113*, 2–20.

Gallagher, R. I., & Espina, V. (2014). Reverse phase protein arrays: Mapping the path towards personalized medicine. *Molecular Diagnosis & Therapy, 18*, 619–630.

Haudenschild, D. R., Eldridge, A., Lein, P. J., & Chromy, B. A. (2014). High abundant protein removal from rodent blood for biomarker discovery. *Biochemical and Biophysical Research Communications, 455*, 84–89.

Hunter, J. M., & Paramithiotis, E. (2010). Protein biomarker quantification by mass spectrometry. *Expert Opinion on Medical Diagnostics, 4*, 11–20.

Jain, K. K. (2010). *A handbook of biomarkers*. New York: Springer.
Jain, K. K. (2012). *A handbook of nanomedicine* (2nd ed.). New York: Springer.
Jain, K. K. (2015). *Textbook of personalized medicine* (2nd ed.). New York: Springer.
Jain, K. K. (2015a). *Proteomics*. Basel: Jain PharmaBiotech Publications.
Jourdain, S., Bulman, A., & Dalmasso, E. (2010). The lucid proteomics system for top-down biomarker research. *Archives of Physiology and Biochemistry, 116*, 158–162.
Katafuchi, T., Esterházy, D., Lemoff, A., et al. (2015). Detection of FGF15 in plasma by stable isotope standards and capture by anti-peptide antibodies and targeted mass spectrometry. *Cell Metabolism, 21*(6), 898–904.
Kobeissy, F. H., Gulbakan, B., Alawieh, A., et al. (2014). Post-genomics nanotechnology is gaining momentum: Nanoproteomics and applications in life sciences. *OMICS, 18*, 111–131.
Nicolini, C., Bragazzi, N., & Pechkova, E. (2012). Nanoproteomics enabling personalized nanomedicine. *Advanced Drug Delivery Reviews, 64*, 1522–1531.
Ning, M. M., Lopez, M., Sarracino, D., et al. (2013). Pharmaco-proteomics opportunities for individualizing neurovascular treatment. *Neurological Research, 35*, 448–456.
Randall, S. A., McKay, M. J., Pascovici, D., et al. (2012). Remarkable temporal stability of high-abundance human plasma proteins assessed by targeted mass spectrometry. *Proteomics. Clinical Applications, 6*, 626–634.
Wilkins, M. R., Sanchez, J. C., Gooley, A. A., et al. (1995). Progress with proteome projects: Why all proteins expressed by genome should be identified and how to do it. *Biotechnology & Genetic Engineering Reviews, 13*, 19–50.
Xu, W., Hu, Y., He, X., et al. (2015). Serum profiling by mass spectrometry combined with bioinformatics for the biomarkers discovery in diffuse large B-cell lymphoma. *Tumour Biology, 36*, 2193–2199.

CHAPTER THREE

Metabolomics and Personalized Medicine

Nadia Koen[1], Ilse Du Preez[1], Du Toit Loots[1,2]
School for Physical and Chemical Sciences, Human Metabolomics, North-West University, Potchefstroom, South Africa
[2]Corresponding author: e-mail address: dutoit.loots@nwu.ac.za

Contents

Abstract

Current clinical practice strongly relies on the prognosis, diagnosis, and treatment of diseases using methods determined and averaged for the specific diseased cohort/population. Although this approach complies positively with most patients, misdiagnosis, treatment failure, relapse, and adverse drug effects are common occurrences in many individuals, which subsequently hamper the control and eradication of a number of diseases. These incidences can be explained by individual variation in the genome, transcriptome, proteome, and metabolome of a patient. Various "omics" approaches have investigated the influence of these factors on a molecular level, with the intention of developing personalized approaches to disease diagnosis and treatment. Metabolomics, the newest addition to the "omics" domain and the closest to the observed phenotype, reflects changes occurring at all molecular levels, as well as influences resulting from other internal and external factors. By comparing the metabolite profiles of two or more disease phenotypes, metabolomics can be applied to identify biomarkers related to the perturbation being investigated. These biomarkers can, in turn, be used to develop personalized prognostic, diagnostic, and treatment

[1] All authors contributed equally to the writing of this manuscript.

Advances in Protein Chemistry and Structural Biology, Volume 102
ISSN 1876-1623
http://dx.doi.org/10.1016/bs.apcsb.2015.09.003

approaches, and can also be applied to the monitoring of disease progression, treatment efficacy, predisposition to drug-related side effects, and potential relapse. In this review, we discuss the contributions that metabolomics has made, and can potentially still make, towards the field of personalized medicine.

1. INTRODUCTION

Conventional disease diagnostics generally entails a physician identifying a disease or abnormality on the basis of a physical examination of the symptomatic patient, with (or without) the additional use of standard diagnostic tests. A positive diagnosis is normally followed by treatment using drugs produced on a large scale and administered at a standardized and universally accepted dosage. These conventional drugs are developed to treat general symptoms or the disease as determined by the mean results obtained over large population groups (Debas, Laxminarayan, & Straus, 2006). However, it is crucial to understand that due to, for example, genetics and a variety of other factors such as individual diet, habits (e.g., smoking), gender etc., not all diseases affect all individuals in the same manner (Jirtle & Skinner, 2007), and neither do all individuals respond to treatment in the same way (Dworkina, McDermottb, Farrarc, O'Connord, & Senn, 2014). This occurrence is clearly reflected by the increasing incidence of treatment failure and relapse, which is especially disturbing when considering their prevalence in life-threatening diseases such as tuberculosis (TB) and acquired immune deficiency syndrome (AIDS).

Although this variation between individuals might not be obvious in the initial clinical presentation of the disease, it is most likely still detectable on a molecular scale. Several researchers groups have subsequently shifted their focus to the development of personalized medicine, which uses the molecular information of an individual, as dictated by his or her genome, transcriptome, proteome, and metabolome (Redekop & Mladsi, 2013) to develop patient-specific diagnostics and drugs. This information can also be used to determine/predict individual treatment response, prior to and during the treatment regimen, in an attempt to lower the incidence of treatment failure or relapse (Salari, 2009), and also to optimize drug dosages, in an attempt to prevent or lessen the severity of the drug-related side effects (Lecea & Rossbach, 2012).

Personalized medicine has undoubtedly been one of the most topical issues over the past decade and several research fields have joined hands in using this as a tool for optimizing disease management, as it has the potential to transform clinical practice and treatment efficacy.

2. THE VALUE OF "OMICS" TECHNOLOGIES IN THE DEVELOPMENT OF PERSONALIZED MEDICINE

Traditionally, genomics was considered the most important approach for determining individual variation and the development of personalized medicine (Jain, 2009). However, several intermediate processes occur between the genotype and disease phenotype in the "omics" cascade, which may influence disease outcome or treatment response, and includes transcription, translation, and metabolism. Furthermore, various other factors, such as environmental influences and age, may also play a role in the disease phenotype, a phenomenon which genotyping is not able to characterize or explain. The development of personalized medicine requires a holistic view of all molecular variation that may differentiate individuals, and researchers are therefore shifting their focus from using exclusively genetics methodologies for instance, to a systematic/integrative "omics" approach. "Omics" is a general term used to describe the study of all genes (genomics), transcription of these genes (transcriptomics), translation into their respective proteins (proteomics), and all the resulting metabolite changes (metabolomics), and is aimed at acquiring large-scale data sets from a single and/or multiple samples (Wheelock et al., 2013). These "omics" research fields, alone or in combination, have shown to be valuable for the identification of new disease biomarkers for the purpose of elucidating disease mechanisms and the development of personalized diagnostic and treatment regimes. The successful completion of the human genome project was considered a major breakthrough in science, especially for those involved in genomics and personalized medicine, as any variation in a person's genome at a specific locus, could potentially have a direct or indirect effect on gene expression, thereby influencing disease susceptibility (Ginsburg & Willard, 2009). Genomics has to date identified various genetic risk factors, for a number of chronic diseases, including heart disease, diabetes, and several form of cancers (Ginsburg & Willard, 2009). For example, it is now well known that women with mutations in BRCA1 or BRCA2 are at risk for developing breast and ovarian cancer (Schwartz, Hughes, & Lynch, 2008), and similarly, individuals with a mutation in MLH1 and MSH2 are at risk for developing colon cancer (Wiesner, Slavin, & Barnholtz-Sloan, 2009). In a similar fashion, genomics can be used to predict drug response, by identifying the effect of drugs on gene expression, as well as the effect of various genes on the variable response of the patient to a drug (Ginsburg & Willard, 2009). One example of this, are the two genes, CYP2C9 and VKORC1, identified using genomics methodologies, which are used to determine/predict the effective maintenance/chronic treatment dose of warfarin, a widely used anticoagulant. CYP2C9 has been shown to be responsible for the metabolic clearance of the *S*-enantiomer of the drug and various genotypic alterations have been associated with a 10-fold variation in *S*-warfarin clearance. It has additionally been estimated that the genotypic variation of CYP2C9, together with that of VKORC1 (which is targeted by warfarin), and other factors including body weight and age, accounts for 35–60% of the variation seen in individual drug response to warfarin (Voora, McLeod, Eby, & Gage, 2005). An additional example of the use of genomics for the purpose of personalized treatment approaches was the development of a blood-based test for the prediction of graft-rejection (Deng, Eisen, & Mehra, 2006). These DNA assessments or "stable" genomics approaches can also be done at birth, to predict a predisposition to a disease or treatment. The other "omics" approaches described above, however, detect "dynamic" disease markers also, which have the capacity to show individual variation due to other factors including diet, individual habits, comedication, coinfection with other pathogens, etc. These biomarkers can therefore be used to characterize diseases and drug action on an even more detailed individualized level (Ginsburg & Willard, 2009).

Several biomarkers identified using transcriptomics approaches have also been validated and used clinically for disease diagnosis and risk assessment. One example of this is the DNA microarray-based diagnostic kit, MammaPrint®, which can be used for breast cancer prognosis and measures the degree of transcription of 70 specified genes (Hong et al., 2013). Oncy*type*DX® is another example of such an assay, which measures the expression of 21 genes for the prediction of recurrences, death, and response to therapy, in patients with estrogen-positive breast cancer (Hong et al., 2013). Heidecker et al. (2008) additionally investigated the potential use of transcriptomics to identify prognostic biomarkers for heart failure, as several genes were identified to be overexpressed in patients with a good versus a bad prognosis, and when the predictive value of these biomarkers were tested, a 74% sensitivity and 90% specificity were obtained.

In addition to the transcriptome, posttranslational modification of proteins has also been shown to play a major role in various biological processes and disease states. Using proteomics, Ono et al. (2006) identified a posttranslational modification of alpha-fibrinogen (4-hydroxylated alpha-fibrinogen) in pancreatic cancer patient plasma and subsequently constructed an antibody and validated the assay on the basis of this for the better diagnosis of the disease. The same group also used proteomics to investigate the life-threatening side effects related to gemcitabine monotherapy, a standard treatment used for pancreatic cancer, and subsequently identified haptoglobin as a biomarker for the prediction of this drug's hematological toxicities.

These studies show the value of genomics, transcriptomics, and proteomics, in all fields of personalized medicine. The contribution of metabolomics, the most recent addition to the "omics" revolution, will subsequently be discussed in detail in the remainder of this review.

3. WHAT IS METABOLOMICS?

As per definition, metabolomics is the nonbiased identification and characterization of "all" the small molecular compounds (metabolites) in a biological system, using highly sensitive analytical techniques, in combination with bioinformatics (Dunn, Bailey, & Johnson, 2005). The metabolome, which is a collective term for all the metabolites in a specific biological system/sample, is the ultimate downstream result of genes, transcription, and translation, and will therefore reflect changes to the genome, transcriptome, and proteome, in addition to that caused by a disease state or other environmental factors. The identification of the main differences between the metabolomes of two sample cohorts (disease vs. healthy controls, for example) is a starting point for the discovery of new metabolite biomarkers. Additionally, a comparison of various cohorts with individuals showing variation to disease or response to treatment can also be done in order to identify markers associated with this type of variation, in the context of individualized medicine.

The extraction and analysis of metabolites from a sample or sample group can be done in an untargeted or semitargeted manner. Untargeted metabolomics aims to extract and detect all metabolites (known and unknown, from all metabolite classes), i.e., the total metabolome, as per definition of metabolomics. Semitargeted metabolomics approaches, however, are focused on the analysis of specific fractions of the metabolome or a subclass of metabolites, such as only the lipids or organic acids, for instance. Sample preparation methods for untargeted metabolome analyses are simple, and the generated metabolite profiles can provide researchers with a good general picture of the effect of the investigated perturbation on the overall metabolome. However, these methods tend to have a lower sensitivity and detection limit when compared to that of the semitargeted approach, which provides simpler metabolite profiles, representing specific metabolic pathways (Wishart, 2010). The choice of the sample preparation method will also depend on the analytical apparatus selected and whether an untargeted or semitargeted approach is required. When using nuclear magnetic resonance (NMR) spectrometry for instance, chemicals such as ethanol and hexane should be avoided, as these solvents are also deuterated and will therefore result in multiple resonances and subsequently interference (Dunn et al., 2005). Currently, there is no single analytical apparatus available with the capacity to identify all the metabolites extracted from a sample, and therefore, when doing untargeted metabolomics, a combination of a number of different analytical approaches is recommended, however, this may not always be a viable option in a particular laboratory as it is dependent on instrument availability. In instances with limited analytical capacity, a lot can still be done in the context of untargeted metabolomics, for instance, derivatization of a sample prior to gas chromatography–mass spectrometry (GC–MS) analysis, in additional to appropriate column selection, can serve well in the detection of a large portion of the metabolome during a single analytical run. Each analytical technique comes with its own set of molecular preferences, advantages, and limitations, as will subsequently be discussed below.

3.1 Analytical Methods Most Often Used for Metabolomics

The most commonly used analytical approaches for metabolome data acquisition include the use of various chromatographic techniques, most commonly gas chromatography (GC), or liquid chromatography (LC), coupled to various different options of mass spectrometry (MS) detectors, and NMR. Without prior separation or derivatization, LC–MS is considered to be the apparatus with the potential to detect the largest variety of metabolites present in a specific sample. However, the derivatization of sample extracts makes GC–MS an even, if not better contender, considering the availability of spectral data for GC–MS compound identification. Furthermore, although LC–MS is ideal for the analysis of polar and ionic compounds, it has a lower chromatographic resolution and higher running costs in comparison. Additionally, a great advancement in GC–MS technology was the development of the GCxGC system, which separates metabolites in two dimensions, on the basis of not only volatility but also polarity, thereby reducing the amount of coeluting peaks and enhancing the resolution of the eluting metabolites (Marriott & Shellie, 2002). GC–MS analysis also requires smaller sample volumes when compared to that required for LC–MS and NMR, but because these samples undergo metabolite separation and derivatization, they are nonrecoverable after GC analysis. Another downside to GC is the rather long analysis times required for compound separation, and the identification of "unknown" metabolites (those compounds detected with mass fragment patterns not in the commercial libraries) is rather complex.

NMR spectroscopy is based on the principle of detecting metabolites according to the signals produced by their proton content, allowing for straightforward metabolite identification (Bonhommea, Gervaisa, & Laurencin, 2014). This relatively fast method (2–3 min per sample), is mostly used for the detection of polar metabolites and is nondestructive to the sample. NMR instrumentation is, however, rather expensive, requires large sample volumes and has a lower sensitivity when compared to other techniques (Dunn et al., 2005; Halket et al., 2005).

Subsequent to sample analysis, one of the most important steps for generating data, which can be used for metabolomics, is the extraction of reliable data matrixes from the complex chromatographic and mass spectrometric outputs, for subsequent statistical analyses and biomarker selection. This course of action includes peak detection, peak deconvolution, peak alignment, compound quantification, and identification, among various other steps. Most of the analytical methods described above come with their own software packages, specifically designed for this purpose (such as ChromaTOF for the Leco GCxGCTOFMS), whereas other universal software packages, such as MET-IDEA, are also freely available for use for processing data generated from a variety of different commercially available analytical techniques (Broeckling, Reddy, Duran, Zhao, & Sumner, 2006). However, because each of these packages comes with their own advantages and limitations, most researchers prefer to use a combination of software packages, in addition to manual inspection, in order to obtain the optimum data matrix for statistical data analysis and biomarker identification.

3.2 Statistical Approaches

The increasing complexity of the data matrixes obtained from the analytical equipment used in metabolomics studies has led to the use of various multivariate chemometric data analysis methods for biomarker identification/extraction from these data sets. In order to get an overview of the data, certain unsupervised methods can be used to highlight trends in the data and grouping or differentiation of various sample sets, and to additionally identify potential outlier samples and batch effects. When employing these unsupervised methods, samples are not assigned to specific groups (for example, disease and control) prior to the statistical analysis, allowing the analyst to determine whether or not the samples are naturally differentiated or grouped based on their analyzed metabolite profiles. For this purpose, principle component analysis (PCA) is the method most commonly used. PCA reduces the dimension of the input data matrix by calculating a weighted sum (score) of the compound (metabolite) concentrations detected in each sample and expresses these in terms of principal components (PCs), with PC1 describing the most variation in the data, PC2 the next highest variation, etc. These PCs subsequently serve as coordinates on a scatter plot and provide an overview of the samples and how they relate to each other on the basis of their analyzed metabolomes. Other chemometric methods, such as self-organizing maps, hidden Markov models, and canonical correlation, can also be implemented in this initial, exploratory stage for the same purpose (Madsen, Lundstedt, & Trygg, 2010; Trygg, Holmes, & Lundstedt, 2007). If those samples, belonging to a specific group, do in fact assemble and group together, supervised methods where individual samples are allocated to their respective sample groups before the analyses, can be applied for the purpose of identifying potential biomarkers best describing the variation detected. Partial least squares discriminant analysis (PLS-DA) is one such method, which uses group membership information to build a discrimination model. The variable influence on the projection (VIP) parameter, which is a weighted sum of the squares of the PLS-DA weights, gives an indication of the importance of the metabolite to the prediction model, and can therefore be used to identify those metabolites which are most characteristic of a specific sample group, or those metabolites which vary the most between the specified groups. The metabolites with the highest VIP scores are then ranked and can be used to identify potential biomarkers. Similar supervised classification models also used for biomarker identification, include, but are not limited to, soft independent modeling of class analogy and support vector machines. The technical details of these chemometric methods fall beyond the scope of this review, but the authors suggest the review by Trygg et al. (2007) for a more detailed description of these.

Various software packages and Web-servers, such as MetaboAnalyst (Xia, Psychogios, Young, & Wishart, 2009), have been developed specifically for researchers with limited statistical knowledge, to perform these essential chemometric analysis on metabolomics data. Although these tools are helpful, most metabolomics research groups still prefer to use qualified biostatisticians, with knowledge of the underlying mathematical programming, for mining the relevant biomarkers form these complex data sets. In these instances, more traditional statistical packages such as Statistica and "R" are used for the analysis of the data in the context of the specific biological question. Identified biomarkers can subsequently be used to explain individual variation in disease and treatment response, by interpreting this as the context of known metabolic pathways, and/or prior genomic, proteomic and transcriptomic data. Furthermore, individual biomarkers or combinations thereof (biosignatures) can be used for diagnostic purposes, the latter of which can be achieved by building a prediction model, such as a classification tree. In the light of this, metabolomics is considered an important tool for personalized medicine.

4. THE APPLICATION OF METABOLOMICS TOWARD PERSONALIZED MEDICINE

The ongoing advancements in technology and subsequent availability of more sensitive analytical equipment for application to metabolomics studies, has made it possible to detect even subtle variations in the metabolomes of even healthy individuals. These variations are now known to be a consequence of a variety of factors, including environment (such as habits, lifestyle differences, and stress), age, genetics (mutations, epigenetics), gender and race (Hernandez & Blazer, 2006).

By implementing various different GC–MS and LC–MS metabolomics techniques, Lawton et al. (2008) investigated the influence of age, gender and race, on the plasma metabolite profiles of 269 healthy individuals. Of these three variables, age was considered to have the most significant influence on the metabolite concentrations of these individuals, and the identification of elevated oxidative stress markers and changes in protein, energy and lipid metabolism could strongly be associated with age. The proposed antiaging androgen, dehydroepiandrosterone-sulfate, additionally showed inverse correlations with age, whereas certain xenobiotics (e.g., caffeine) positively correlated, potentially indicating decreased activity of cytochrome P450. A significant interaction between age and gender was also identified and a subsequent positive correlation between age and urea concentrations was shown, which is more pronounced in females than in males. When investigating the metabolome-specific gender-related differences, 35 metabolite markers were detected, indicative of an increased amino acid, energy and nucleotide metabolism, in the males comparatively. When investigating metabolome differences in race, caffeine, paraxanthine and theobromine were detected in far lower concentrations, and hydroxyproline, glycochenodeoxycholate and glycolate were detected in comparatively higher concentrations in African-Americans versus Caucasians. Although variations were also determined when comparing metabolite profiles obtained of Hispanics and African-Americans or Hispanics and Caucasians, the identity of these compounds could not be identified.

In a similar study, Urban, Kavvadiasa, Riedela, Scherera, and Trickerb (2006) investigated variations in the metabolite profiles of smokers versus nonsmokers using LC–MS/MS and GC–MS metabolomics. Various acrylamide (a neurotoxin known to be present in tobacco smoke) metabolites (*N*-acetyl-*S*-(2-carbamoylethyl)-L-cysteine, glycidamide *N*-(R/S)-acetyl-*S*-(2-carbamoyl-2-hydroxyethyl)-L-cysteine, and *N*-2-carbamoylethylvaline) were detected in significantly higher concentrations in the urine and blood samples of the 60 smokers and strongly correlated to the number of cigarettes they smoked daily. These compounds are therefore considered valid biomarkers of acrylamide exposure.

From these and many other related studies, it is evident that metabolomics can be useful for identifying biomarkers explaining individual variation in the metabolome profiles of healthy individuals, related to various "natural" or "habitual" factors. The identification of the metabolite biomarkers related to the variation detected in various disease states and/or the xenobiotic treatment of these diseases, and/or the influence of various other environmental factors on this, can assist in explaining the underlying mechanisms associated with individual variation in disease diagnosis, disease susceptibility, treatment response, and drug-related side effects (Lawton et al., 2008). These findings will consequently contribute to the development of personalized medicine, a concept which not only involves personalized diagnosis and treatment but also personalized disease susceptibility assessment, health monitoring, and preventative medicine.

4.1 Identification of Disease Biomarkers for Metabolomics-Based Diagnostics

Various metabolomics studies have been performed for the purpose of discovering new diagnostic biomarkers for a variety of diseases. These biomarkers are identified as those metabolites with the highest intergroup variation after comparing metabolome profiles obtained from healthy or nonspecific-diseased sick individuals with similar symptoms, to those obtained from the specific/investigated diseased patients. One such a study was done with the aim of identifying diagnostic breath biomarkers for pediatric asthma, using NMR metabolomics. By applying linear discriminant analysis and PLS-DA to the metabolite profiles obtained from exhaled breath condensate of 25 children with asthma and 11 healthy controls, Carraro et al. (2008) were able to differentiate between the two groups with a success rate of 86%, which is slightly better when compared to the conventionally used methods (81%). The metabolites contributing most to this differentiation were identified as oxidized and acetylated compounds, indicative of increased oxidative stress and inflammation in the airways of the lungs. Considering this, despite their superior diagnostic ability and capacity to explain disease mechanisms, these compounds are not necessarily specific to asthma, but are rather indicative of a general disease state. For this reason, biomarkers identified in diseases, using healthy subjects as controls, although for the purpose of describing diseased mechanisms, have serious limitations for diagnostic applications. Hence, in order to validate diagnostic markers, comparison to other diseased states with similar symptoms is a true test for potential application in a clinical diagnostic scenario.

A good example of this is a study done by Sugimoto, Wong, Hirayama, Soga, and Tomita (2010), who analyzed saliva metabolites obtained from three different groups of cancer patients (69 individuals with oral cancer, 18 with pancreatic cancer, and 30 with breast cancer), 11 periodontal disease patients, and 87 healthy controls, using capillary electrophoresis time-of-flight mass spectrometry (CE-TOF-MS). Different sets of metabolites were subsequently identified which could be used to differentiate each individual disease state from the healthy controls, with the area under the receiver operating characteristic curves (AUCs) of 0.865 for oral cancer, 0.973 for breast cancer, 0.993 for pancreatic cancer, and 0.969 for periodontal diseases. Although these diseases could not be differentiated from each other using PCA, three metabolites were identified specific for oral cancer (detected in concentrations significantly different to all other groups) and eight compounds specific to pancreatic cancer. No biomarkers could, however, be uniquely identified for breast cancer. The authors hypothesized that the reason for this may be due to the fact that breast cancer has various structurally different forms, dependant on the expression of hormone receptors, and they added that additional factors such as age and menopausal status can additionally contribute to the diverse metabolite profiles detected in these patients. Considering this, when identifying diagnostic biomarkers for a particular disease, it is crucial to consider all the contributing factors, such as the various stages of the disease, characteristics of the pathogen (drug resistance, pathogen subspecies, active vs. latent infection), and individual variation due to genetic and environmental factors, which could have an influence on the detected metabolite profiles. Various metabolomics studies have investigated the influence of these and other factors with the aim of developing personalized diagnostic approaches.

In a semitargeted metabolomics study, Zhang et al. (2006) investigated the influence of different stages of hepatitis C virus (HCV)-induced liver fibrosis on plasma amino acid profiles. The disease progression, from early to advanced fibroses, correlated with a decrease in branched-chain amino acids and a simultaneous increase in aromatic amino acids. These amino acid profiles were subsequently used to develop a diagnostic classification model which was able to differentiate between early and advanced fibrosis ($AUC = 0.9 \pm 0.04$) and cirrhosis ($AUC = 0.99 \pm 0.01$). Similarly, biomarkers related to disease progression have also been identified for colorectal cancer (Nishiumi et al., 2012) and Parkinson's disease (Bogdanov et al., 2008), thereby proving that metabolomics not only has the capacity to identify markers specific to a disease but also to identify the specific disease stages. Such approaches can not only be used to identify biomarkers for use in less invasive diagnostic methods (for example, identifying markers in plasma, as opposed to performing a colonoscopy in colorectal cancer patients for instance) but also to assist clinicians in determining the most appropriate treatment protocol, with regards to the patient's individualized disease stage or severity.

When investigating the metabolome of patients with disease resulting from an infectious pathogen, the individual characteristics of the pathogen can also potentially add further interindividual variation to the presentation and treatment response of the patient. Olivier and Loots (2012) used GC–MS metabolomics to identify biomarkers classifying the various *Mycobacterium* species forming part of TB complex, which are known to cause infections with symptoms closely related to that caused by *Mycobacterium tuberculosis* in pulmonary TB patients, but respond differently to standard TB treatment. Based on the lipid metabolite profiles, they were able to differentiate between *M. tuberculosis*, *M. avium*, *M. bovis*, and *M. kansasii*, and they furthermore identified the 12 compounds which best describe the variation between these disease causing *Mycobacterium* species. These potential biomarkers were then used to build a multivariate diagnostic model, which was able to correctly classify unknown samples with probabilities ranging from 72% to 100%, at an exceedingly better detection limit as compared to current speciation methods. By using the same research approach, this group was additionally able to differentiate various different drug-resistant strains from each other (Du Preez & Loots, 2012; Loots, 2014). In another example of such an application, Denery, Nunes, Hixon, Dickerson, and Janda (2010) used a LC–MS metabolomics approach to differentiate serum and plasma samples collected from patients with active onchocerciasis (*Onchocerca volvulus* positive) and controls (*O. volvulus* negative), and subsequently identified 14 compounds as biomarkers for this infection. They subsequently applied this biomarker set to samples collected from individuals living in onchocerciasis endemic areas, where ivermectin was given as a chronic medication, and were able to distinguish between those subjects with compromised worm viability due to treatment and those still with an active infection, and hence these markers could subsequently be used not only for disease diagnostics but also to determine disease severity and to predict treatment outcome or monitor individual treatment response. These studies subsequently show the capacity of metabolomics for identifying markers which not only accurately diagnose disease but also have the capacity to detect individual variation not necessarily evident from the clinical symptoms, such that an individualized treatment regimen can be given. These markers can additionally be used to monitor treatment response or predict treatment outcome (Redekop & Mladsi, 2013).

Lastly, despite the actual metabolome changes induced by a disease or pathogen, individual variation due to genetic and environmental factors can also influence the identified diagnostic biomarkers, in a similar fashion to that previously described in healthy individuals. Szymanska et al. (2012), for example, showed that metabolic biomarkers used for the diagnosis of obesity are strongly influenced by gender. When investigating the plasma and serum metabolite profiles of healthy overweight men and women, central obesity, insulin, cholesterol, VLDL, and certain triacylglycerols, were identified as potential biomarkers for abdominal visceral (VAT) fat distribution in women, whereas in men, the identified VAT-associated biomarkers included different triacylglycerols as compared to women, phosphatidylcholine, and VLDL. These results highlight the fact that individual variation, such as gender, can be detected via metabolomics, and that these variations should be considered when identifying diagnostic biomarkers.

4.2 Pharmacometabonomics

Apart from the contribution that metabolomics has made to the early detection and characterization of a disease state in the context of personalized clinical care, it has also shown promise in the domain of pharmacokinetics, and subsequently to the new research domain termed pharmacometabonomics. This research field investigates the response of a patient to xenobiotics from a metabolism perspective (Baraldi et al., 2009). Xenobiotics include those substances foreign to the biological system, which may originate from the individual's environment, including diet and medication. When these xenobiotic substances are ingested, the first reaction of the body is to eliminate these compounds, by modifying their chemical characteristics by means of any number of enzymatic reactions (xenobiotic metabolism). This response also forms the basis for the activation of various medications, the products of which act to alleviate symptoms or treat the respective disease. Individual variation in xenobiotic metabolism is a concept known for quite some time now and plays an important role in individual response to treatment, and hence treatment outcome, relapse, and side effects (de Villiers & Loots, 2013). It has been shown that various innate factors including race, ethnicity, and gender, in addition to the previously mentioned environmental factors or social habits, such as diet, alcohol consumption, and smoking, can influence individual variation to xenobiotic metabolism and response (Matthews, 1995). Identifying these factors and determining their influence or correlations to individual treatment response, forms the basis of personalized disease treatment and will subsequently be discussed in the following sections, in the context of metabolomics.

4.2.1 Metabolomics and the Elucidation of Drug Mechanisms

Before treatment strategies can be tailored to an individual's unique response to therapy, it is important to understand the general xenobiotic metabolism and underlying mechanisms of the proposed drug. For this purpose, pharmacometabonomics can be applied in a number of ways: (1) the comparison of the metabolite profiles of a patient group receiving the vehicle/placebo versus those receiving the xenobiotic treatment, (2) the comparison of the metabolite profiles of patient groups receiving unlabeled xenobiotic treatment versus stable isotope-labeled xenobiotic treatment, and (3) the comparison of the metabolite profiles of wild-type and genetically modified animals receiving the investigated xenobiotic (Lan & Jia, 2010). These methodologies have been implemented to investigate the metabolism of various nutrients, drugs, and other xenobiotics, using a variety of analytical equipment and bioinformatics strategies (Lan & Jia, 2010). In addition to drug-derived metabolites originating directly from xenobiotic metabolism, these drug exposure signatures will also include drug-induced alterations to normal metabolism, representing the patient's altered metabolic state in response to the treatment. In one such instance, Rozen et al. (2005) were able to differentiate between patients with motor neuron diseases (MND) and healthy controls, in addition to MND patients with and without Riluzole treatment, using the plasma metabolite profiles of these individuals. Two fatty acid-based compounds were uniquely associated with patients on Riluzole, and although these compounds could not be identified completely, they did not appear to be derivatives of the drug, but rather reflected changes to the patient's normal metabolome. Absolute identification of these compounds might help to elaborate on the still poorly understood drug mechanism of Riluzole and can assist with the deciphering its related side effects.

In a similar fashion, pharmacometabonomics can be implemented to verify or complement drug mechanisms proposed by other omics approaches. Tiziani et al. (2009) applied this approach to investigate the antileukemic activities of the combination of bezafibrate (BEZ) and medroxyprogesterone acetate (MPA), against acute myeloid leukemia (AML). Previous studies have linked the xenobiotic metabolism of these drugs to the production of reactive oxygen species (ROS), and therefore this group proposed that this is responsible for the drug action (Tiziani et al., 2009). In order to prove this hypothesis, and potentially other previously unknown drug mechanisms, the group applied NMR metabolic profiling to three AML cell lines treated with BEZ and MPA. From the generated biosignatures, they were able to identify changes to TCA cycle intermediates (more specifically alterations to the conversion of α-ketoglutarate to succinate), which are consistent with ROS action.

In order to minimize the influence of individual variation on the resultant metabolite profiles, however, many of these studies were done on samples collected from homogeneous patient populations or animal models kept under identical conditions, thereby ensuring that the metabolites emanating as biomarkers are, in fact, related to the xenobiotic metabolism exclusively, with little or no individual variation due to either genetic or environmental factors. Various external factors such as age, stress, diet, gut microbes (microbiome), medication, lifestyle, and disease, in addition to genetic factors including gender, epigenetics, and polymorphisms in genes encoding for xenobiotic-metabolizing components such as enzymes, transporters, receptors, and ion channels, can also influence xenobiotic metabolism and account for individual variation (Johnson, Patterson, Idle, & Gonzalez, 2012). Clayton, Baker, Lindon, Everett, and Nicholson (2009) demonstrated the application of pharmacometabonomics toward a better understanding of these variables by analyzing pre- and postdose urinary metabolites of patients on acetaminophen (paracetamol), using NMR spectrometry. When analyzing the predose profiles of these individuals, considering the levels of the excreted drug-derived metabolites, they found high levels of predose *p*-cresol sulfate, which correlated with low postdose ratios of acetaminophen sulfate to acetaminophen glucuronide. The ratio of these derivatives, which indicate the extent to which acetaminophen is metabolized through two major phase 2 conjugating processes (O-sulfonation and glucuronidation), is known to be a site and indicator of individual variation in response to paracetamol. The group finally concluded that in patients with high levels of gut microbiome-mediated *p*-cresol generation, competitive *p*-cresol O-sulfonation reduces the capacity to sulfonate acetaminophen, which in turn results in an increased likelihood of drug-induced hepatotoxicity. This study subsequently proves the capacity of metabolomics to identify individual variation in xenobiotic metabolism, related to a variation to individual patient environment.

Pharmacometabonomics can also be implemented as an informative tool, assisting pharmacogenomics in the investigation of genome-related variation in drug metabolism (Johnson et al., 2012). With this goal, Ji et al. (2011) investigated urine metabolites of individuals with major depressive disorder undergoing therapy with selective serotonin reuptake inhibitors (SSRIs), citalopram and escitalopram. On average, 40% of patients generally do not respond to this treatment and previous pharmacogenomic studies failed to identify potential polymorphisms which could be used for the prediction of a SSRI treatment outcome. From the obtained metabolite profiles, elevated glycine levels were identified to be associated with a decreased treatment response, and after subsequent pharmacogenomics studies, polymorphisms in the glycine dehydrogenase gene were identified. These results show that metabolomics can additionally play a significant role in supporting or initiating pharmacogenomics studies, with the intention of identifying genetic factors related to individual variation in drug metabolism.

Considering these and several other studies on the topic (de Carvalho, Darby, Rhee, & Nathan, 2011; Halouska et al., 2007; Lu, Deng, Li, Wang, & Li, 2014; Wang et al., 2013), it is evident that the identification of drug exposure metabolites can play a significant role in the elucidation of drug mechanisms and the influence of individual variation on these, which in turn, can contribute to the development of more effective drugs, or the positive adjustment of drug dosages and treatment schedules, on the basis of the individual differences detected and explained.

4.2.2 Metabolomics and Understanding Response to Treatment

Despite the elucidation of drug mechanisms, pharmacometabolomics can also be used to investigate and predict an individual's response to treatment. Statins, for example, are commonly prescribed for patients with increased levels of LDL-cholesterol and risk for cardiovascular disease, despite the substantial individual variation in response to this therapy. Trupp et al. (2012) investigated this occurrence using GC–MS metabolomics analyses of patient plasma samples, prior to, and 6 weeks after simvastatin treatment onset. A number of metabolites were identified implicating genetic, gut microbiome and various environmental factors, contribute to the variation in simvastatin response. Additionally, responders and nonresponders to the drug could be differentiated based on their baseline metabolite profiles and the most significant compounds responsible for this differentiation were identified. These metabolite variations could be correlated to different treatment responses and subsequently described the mechanisms related to the individual variation to this therapy. These markers additionally have the potential to be implemented preclinically, to identify those patients who would/would not benefit from simvastatin treatment, prior to commencing treatment. Using the same approach, Wei et al. (2013) built a prediction model, based on four identified serum metabolite biomarkers, in order to predict the outcome of breast cancer neoadjuvant chemotherapy. The model was able to predict complete response (disappearance of all tumor deposits) versus stable disease (tumor reduction less than 50%) with 100% specificity and 80% sensitivity (AUC of 0.95). Despite their prognostic value, the biomarkers identified in these studies also show promise in the development of new, more efficient drugs, and also to subclassify patients during clinical trials.

The same approach has been used to explain and predict variation to treatment response in patients diagnosed with various infectious diseases. Das et al. (2015) investigated the influence of anti-TB drug treatment on the urine metabolic profiles of TB patients at various treatment intervals. A clear treatment-dependent trend could be seen on the PCA, as the metabolite profiles of each consecutive treatment interval shifted closer to that of healthy controls, with profiles of clinically cured patients very closely resembling that of the control group. With these profiles, one might be able to build a prediction model for treatment outcome, provided that profiles of patients with failed treatment outcomes are also incorporated. These profiles may additionally give clues as to why certain patients fail to respond to TB treatment within the recommended 6-month therapy regime, and by using a similar approach, metabolite biomarkers predicting treatment failure or relapse might also be identified.

When investigating infectious diseases in this manner, it is important to bear in mind that individual variation to drug treatment outcomes can be a result of the host's variable response to the drug, or the pathogen's resistance to the drug, and therefore, both factors should be considered when identifying biomarkers reflecting treatment response. In a GC–MS metabolomics study, Du Preez and Loots (2012) investigated rifampicin resistance in pulmonary TB by comparing the fatty acid metabolomes of two *M. tuberculosis* strains, with resistance conferring mutations on different locations on the *rpoB* gene, to a fully susceptible wild-type parent strain. All three groups showed a clear differentiation when doing PCA, and a number of metabolites indicating a decreased synthesis of various 10-methyl branched-chain fatty acids and cell wall lipids, and an increased use of the shorter-chain fatty acids as carbon sources, were identified as markers in the drug-resistant strains. In addition, the *rpoB* S531L mutant, previously reported to occur in well over 70% of all clinical rifampicin-resistant *M. tuberculosis* strains, showed a better capacity for using these alternative energy sources, compared to the less frequently detected *rpoB* S522L mutant. This study therefore shows that pharmacometabonomics has the power to not only detect metabolome changes related to pathogen-induced drug resistance, but it can also differentiate between the various genotypes leading to the observed phenotype. The clinical use of these identified markers can significantly contribute to the development of improved treatment approaches, thereby bettering treatment outcomes in patients with drug-resistant TB. When developing such personalized therapeutic approaches, it is also important to realize that not all individuals will respond to an infection in a similar fashion, and although infected with identical pathogen strains, treatment outcomes may vary, and through metabolomics, markers can be identified in order to explain and predict this phenomenon.

4.2.3 Metabolomics and Understanding Drug Toxicology

Several metabolomics studies have identified universal, nonspecific biomarkers, related to general toxicity of various different drugs, which are unrelated to the specific drug type, drug mechanism, or site of action. These metabolites, including decreased concentrations of TCA cycle intermediates and hippurate, are a reflection of secondary side effects related to the ingestion of these drugs and include general changes to energy demand or energy metabolism, and changes to gut microbiota (Keun, 2006). Additionally, many metabolomics studies have also identified general markers related to hepatotoxicity (Holmes et al., 1992; Robertson et al., 2000; Schnackenberg, Dragan, Reily, Robertson, & Beger, 2007) and nephrotoxicity (Garrod et al., 2005; Lenz, Nicholson, Wilson, & Timbrell, 2000). These markers are especially important in the preclinical drug development phase, considering that liver and kidney damage are the two major reasons for drug withdrawal from the market. Although knowledge of these general toxicity markers is of importance, metabolomics also has the capacity to identify markers related to specific drug mechanisms and side effects to vital organs also, and by using blood and urine for biomarker detection, eliminates the need for intrusive procedures for monitoring these outcomes, such as tissue biopsy for instance. Sumner, Burgess, Snyder, Popp, and Fennell (2010) detected urinary markers related to abnormalities in inositol, carbohydrate, glycerolipid, and glyoxylate metabolism, correlating to hepatic microvesicular lipid accumulation (MVLA), a histopathological side effect related to the treatment of TB with isoniazid. They propose that, if validated, these metabolite changes can be used to develop a noninvasive method for the early detection of MVLA.

A study entitled Consortium for Metabonomic Toxicology (COMET) was one of the most extensive pharmacometabolomics investigations conducted to date. The aim of COMET was to build models for the prediction of organ toxicity (mainly liver and kidney), from NMR spectra obtained from rodent urine and serum, from multiple toxicity studies (Lindon et al., 2005, 2003). A validation of these markers indicated that these methods could predict liver and kidney toxicity with specificities of 100% and 77% and sensitivities of 41% and 67%, respectively (Ebbels et al., 2007). These models are now being implemented by the six pharmaceutical companies involved in the consortium, in preclinical studies.

Another important general drug-related adverse effect to be considered is the change in the patient's intestinal microbiota composition, in response to antibiotic treatment. Pharmacometabonomics has also contributed significantly to this intensely studied research area, indicating that 87% of all metabolites in the intestinal metabolome are influenced by antibiotic treatment. Furthermore, a disturbance in a number of metabolic pathways, including bile acid, eicosanoid, and steroid hormone synthesis, were identified subsequent to antibiotic treatment (Antunes et al., 2011).

In addition to better describing the molecular mechanisms resulting in general adverse effects such as hepatotoxicity, nephrotoxicity, and intestinal microbiota response to antibiotic treatment, metabolomics has also been used to investigate the mechanisms of more specific drug-related side effects associated with specific treatment regimens. To this end, Loots, Wiid, Page, Mienie, and Helden (2005) evaluated the effects of the combined anti-TB drug, Rifater, on the metabolic profiles of Sprague-Dawley rats. The identified biomarkers indicated elevated oxidative stress levels in the animal models receiving treatment, and the metabolite profiles closely resembled that of human patients suffering from a multiple acyl-CoA dehydrogenase deficiency (MADD). These findings indicated that Rifater treatment could be linked to an inhibition of the electron transport chain flavoproteins, and the group subsequently indicated that this MADD metabolite profile, and hence the associated drug-related side effects, could be corrected for by the coadministration of melatonin. From these and other studies, it is evident that metabolomics can assist in a better description and understanding of the general and specific side effects related to various drugs, which may contribute to the development of safer treatment approaches, or the expansion of therapeutic strategies for the prevention or early management of these occurrences. Once again, it is important to consider all factors which may influence these metabolite changes, including the previously mentioned environmental and genetic factors, when identifying those biomarkers related to toxic insult (Beger, Sun, & Schnackenberg, 2010). Although adverse side effects are not uncommon for many prescription drugs, especially those targeting life-threatening disease states such as cancer for instance, it can also occur in drugs which are widely recognized as safe and are normally well tolerated by most patients. As discussed earlier, the largest contributor to the variation in individual patient xenobiotic metabolism, and hence their toxicology phenotypes, are polymorphisms in genes encoding for xenobiotic-metabolizing enzymes. However, various other factors have been identified to contribute to this, including the patient's health status (hepatic dysfunction, inflammation, infection, and cancer), drug–drug interactions, and exposure to contaminants such as heavy metals, all of which can influence the activities of these drug-metabolizing enzymes (Chen, Gonzalez, & Idle, 2007). Although pharmocogenomics is the primary approach used for the investigation of such variation, metabolomics studies have also shown to be useful in the elucidation of the mechanisms leading to these phenotypes. For example, a number of genetics studies have linked a polymorphism in the CYP2D6 with an excessive hypotensive response to the antihypertensive drug; debrisoquine. When investigating this phenomenon using LC–MS metabolomics, metabolite profiles from urine could differentiate the treatment response phenotypes (poor metabolizers vs. extensive metabolizers). As expected, debrisoquine was significantly higher in the profiles of poor responders, whereas the products of drug metabolism: 4-hydroxy-debrisoquine, and two-open ring products of debrisoquine: 2-(guanidinoethyl)benzoic acid (I) and 2-(guanidinomethyl) phenyl acetic acid (II), where detected in higher levels in the profiles of the good responders. The two CYP2D6 genotypes could thus be identified using biomarkers determined by metabolomics, proving that pharmacometabonomics has the capacity to identify individual variation in drug-related side effects, originating from these polymorphisms (Zhen et al., 2006). In a similar fashion, metabolomics can also be implemented to investigate other factors leading to this variation, thereby paving the way for the development of more effective, personalized medicines, with a lowered incidence of drug-induced side effects.

5. CONCLUDING REMARKS

The leap in technology over the past few decades has significantly contributed to our knowledge of biological systems on a molecular level, and the newly gained information has subsequently shed light on the variation that exists in the genetic, protein, and metabolite compositions of different individuals. The application of this knowledge to the development and implementation of personalized medicine is currently a hot topic among clinical researchers, and as described above, metabolomics has been a valuable contributor to this. However, before the identified metabolite biomarkers (or any other marker for that matter, identified via genomic, proteomic, or transcriptomic approaches) can be applied toward personalized clinical care, they should be validated using large sample cohorts, with all influencing variables, such as age, gender, habits, and diet considered. It should additionally be noted that current metabolomics, genomics, transcriptomics, and proteomics techniques require highly trained technical competence and infrastructure, and hence, is regarded to be rather expensive to perform, especially if these methods are being considered for use directly for clinical application to diagnostics. For this reason, various research efforts have been made toward using these newly identified diagnostic and prognostic markers for clinical application, through the development of simple, cheap tests, targeted for identifying these specific compounds/biomarkers previously identified, and validated using these "omics" approaches. The use of genomics, transcriptomics, proteomics, and also more recently, metabolomics data, in large databases such as DrugBank (Knox et al., 2011) (which is a comprehensive collection of drug mechanism and drug target information), proves that despite challenges in these fields, the contribution of all of these technologies to the development of personalized medicine is invaluable.

ACKNOWLEDGMENTS

The financial assistance of the National Research Foundation (NRF) of South Africa for this research is gratefully acknowledged. The opinions expressed and conclusions derived are those of the authors and are not necessarily those of the NRF.

REFERENCES

Antunes, C., Han, J., Ferreira, R., Lolic, P., Borchers, C., & Finlay, B. B. (2011). Effect of antibiotic treatment on the intestinal metabolome. *Antimicrobial Agents and Chemotherapy*, *55*(4), 1494–1503.

Baraldi, E., Carraro, S., Giordano, G., Reniero, F., Perilongoand, G., & Zacchello, F. (2009). Metabolomics: Moving towards personalized medicine. *Italian Journal of Pediatrics*, *35*(30), 12–16.

Beger, R., Sun, J., & Schnackenberg, L. (2010). Metabolomics approaches for discovering biomarkers of drug-induced hepatotoxicity and nephrotoxicity. *Toxicology and Applied Pharmacology*, *243*, 154–166.

Bogdanov, M., Matson, W., Wang, L., Matson, T., Saunders-Pullman, R., Bressman, S., et al. (2008). Metabolomic profiling to develop blood biomarkers for Parkinson's disease. *Brain*, *131*, 389–396.

Bonhommea, C., Gervaisa, C., & Laurencin, D. (2014). Recent NMR developments applied to organic–inorganic materials. *Progress in Nuclear Magnetic Resonance Spectroscopy*, *77*, 1–48.

Broeckling, C., Reddy, I., Duran, A., Zhao, X., & Sumner, L. (2006). MET-IDEA: Data extraction tool for mass spectrometry-based metabolomics. *Analytical Chemistry*, *78*, 4334–4341.

Carraro, S., Andreola, B., Alinovi, R., Corradi, M., Freo, L., & Da Dalt, L. (2008). Exhaled leukotriene B4 in children with community acquired pneumonia. *Pediatric Pulmonology*, *43*, 982–986.

Chen, C., Gonzalez, F., & Idle, J. R. (2007). LC-MS-based metabolomics in drug metabolism. *Informa Healthcare*, *39*, 581–597.

Clayton, A., Baker, D., Lindon, J., Everett, J., & Nicholson, J. (2009). Pharmacometabonomic identification of a significant host-microbiome metabolic interaction affecting human drug metabolism. *PNAS*, *106*(34), 14728–14733.

Das, M. K., Bishwal, S., Das, A., Dabral, D., Badireddy, V., Pandit, B., et al. (2015). Deregulated tyrosine – phenylalanine metabolism in pulmonary tuberculosis patients. *Journal of Proteome Research*, *14*, 1947–1956.

de Carvalho, L. P., Darby, C., Rhee, K., & Nathan, C. (2011). Nitazoxanide disrupts membrane potential and intrabacterial ph homeostasis of mycobacterium tuberculosis. *ACS Medicinal Chemistry Letters*, *2*, 849–854.

de Villiers, L., & Loots, D. T. (2013). Using metabolomics for elucidating the mechanisms related to tuberculosis treatment failure. *Current Metabolomics*, *1*(4), 306–317.

Debas, H., Laxminarayan, R., & Straus, S. (2006). *Complementary and alternative medicine. In D. Jamison (Ed.), Disease control priorities in developing countries* (2nd ed.). Washington, DC, USA: World Bank. http://www.ncbi.nlm.nih.gov/books/NBK11796/.

Denery, J., Nunes, A., Hixon, M., Dickerson, T., & Janda, K. (2010). Metabolomics-based discovery of diagnostic biomarkers for onchocerciasis. *PLoS Neglected Tropical Diseases, 4*(10). http://dx.doi.org/10.1371/journal.pntd.0000834.

Deng, M., Eisen, H., & Mehra, M. (2006). Noninvasive discrimination of rejection in cardiac allograft recipients using gene expression profiling. *American Journal of Transplantation, 6*, 150–160.

Du Preez, I., & Loots, D. T. (2012). Altered fatty acid metabolism due to rifampicin-resistance conferring mutations in the rpoB gene of mycobacterium tuberculosis: Mapping the potential of pharmaco-metabolomics for global health and personalized medicine. *A Journal of Integrative Biology, 16*(11), 596–603.

Dunn, W., Bailey, N., & Johnson, H. (2005). Measuring the metabolome: Current analytical technologies. *Analyst, 130*, 606–625.

Dworkina, R., McDermottb, M., Farrarc, J., O'Connord, A., & Senn, S. (2014). Interpreting patient treatment response in analgesic clinical trials: Implications for genotyping, phenotyping, and personalized pain treatment. *PAIN, 155*, 457–460.

Ebbels, T., Keun, H., Beckonert, O., Bollard, M., Lindon, J., Holmes, E., et al. (2007). Prediction and classification of drug toxicity using probabilistic modeling of temporal metabolic data: The consortium on metabonomic toxicology screening approach. *Journal of Proteome Research, 6*, 4407–4422.

Garrod, S., Bollard, M., Nicholls, A., Connor, S., Connelly, J., Nicholson, J., et al. (2005). Integrated metabonomic analysis of the multiorgan effects of hydrazine toxicity in the rat. *Chemical Research in Toxicology, 18*(18), 115–122.

Ginsburg, G. S., & Willard, H. F. (2009). Genomic and personalized medicine: Foundations and applications. *Translational Research, 154*, 277–287.

Halket, J. M., Waterman, D., Przyborowska, A. M., Patel, R. K., Fraser, P. D., & Bramley, P. M. (2005). Chemical derivatization and mass spectral libraries in metabolic profiling by GC/MS and LC/MS/MS. *Journal of Experimental Botany, 56*, 219–243.

Halouska, S., Chacon, O., Fenton, R., Zinniel, D., Barletta, R., & Powers, R. (2007). Use of NMR metabolomics to analyze the targets of D-cycloserine in mycobacteria: Role of D-alanine racemase. *Journal of Proteome Research, 6*, 4608–4614.

Heidecker, B., Kasper, E., Wittstein, I., Champion, H., Breton, E., Russell, S., et al. (2008). Transcriptomic biomarkers for individual risk assessment in new-onset heart failure. *Circulation, 118*, 238–246.

Hernandez, L., & Blazer, D. (2006). *Genetics and Health. In Institute of Medicine (US) Committee on Assessing Interactions Among Social, Genes, Behavior, and the Social Environment (Ed.), Moving beyond the nature/nurture debate.* Washington, DC: National Academies Press (US). http://www.ncbi.nlm.nih.gov/books/NBK19932/.

Holmes, E., Bonner, F., Sweatman, B., Lindon, J., Beddell, C., Rahr, E., et al. (1992). Nuclear magnetic resonance spectroscopy and pattern recognition analysis of the biochemical processes associated with the progression of and recovery from nephrotoxic lesions in the rat induced by mercury(II) chloride and 2-bromoethanamine. *Molecular Pharmacology, 42*, 922–930.

Hong, H., Zhang, W., Su, Z., Shen, J., Ge, W., Ning, B., et al. (2013). Next-generation sequencing (NGS): A revolutionary technology in pharmacogenomics and personalized medicine. In *Omics for personalized medicine* (pp. 39–61). India: Springer.

Jain, K. K. (2009). *Textbook of personalized medicine.* New York, NY: Springer Science +Business Media.

Ji, Y., Hebbring, S., Zhu, H., Jenkins, G., Biernacka, J., Snyder, K., et al. (2011). Glycine and a glycine dehydrogenase (GLDC) SNP as citalopram/escitalopram response biomarkers in depression: Pharmacometabolomics-informed pharmacogenomics. *Clinical Pharmacology & Therapeutics*, *89*(1), 97–104.

Jirtle, R. L., & Skinner, M. K. (2007). Environmental epigenomics and disease susceptibility. *Nature Reviews*, *8*, 253–262.

Johnson, C., Patterson, A., Idle, J., & Gonzalez, F. (2012). Xenobiotic metabolomics: Major impact on the metabolome. *Annual Review of Pharmacology and Toxicology*, *52*, 37–56.

Keun, H. C. (2006). Metabonomic modeling of drug toxicity. *Pharmacology & Therapeutics*, *109*, 92–106.

Knox, C., Law, V., Jewison, T., Liu, P., Ly, S., Frolkis, A., et al. (2011). DrugBank 3.0: A comprehensive resource for 'Omics' research on drugs. *Nucleic Acids Research*, *39*, D1035–D1041.

Lan, K., & Jia, W. (2010). An integrated metabolomics and pharmacokinetics strategy for multi-component drugs evaluation. *Current Drug Metabolism*, *11*(1), 105–114.

Lawton, K., Berger, A., Mitchell, M., Milgram, E., Evans, A., Guo, L., et al. (2008). Analysis of the adult human plasma metabolome. *Future Medicine*, *9*(4), 383–397.

Lecea, M. G., & Rossbach, M. (2012). Translational genomics in personalized medicine–scientific challenges en route to clinical practice. *The HUGO Journal*, *6*(2), 1–9.

Lenz, E., Nicholson, J., Wilson, I., & Timbrell, J. (2000). A 1H NMR spectroscopic study of the biochemical effects of ifosfamide in the rat: Evaluation of potential biomarkers. *Biomarkers*, *5*, 424–435.

Lindon, J., Keun, H., Ebbels, T., Pearce, J., Holmes, E., & Nicholson, J. (2005). The consortium for metabonomic toxicology (COMET): Aims, activities and achievements. *Pharmacogenomics*, *6*, 691–699.

Lindon, J., Nicholson, J., Holmes, E., Antti, H., Bollard, M., Keun, H., et al. (2003). Contemporary issues in toxicology the role of metabonomics in toxicology and its evaluation by the COMET project. *Toxicology and Applied Pharmacology*, *187*, 137–146.

Loots, D. T. (2014). An altered mycobacterium tuberculosis metabolome induced by katG mutations resulting in isoniazid resistance. *Antimicrobial Agents and Chemotherapy*, *58*(4), 2144–2149. http://dx.doi.org/10.1128/AAC.02344-13.

Loots, D. T., Wiid, I. J., Page, B. J., Mienie, L. J., & Helden, P. D. (2005). Melatonin prevents the free radical and MADD metabolic profiles induced by antituberculosis drugs in an animal model. *Journal of Pineal Research*, *38*(2), 100–106.

Lu, C., Deng, J., Li, L., Wang, D., & Li, G. (2014). Application of metabolomics on diagnosis and treatment of patients with psoriasis in traditional Chinese medicine. *Biochimica et Biophysica Acta*, *1844*, 280–288.

Madsen, R., Lundstedt, T., & Trygg, J. (2010). Chemometrics in metabolomics—A review in human disease diagnosis. *Analytica Chimica Acta*, *659*, 23–33.

Marriott, P., & Shellie, R. (2002). Principles and applications of comprehensive two-dimensional gas chromatography. *Trends in Analytical Chemistry*, *21*, 573–583.

Matthews, H. (1995). Racial, ethnic and gender differences in response to medicines. *Drug Metabolism and Personalized Therapy*, *12*(2), 77–92.

Nishiumi, S., Kobayashi, T., Ikeda, A., Yoshie, T., Kibi, M., Izumi, Y., et al. (2012). A novel serum metabolomics-based diagnostic approach for colorectal cancer. *PLoS One*, *7*(7). http://dx.doi.org/10.1371/journal.pone.0040459.

Olivier, I., & Loots, D. (2012). A metabolomics approach to characterise and identify various Mycobacterium species. *Journal of Microbiological Methods*, *88*(3), 419–426.

Ono, M., Shitashige, M., Honda, K., Isobe, T., Kuwabara, H., Matsuzuki, H., et al. (2006). Label-free quantitative proteomics using large peptide data sets generated by nanoflow liquid chromatography and mass spectrometry. *Molecular & Cellular Proteomics*, *5*, 1338–1347.

Redekop, K., & Mladsi, D. (2013). The faces of personalized medicine: A framework for understanding its meaning and scope. *Value in Health*, *16*, S4–S9.

Robertson, D., Reily, M., Sigler, R., Wells, D., Paterson, D., & Braden, T. (2000). Metabonomics: Evaluation of nuclear magnetic resonance (NMR) and pattern recognition technology for rapid in vivo screening of liver and kidney toxicants. *Toxicological Sciences*, *57*, 326–337.

Rozen, S., Cudkowicz, M., Bogdanov, M., Matson, W., Kristal, B., Beecher, C., et al. (2005). Metabolomic analysis and signatures in motor neuron disease. *Metabolomics*, *1*(2), 101–108.

Salari, K. (2009). The dawning era of personalized medicine exposes a gap in medical education. *PLoS Medicine*, *6*(8), 1–3.

Schnackenberg, L., Dragan, Y., Reily, M., Robertson, D., & Beger, R. (2007). Evaluation of NMR spectral data of urine in conjunction with measured clinical chemistry and histopathology parameters to assess the effects of liver and kidney toxicants. *Metabolomics*, *2*, 87–100.

Schwartz, G., Hughes, K., & Lynch, H. (2008). Proceedings of the international consensus conference on breast cancer risk, genetics, & risk management. *Cancer*, *113*, 2627–2637.

Sugimoto, M., Wong, D., Hirayama, A., Soga, T., & Tomita, M. (2010). Capillary electrophoresis mass spectrometry-based saliva metabolomics identified oral, breast and pancreatic cancer-specific profiles. *Metabolomics*, *6*, 78–95.

Sumner, S., Burgess, J., Snyder, R., Popp, J., & Fennell, T. (2010). Metabolomics of urine for the assessment of microvesicular lipid accumulation in the liver following isoniazid exposure. *Metabolomics*, *6*, 238–249.

Szymanska, E., Bouwman, J., Strassburg, K., Vervoort, J., Kangas, A., Soininen, P., et al. (2012). Gender-dependent associations of metabolite profiles and body fat distribution in a healthy population with central obesity: Towards metabolomics diagnostics. *OMICS A Journal of Integrative Biology*, *16*(12), 652–667.

Tiziani, S., Lodi, A., Khanim, F., Viant, M., Bunce, C., & Günther, U. (2009). Metabolomic profiling of drug responses in acute myeloid leukaemia cell lines. *PLoS One*, *4*(4). http://dx.doi.org/10.1371/annotation/39584d38-04f5-4b37-bfd8-eae2318ec6f9.

Trupp, M., Zhu, H., Wikoff, W., Baillie, R., Zeng, Z.-B., Karp, P., et al. (2012). Metabolomics reveals amino acids contribute to variation in response to simvastatin treatment. *PLoS One*, 7(7). http://dx.doi.org/10.1371/journal.pone.0038386.

Trygg, J., Holmes, E., & Lundstedt, T. (2007). Chemometrics in metabonomics. *Journal of Proteome Research*, *6*, 469–479.

Urban, M., Kavvadiasa, D., Riedela, K., Scherera, G., & Trickerb, A. (2006). Urinary mercapturic acids and a hemoglobin adduct for the dosimetry of acrylamide exposure in smokers and nonsmokers. *Inhalation Toxicology: International Forum for Respiratory Research*, *18*(10), 831–839.

Voora, D., McLeod, H., Eby, C., & Gage, B. (2005). The pharmacogenetics of coumarin therapy. *Pharmacogenomics*, *6*, 503–513.

Wang, X., Zhang, A., Wang, P., Sun, H., Wu, G., Sun, W., et al. (2013). Metabolomics coupled with proteomics advancing drug discovery toward more agile development of targeted combination therapies. *Molecular & Cellular Proteomics*, *12*, 1226–1238.

Wei, S., Liub, L., Zhang, J., Bowers, J., Gowda, G. N., Seeger, H., et al. (2013). Metabolomics approach for predicting response to neoadjuvant chemotherapy for breast cancer. *Molecular Oncology*, 7, 297–307.

Wheelock, C., Goss, V., Balgoma, D., Nicholas, B., Brandsma, J., Skipp, P., et al. (2013). Application of 'omics technologies to biomarker discovery in inflammatory lung diseases. Omics technologies to biomarker discovery. *The European Respiratory Journal*, *42*, 802–825.

Wiesner, G., Slavin, T., & Barnholtz-Sloan, J. (2009). Colorectal cancer. In H. Willard & G. Ginsburg (Eds.), *Genomic and personalized medicine* (pp. 879–897). Durham: Elsevier.

Wishart, D. (2010). Computational approaches to metabolomics. In R. Matthiesen (Ed.), *Bioinformatics methods in clinical research* (pp. 283–313). Hatfield, Hertfordshire, UK: Humana Press.

Xia, J., Psychogios, N., Young, N., & Wishart, D. (2009). MetaboAnalyst: A web server for metabolomic data analysis and interpretation. *Nucleic Acids Research*. Web Server issue: W652e60.

Zhang, M., Rosenberg, P., Brown, D., Preiss, L., Konkle, B., Eyster, E., et al. (2006). Correlates of spontaneous clearance of hepatitis C virus among people with hemophilia. *Blood*, *107*, 892–897.

Zhen, Y., Slanar, O., Krausz, K. W., Chen, C., Slavik, J., McPhail, K. L., et al. (2006). 3,4-Dehydrodebrisoquine, a novel debrisoquine metabolite formed from 4-hydroxydebrisoquine that affects the CYP2D metabolic ratio. *Drug Metabolism and Disposition*, *34*, 1563–1574.

CHAPTER FOUR

Clinical Perspectives on Targeting Therapies for Personalized Medicine

Donald R.J. Singer*[,1], Zoulikha M. Zaïr[†]
*Fellowship of Postgraduate Medicine, London, United Kingdom
[†]Warwick Medical School, University of Warwick, Coventry, United Kingdom
[1]Corresponding author: e-mail address: fpm.chandos@gmail.com; drjsinger@gmail.com

Contents

Advances in Protein Chemistry and Structural Biology, Volume 102
ISSN 1876-1623
http://dx.doi.org/10.1016/bs.apcsb.2015.11.003

Abstract

Expected benefits from new technology include more efficient patient selection for clinical trials, more cost-effective treatment pathways for patients and health services and a more profitable accelerated approach for drug developers. Regulatory authorities expect the pharmaceutical and biotechnology industries to accelerate their development of companion diagnostics and companion therapeutics toward the goal of safer and more effective personalized medicine, and expect health services to fund and prescribers to adopt these new therapeutic technologies.

This review discusses the importance of a range of new approaches to developing new and reprofiled medicines to treat common and serious diseases, and rare diseases: new network pharmacology approaches, adaptive trial designs with enriched populations more likely to respond safely to treatment, as assessed by companion diagnostics for response and toxicity risk and use of "real world" data.

Case studies are described of single and multiple protein drug targets in several important therapeutic areas. These case studies also illustrate the value and complexity of use of selective biomarkers of clinical response and risk of adverse drug effects, either singly or in combination.

1. INTRODUCTION

Francis Collins, Director of the US National Institutes of Health, has stated that "the 21st century is the century of biology. The nation that invests in biomedical research will reap untold rewards in its economy and the health of its people" (Collins, 2015). This assertion reflects optimism on two fronts: that better characterization of disease will lead to more effective and safer use of current treatments for common long-term medical disorders; and that new biomedical technology and expertise will accelerate development of improved treatments or new treatments where none currently exist. Of particular interest are new approaches to currently refractory cancers, resolving acquired resistance to cancer chemotherapy and resistance to antimicrobial chemotherapy, the ability to deal rapidly with epidemic viral diseases such as Ebola hemorrhagic fever, SARS (Severe Acute Respiratory Syndrome), and MERS (Miiddle Eastern Respiratory Syndrome), developing disease-modifying treatment for rare inherited diseases, and improving treatment for endemic infections in the developing world for which current treatments are toxic, poorly effective, or not available.

Personalizing use of medicines needs to involve development of the right drugs in parallel with biomarkers of response, both for disease outcomes and for adverse drug reactions (ADRs). Accelerated drug development usually means more dependence on surrogate markers of response. Regulators need to ensure that long-term monitoring and reporting is systematic for clinically important endpoints and coupled to the right to continued licensing of medicines for clinical use.

This review will consider the extent to which this promise may be met with respect to precision medicine—the application of molecular approaches to improving effectiveness and safety in the use of medicines within stratified groups of patients with well-defined genotypes and phenotypes. This review also discusses the importance of timely development of companion diagnostics to improve treatment planning in selected patients by increasing the likelihood of effectiveness and reducing the risk of adverse affects from an associated companion therapeutic.

2. WHAT ARE PERSONALIZED MEDICINES?

The USA drugs regulatory authority the Food and Drugs Administration, when describing personalized medicines (Marburger & Kvamme, 2008), notes that this does not mean "the creation of drugs or medical devices that are unique to a patient but rather the ability to classify individuals into subpopulations that differ in their susceptibility to a particular disease or their response to a specific treatment. Preventive or therapeutic interventions can then be concentrated on those who will benefit, sparing expense and side effects for those who will not."

This reflects the clinical practicality of delivering "personalizing" medicines, as precision medicine: a stratified approach to treatment of smaller subgroups of patients, as technology to test for biomarkers of drug effectiveness and toxicity becomes increasingly affordable.

3. RARE DISEASES

Around 7300 rare genetic diseases have been included in the registry prepared by the US National Organization for Rare Disorders (NORD, 2015). NORD has also provided support for the development of rare disease registries within the ambit of the European Union. These registries will play a vital role in identifying patients and their families willing to take part in research into new and improved treatments for serious rare diseases. However, even when these registries are pooled, new approaches will be needed to obtain clinical data on the safety and effectiveness of treatments in very small groups of patients.

4. EVIDENCE FOR PRECISION MEDICINES FROM REAL WORLD DATA

A major challenge in developing drugs for precision medicine concerns achieving adequate sample size both for insight into efficacy and safety of medicines: obtaining sufficient statistical power for understanding use of medicines in smaller subsets of patients with heterogeneous variants of common diseases and in the often very small numbers of patients with individual rare diseases. Interest is rapidly growing in the idea of complementing results of "gold standard" formal randomized controlled trials (RCTs) with use of "real world data" (Sun et al., 2015). To be as effective as possible, accessed data should be homogeneous in content and standard in use of taxonomy within health system electronic health records, within disease registries, observational cohorts, and other relevant health data sets.

The quality of "big data" obtained from within electronic health records will need to be improved, supported by new ethical approaches to patient confidentiality to ensure as far as possible completeness of the information in EHRs. In addition, new statistical approaches will be needed to analyze these complex real world data sets.

To these ends, public–private partnerships supported by the European Union's Innovative Medicines Initiatives are currently exploring new approaches that will be needed within two major programmes: Accelerated Development of Appropriate Patient Therapies (ADAPT SMART, 2015) and IMI Get Real (Schmidt, Klungel, & Groenwold, 2016), both projects funded by the EU Innovative Medicines Initiative. A current limiting factor is the estimated shortfall of 125,000 data analysts needed across the EU alone for effective development, maintenance, and analysis of current and envisaged very large clinical datasets.

In the United States, the 2007 Food and Drugs Administration Amendment Act provided for instruments toward clinical database development, including with regard to informed patient consent, with the aim of providing access to health data on 100 m individuals, by using systems to be supported by 18 data-partners and 35 subcontractors.

5. USING "REAL WORLD" DATA

The Sentinel Initiative (Southworth, Reichman, & Unger, 2013) is a Harvard-based database sourced from patient data derived from insurance claim reports and other administrative data. The Sentinel Initiative (Southworth et al., 2013) is already beginning to be used to assess "in world" risk from selected drugs and vaccines, providing reports on issues of drug safety concern within months, rather than the years needed for RCTs. Outcomes for the FDA regulators from projects arising from this initiative have included questions on a trivalent flu vaccine and febrile seizures, with no increase in risk observed. Sentinel and other international databases have also identified the need for new warnings in patient information leaflet that the live rotavirus vaccine Rotarix may cause intussusception (Rha et al., 2014) and that the angiotensin Type 1 receptor olmesartan may cause sprue as an important adverse effect (FDA, 2013).

A study of the Sentinel database to assess links between dabigatran and bleeding risk illustrates some of the challenges to use of the data. Dabigatran was licensed as an orally active alternative to reducing risk of embolic stroke in patients with atrial fibrillation due to nonvalvular causes. RE-LY, the initial RCT, indicated that dabigatran was better than warfarin at preventing stroke, with similar overall bleeding risk (Connolly et al., 2009). Following the market authorization for dabigatran, the FDA received an expectedly large number of reports of serious bleeding episodes associated with the drug. This was interpreted as possibly due to underreporting about bleeding due to the much older drug warfarin and overreporting for the new drug dabigatran. Other possible reasons included inability to obtain sufficient personal data on individuals featured in the ADR reports, including for details of patient populations, comorbidity, concurrent medications, drug dose and whether it had been adjusted for renal function (Southworth et al., 2013). The Sentinel database suggested that in fact, as in the RE-LY RCT, bleeding risk from intracranial and gastrointestinal hemorrhages for dabigatran was similar to that from warfarin (Southworth et al., 2013). However, the authors of the Sentinel report noted that its limitations also included lack of adjustment for some key variables, including confounding by indication for the treatment (Freemantle et al., 2013), and lack of validation about whether an insurance claim code reflected an actual bleeding episode (Southworth et al., 2013). Further analyses are planned to aim to adjust as far as possible for these key factors.

6. ETHICAL CONCERNS

Barriers to patients' willingness for their data to be made available for clinical researchers have to be anticipated and resolved so that patients will as far as possible not to be deterred from allowing their clinical data to be used. Patients' concerns include that by agreeing to testing with regard to their personal risk of subtypes of serious diseases and their likelihood of responding to treatment, there may be an adverse impact on their financial position, e.g., on employability, eligibility for a mortgage and other loans, and obtaining insurance cover.

By voluntary agreement, insurance companies in some jurisdictions have previously waived access to knowledge of the results of testing for biomarkers of risk of serious disease and by implication of reduced likelihood of optimal response to treatment. This area may need statutory protection if these voluntary agreements prove insufficient to protect patients and the public from the impact of testing for personal risk, when serious and/or currently untreatable disease risks are identified. Of note, the genetic testing company 23andMe reports having genotyped 1 million people worldwide, of whom about 80% are reported to have consented to let their data be used for research (Chen, 2015).

Precision medicine testing may result in depriving patients of treatments, when in clinical practice the benefits of these treatments may be better than anticipated from formal clinical trials (Singer & Watkins, 2012). The response to treatment may be influenced by compensatory effects of many factors, including other treatments that affect drug-metabolizing enzymes or drug binding, effects of comorbidity and lifestyle factors on pharmacokinetics, single-nucleotide polymorphisms (SNPs) or copy number polymorphisms (CNPs) in genes for other metabolic pathways, and differences in alternative signaling pathways (Singer & Watkins, 2012).

This raises the major ethical question: what level of certainty about the potential benefit of a "precision medicine" is acceptable in patients with no other treatment options, or much less effective or toxic options within current therapy?

7. ADAPTIVE TRIAL DESIGN

Growing interest in development of stratified medicines is leading to the increasing importance of subgroup analyses in confirmatory clinical trials and in particular achieving adequate statistical power in relatively small subgroups of patients (Ogbagaber, Karp, & Wahed, 2015; Shan, Wilding, Hutson, & Gerstenberger, 2015; Stallard, Hamborg, Parsons, & Friede, 2014).

For example, the B-Raf protein regulates normal cell growth and the *BRAF V600E*-activating gene mutation is important for tumor growth in late-stage malignant melanoma in around 50% of patients with this aggressive tumor type (Chapman et al., 2011; McArthur et al., 2014). As an example of adaptive trial design, the B-Raf enzyme inhibitor vemurafenib has only been assessed for treating malignant melanoma in patients who have the above mutation (Chapman et al., 2011; McArthur et al., 2014), reducing costs of developing the drug. This mutation occurs in around 8% of colorectal and other solid cancers (Yang et al., 2012), suggesting that the *BRAF V600E* companion diagnostic could be used for adaptive trial design when studying treatment with inhibitors of B-Raf for these additional types of cancer.

New adaptive designs for clinical trials include monitoring the relative effectiveness of treatments in subgroups of interest, with the option then to alter the recruitment strategy for future patient entry into trials. These adaptive approaches need new analytical methods and resolution of discussion with regulators about use of resulting data in applications for licensing new drugs and devices evaluated in this way (Stallard et al., 2014).

These medicine adaptive pathways to patients (MAPPs) aim to support new types of clinical trials that adapt to a given patient's response (Schulthess, Chlebus, Bergstrm, & Baelen, 2014). These MAPPs are intended to lead to accelerated but limited commercial marketing authorization for a patient group to benefit from a new therapeutic agent, while additional clinical endpoints are validated.

8. COMPANION DIAGNOSTICS

For precision medicine, diagnostics are needed both for the initial decision to prescribe and, if so, whether dose modification is needed in the light of genetic or functional information on individual variability in pharmacokinetics. Further tests are needed after a drug treatment has been started to monitor disease outcome and to monitor for adverse effects.

The concept of combination diagnostics (Singer & Watkins, 2012) was applied initially in the 1980s, mainly in veterinary and dental practice. The further term theranostics—used interchangeably with the phrase Companion Diagnostics—describes mandatory safety or efficacy diagnostic tests to be used before a companion drug can be prescribed (Singer & Watkins, 2012).

The US drugs regulatory authority the FDA defines a companion diagnostic device as an "*in vitro* diagnostic device or an imaging tool that provides information that is essential for the safe and effective use of a corresponding therapeutic product" (FDA, 2015a). The therapeutic and its companion diagnostic are defined as a combination product (FDA, 2015b).

Within the jurisdictions of the FDA and other national regulatory authorities (Crabb, Marlow, Bell, & Newland, 2012), a companion diagnostic device with a particular therapeutic product is "stipulated in the instructions for use in the labeling of both the diagnostic device and the corresponding therapeutic product, as well as in the labeling of any generic equivalents and biosimilar equivalents of the therapeutic product" (FDA, 2015a). This means that the Companion Diagnostic test must be performed before a new partner therapeutic can be prescribed.

Understanding whether these tests will improve cost-effectiveness needs consideration of the costs of performing and interpreting assays for these biomarkers of drug transport, drug costs compared to other treatment options if they exist, the epidemiology of these biomarkers, and whether there is heterogeneity in their tissue expression (Singer & Watkins, 2012).

Licensing by national regulators of pharmacogenetic companion diagnostics to coincide with approval for a new drug has the aims of both improving safety in use of potentially toxic drugs and minimizing future treatment costs resulting from better treatment effectiveness. The USA FDA has already licensed several companion diagnostic devices for use with a corresponding therapeutic product (FDA, 2015a).

It is ideal on clinical and cost grounds that effective companion diagnostic devices are identified prior to licensing of a treatment (Fig. 1). This now forms part of strategy for major pharmaceutical companies with, for example, ~50% of new drugs under development by Roche having diagnostics developed in parallel.

This is important both to ensure that effective treatments can be selected and that patients can be protected from predictable significant ADRs. It is also important for effective planning of the costs of a new treatment where those costs will include additional mandated testing. Agreeing reimbursement of the combined costs of testing and treatment is much more difficult when genetic biomarkers and other companion diagnostic devices are identified after initial drug licensing.

9. BIOLOGICAL TREATMENTS

Biologics as biopharmaceuticals are typically a class of therapeutic proteins synthesized by biological processes enabled by recombinant DNA technology. Monoclonal antibodies (MCA) are similar to the antibody

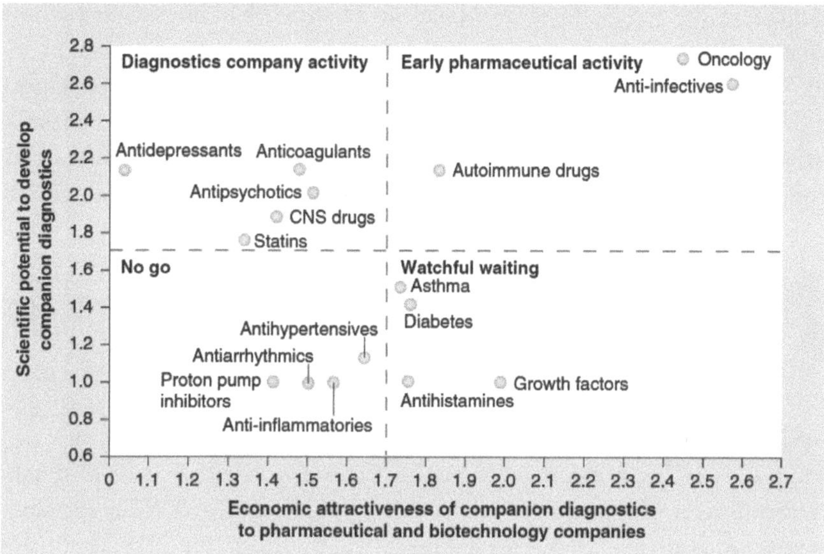

Figure 1 Economic attractiveness of companion diagnostics to pharmaceutical and biotechnology companies. *Reproduced with permission from Davis, Furstenthal, and Desai (2009).*

proteins that the human immune system uses but MCA are designed specifically to block a biochemical or specific cellular target in the body. Examples of MCA for therapeutic use are given in the case studies below. Receptor constructs (fusion proteins) are usually based on a naturally occurring receptor linked to an immunoglobulin frame. The receptor provides the construct with specificity with the immunoglobulin-structure conferring stability.

Nucleic acid biologics are also attractive as potential new therapeutic agents, based on the ability of current technology to achieve rapid synthesis of small interfering RNAs to inhibit gene expression. For example, the public–private RNAi consortium based at the Broad Institute in Boston has synthesized 160,000 RNAi constructs for evaluation for potential therapeutic use, targeted against 15,000 human and 15,000 mouse genes (Broad, 2015). Noncoding microRNAs may also provide biological options for treatment targets (Pichler & Calin, 2015). However, delivery to tissues of interest remains a major limiting factor.

The biologics class of medicines has clinical indications for treating a wide range of organ-based and systemic diseases. Their high cost has led to interest in the development of biosimilars (Ramanan & Grampp, 2014). These can be licensed at much lower cost, however, differences in manufacturing and quality control may lead to the risk of unexpected serious adverse effects, for example, pure red cell aplasia resulting from patient antibodies being formed to a biosimilar to erythropoietin (EPO) (Macdougall et al., 2012). These antibodies to EPO not only neutralize the biological activity of the administered biological EPO but also endogenous EPO, thus stopping red cell production by the bone marrow (Macdougall et al., 2012).

10. CASE STUDIES: TARGETS FOR PRECISION MEDICINES AND COMPANION DIAGNOSTICS

10.1 Drug Transporter Proteins

Drug transporter proteins are involved in the pharmacokinetics of many drugs and are categorized according to their role in either cellular uptake or elimination of transported molecules (Wen et al., 2015; Zair, Eloranta, Stieger, & Kullak-Ublick, 2008). Influx transporters (Fig. 2), comprising organic anion transporter proteins (OATPs), organic anion transporters (OATs), and organic cation transporters (OCTs), transport drugs across

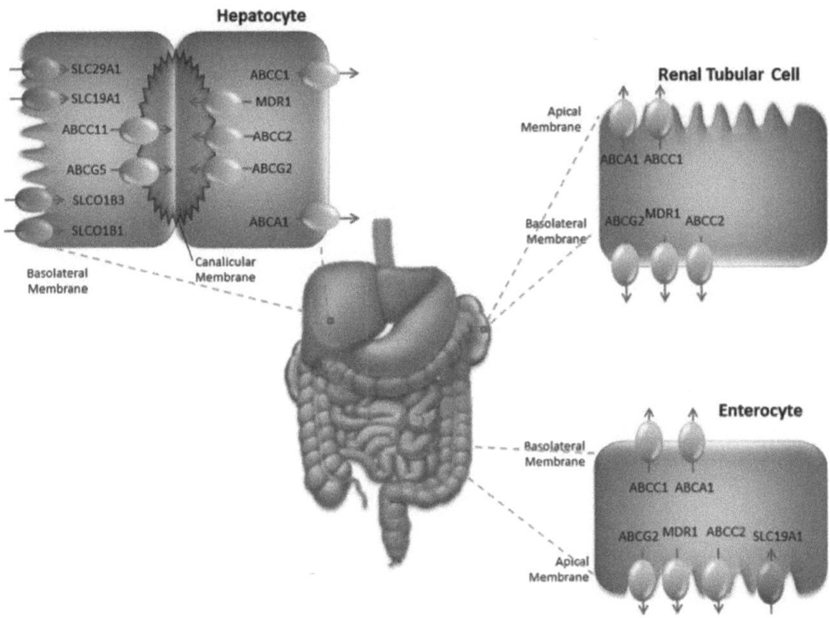

Figure 2 Key influx transporters (represented by red (gray in the print version) ovals) belong to the SLC protein family and major efflux transporters (represented by green (gray in the print version) ovals) belong to the ABC protein family. *Adapted and reproduced with permission from Zair et al. (2008).*

epithelial barriers such as enterocytes, hepatocytes, and renal tubule cells (Hagenbuch & Stieger, 2013; Zair et al., 2008). Efflux transporters (Fig. 2), common among which are members of the ATP-binding cassette (ABC) superfamily, are responsible for cellular efflux and drug resistance (Holland, 2011; Zair et al., 2008). Perturbations in transporter-mediated drug uptake dramatically influence drug pharmacokinetics, manifestations including subtherapeutic drug concentration, drug resistance (Wen et al., 2015), and/or adverse effects.

Particular SNPs and CNPs (Sebat et al., 2004) may alter drug transporter protein expression and function and thus alter drug pharmacokinetics (Franke, Gardner, & Sparreboom, 2010; Megaraj et al., 2011; Niemi, Pasanen, & Neuvonen, 2011). It is thus important that drug design and development take account of potential effects on candidate bioactive compounds of variability in drug transporter activity.

Testing for pharmacogenetic and functional variants in drug transporters also provides opportunities for developing new companion diagnostics. Thus, a patient's drug transporter profile may be useful in predicting treatment response (Zair et al., 2008), as well as risk of adverse effects from treatments of cancer, and a wide range of other therapeutic classes. For instance, increased OATP1B1 protein levels are seen in tumors of the colon, endometrium, esophagus, lung, prostate, stomach, testis, and bladder, and genetic variation both in OATP1B1 and in OATP1B3 contribute to the *in vitro* cytotoxicity of paclitaxel in ovarian cancer cells (Svoboda et al., 2011). *OATP1B1* genetic variants were also associated with increased irinotecan-induced treatment response and survival in colorectal cancer patients in a prospective multicenter study (Huang et al., 2013). Regarding efflux transporters, SNPs in genes expressing ABCB1 multidrug resistance protein variants have been related to substrate and inhibitor-dependent functional modifications in *in vitro* studies and reduced expression in tissues. Furthermore, ABCB1 proteins have been shown to have a potent ability to interact with numerous clinically important kinase inhibitors, including imatinib, gefitinib, and erlotinib, conferring cancer treatment resistance (Gong & Kim, 2013; Ozvegy-Laczka et al., 2004; Shi et al., 2007; Sugimoto, Tsukahara, Ishikawa, & Mitsuhashi, 2005).

10.2 Pleiotropic Effects of Statins

Statins provide an illustrative case study for the pleiotropic effects of drugs and for the potential for development of companion diagnostics. Statins are well-recognized to reduce the incidence and severity of ischaemic heart disease and atheromatous vascular complications in the brain and other circulations by inhibiting the key rate-limiting enzyme for cholesterol synthesis HMG CoA reductase (Goldstein & Brown, 1990), and by upregulation of LDL receptors to increase cholesterol scavenging (Ma et al., 1986). However, statins also have important pleiotropic actions (Koh, Sakuma, & Quon, 2011; Liao & Laufs, 2005), independent of their lipid-lowering properties. These pleiotropic actions may contribute both to the cardiovascular benefits of statins and to their adverse effects. For example, direct (Hauck et al., 2007; Hermida et al., 2013; Rajtik et al., 2012; Thuc et al., 2010; Zheng & Hu, 2005) and indirect evidence (Jenkins, Grieve, Yacoub, & Singer, 1996; Ozawa et al., 2009) in experimental models and in patients suggests that beneficial effects of statins may include improving cardiac function by effects independent of their cholesterol-lowering effects, by modulating a range of signaling pathways, including mitigation of apoptosis (Rajtik et al., 2012).

However, recent secondary research suggests that part of the cardiovascular benefits in some patients on statin treatment may be offset by the increased incidence of Type 2 diabetes mellitus which may be caused by the drug as a pleiotropic effect (Sattar et al., 2010). In a meta-analysis of 13 statin trials (91,140 participants), 2226 assigned statins and 2052 assigned control developed diabetes over an average of 4 years. Statin therapy was associated with a 9% increase in risk for incident diabetes (95% confidence interval (CI) 2–17%) (Sattar et al., 2010).

The most serious adverse affect that limits ability to take long-term statin treatment is rhabdomyolysis—injury to skeletal muscle cells leading to leak of myoglobin into the circulation and risk of acute kidney failure. An SNP with population prevalence of 15% has now been identified linked to serious statin-induced myopathy, the SNP found in 60% of affected patients in the initial report (Ghatak, Faheem, & Thompson, 2010; Link et al., 2008). Findings arose from a genomewide association study (GWAS) in 85 subjects with confirmed or developing myopathy and 90 controls. All were on a high daily dose of simvastatin (80 mg) as part of a trial with ~12,000 participants.

The GWAS showed a single highly significant graded association of myopathy with the rs4363657 SNP within *SLCO1B1* on chromosome 12 (odds ratio for myopathy: 4.5 (95% CI 2.6–7.7) per copy of the C allele; 16.9 (95% CI 4.7–61.1) in CC vs. TT homozygotes). This region encodes the organic anion-transporting polypeptide OATP1B1 that regulates hepatic uptake of statins and is in nearly complete linkage disequilibrium with the rs4149056 SNP that is linked to statin metabolism. The association of rs4149056 with myopathy was replicated in a trial of 40 mg of simvastatin daily, which also showed an association between rs4149056 and the cholesterol-lowering effects of simvastatin. No SNPs in any other region were clearly associated with myopathy.

Drugs metabolized by the OATP1B1 variant accumulate to higher concentrations in blood than with the normally functioning transport protein. Statins affected include simvastatin, pravastatin, and rosuvastatin. Other drugs transported less well and with adverse effects potentially attributable to the abnormal variant OATP1B1 protein include the antimetabolite methotrexate (Buxhofer-Ausch et al., 2013), the SN-38 active metabolite of the cancer chemotherapy irinotecan (Fujita et al., 2014), the antiviral agent lopinavir (Williamson et al., 2013), and the antibiotic rifampicin (Niemi et al., 2006). However, data for rifampicin suggest that other transporters may compensate for altered uptake of the drug by the OATP1B1 protein (Niemi et al., 2006).

10.3 Flecainide: Potential Off-Target Mechanism for Sudden Death

Ventricular arrhythmias are a major cause of sudden cardiac death. The antiarrhythmic drug flecainide blocks the Nav1.5 sodium channel in the heart, slowing conduction of cardiac electrical impulse (Investigators, 1989). However, in a treatment trial with the antiarrhythmic drug flecainide led to a 25% excess mortality compared to placebo (Investigators, 1989). New chemical genomic methods allow unrecognized pleiotropic beneficial and harmful targets for medicines to be identified. If these targets are genetically variable in their expression, potentially beneficial drugs might be made available to selected patients who do not have the high-risk genotype(s) and phenotype(s). Magic Tag® is an example of technology for new target discovery through allowing immobilization of bioactive molecules coupled with bacteriophage display, followed by an ELISA assay, to test for interactions of drugs with new putative protein targets (Dilly et al., 2007; Ladwa, Dilly, Clark, Marsh, & Taylor, 2008). Using these methods, an alternative initiation of translation was exemplified for flecainide as a potential toxicity factor by showing its binding to protein products of genes linked to sudden cardiac death (Taylor, Clark, Marsh, Singer, & Dilly, 2013).

10.4 EGF Receptor 2 Antagonism with Trastuzmab

Human epidermal growth factor 2 (HER2) (Gijsen et al., 2010) is a normal pathway for cell growth and division. These growth receptors may be 100 times overexpressed in cancer cells. The first Companion Diagnostic was the

HercepTest, an immunohistochemistry assay to detect HER2 (Gijsen et al., 2010) receptor positive breast cancers (Hurvitz et al., 2012; Slamon et al., 2011; Stern et al., 2015). Trastuzumab was developed as an antibody against HER2. It binds to domain IV of the extracellular segment of the receptor, leading to reduced cell proliferation due to arrest of cells in G1 phase of the cell cycle (Le, Pruefer, & Bast, 2005). Mechanisms for this include activation of the tumor suppressor p27 through inhibiting at least six major intracellular signaling pathways (Le et al., 2005). Trastuzumab also inhibits angiogenesis (Banerjee, Dowsett, Ashworth, & Martin, 2007), attributable in part to reduction in expression of vascular endothelial growth factor (VEGF) and reduced activity of the PI3 kinase/Akt pathway (Kou, Vatish, & Singer, 2007; Le et al., 2005). Use of trastuzumab to treat breast cancer is conditional on the HER2 companion diagnostic test confirming that a breast cancer expresses EGF receptor 2. Around ~30% of breast cancer patients express EGF receptor 2 and thus may benefit from this biological drug. The HER2 test was approved in 1999 by the FDA (Roche & Ingle, 1999), and then by NICE (Crabb et al., 2012) and other regulatory agencies, as a companion diagnostic for use with trastuzumab as the companion therapeutic agent: the first example of formal regulatory licensing of a companion diagnostic.

The Cochrane Collaboration publishes secondary research on treatment outcomes. A 2014 Cochrane Review reported on the benefits of trastuzumab-containing therapies for HER2-positive metastatic breast cancer (Balduzzi et al., 2014). Hazard ratios for overall survival were 0.82 (95% CI 0.71–0.94, $P=0.004$) and 0.61 (95% CI 0.54–0.70, $P<0.00001$) for progression-free survival. The authors concluded that contamination of control study arms by allowing addition of trastuzumab at the end of a trial may have led to the true effectiveness of the antibody being underestimated (Balduzzi et al., 2014).

Trastuzumab is ineffective against EGF receptor 2 negative cancers and may cause serious adverse effects, including allergic reactions and heart failure (relative risk from the drug: 3.49 (90% CI 1.88–6.47, $P=0.0009$)) (Balduzzi et al., 2014). EGF receptor 2 testing avoids exposing patients to unnecessary risk without benefit and reduces costs of unnecessary treatment.

EGF receptor 2 has been identified as a treatment target in other cancers and trastuzumab is now also indicated for treatment of EGF receptor 2 positive metastatic gastric cancer (Gomez-Martin et al., 2014) and gastroesophageal junction adenocarcinoma (Gomez-Martin et al., 2014). Further randomized controlled studies are assessing the effects of adding trastuzumab to combination chemotherapy against advanced HER2 positive gastric cancer in Japan (Kataoka et al., 2015), a country where the incidence of gastric cancer is very high by international comparison (Charvat et al., 2015).

Cancer chemotherapy trials have evolved to evaluating MCA in combination to target treatment against multiple cancer pathways simultaneously, aiming for additive or synergistic effects, with preclinical or clinical results available for 25 antibody combinations (Henricks, Schellens, Huitema, & Beijnen, 2015). For example, trastuzumab and pertuzumab (an inhibitor of EGF receptor 2 dimerization) have already been shown to be a beneficial combination as double treatment against the EGF receptor 2 receptor for treating HER2-positive metastatic breast cancer, with median progression-free survival increasing from 12 to 18 months (hazard ratio for progression or death, 0.62; 95% CI 0.51–0.75; $P<0.001$) with adding pertuzumab to trastuzumab combined with the microtubule assembly inhibitor paclitaxel (Baselga et al., 2012). There was no significant increase in serious adverse effects with this antibody combination (Baselga et al., 2012).

It was expected that combining antibodies would be less toxic than conventional combination chemotherapy for cancer. However, a meta-analysis of 4 RCTs of 2069 patients with advanced colorectal cancer suggests that the combination of bevacizumab (an antiangiogenic antibody against VEGF) with cetuximab or panitumumab (antibodies against EGF receptor 1) led to poorer survival and worse toxicity (Lv et al., 2015).

Further studies are needed to assess to what extent differences in outcomes are due to the antibodies themselves, characteristics of the cancers treated or patient factors.

10.5 Diagnostic Markers to Guide Selection of Anti-cancer Tyrosine Kinase Receptor Inhibitors

Chronic myeloid leukemia (CML) was the first human cancer linked to an acquired genetic abnormality, the Philadelphia translocation (Quintas-Cardama & Cortes, 2009). This provided the opportunity for developing both new diagnostics and selective drugs to target the resulting mutation. This has been identified as an enzyme with tyrosine kinase activity resulting from fusion of part of the breakpoint cluster region (*bcr*) gene from chromosome 22 with the *abl* gene on chromosome 9 (Quintas-Cardama & Cortes, 2009). The fused Bcr-Abl protein is continuously active and is coupled to the interleukin 3 beta(c) receptor subunit, increasing cell turnover and impairing DNA repair (Quintas-Cardama & Cortes, 2009). Imatinib was developed as the first clinically effective tyrosine kinase receptor inhibitor (TKI) for treating patients with CML who are positive for the *BCR-ABL1* mutation (Hochhaus et al., 2009; Hughes et al., 2010; O'Brien et al., 2003). The disease commonly proceeds to an accelerated phase and then to a blast crisis with very poor clinical outcome. The International Randomized Study of Interferon versus STI571 (IRIS) trial of untreated CML in chronic phase assessed benefits of randomization to imatinib ($n=553$) versus interferon-alpha plus cytarabine ($n=553$) (Hochhaus et al., 2009). In a 6-year update from IRIS in patients receiving imatinib as first-line therapy, there were no reports of disease progression to accelerated phase or blast crisis (Hochhaus et al., 2009). The estimated event-free survival at 6 years was 83%, with a 93% estimated rate of freedom from progression to accelerated phase and then to a blast crisis; estimated overall survival was 88%, 95% survival for CML-related deaths (Hochhaus et al., 2009).

Identifying Bcr-Abl as a treatment target also provided a diagnostic to predict treatment response (Millot et al., 2014). The *BCR-ABL1* transcript level early after Bcr-Abl inhibitory treatment predicts short-term clinical response and longer term outcome. For example, in a French trial of imatinib for CML treatment in 40 children, those with a *BCR-ABL1/ABL* ratio > 10% at 3 months starting the drug had a larger spleen and higher white cell count compared with those with a lower ratio (Millot et al., 2014). Children with *BCR-ABL1/ABL* ≤10% 3 months after starting treatment had higher rates of complete cytogenetic and molecular responses after 12 months and better progression-free survival on median follow-up of 71 months (range, 22–96 months) (Millot et al., 2014).

However, point mutations in the kinase domain of Bcr-Abl are a reason for resistance developing to imatinib, with >90 mutations already known to confer resistance (Soverini et al., 2011). New TKIs (see below) have therefore been developed with different profiles for pathways associated with resistance and adverse effects.

Tyrosine kinases are also targets for treating a range of other cancers. For example, the c-KIT tyrosine kinase promotes growth of gastrointestinal stromal tumors (GIST). Patients who have a c-KIT positive GIST are also responsive to imatinib (Joensuu et al., 2012, 2015). Further inhibitors effective against a multiple range of receptor tyrosine kinases (e.g., sunitinib (Montemurro et al., 2013), nilotinib (Blay et al., 2015; Hughes et al., 2014; Kim et al., 2015; Shimoni et al., 2015), dasatanib (Iriyama et al., 2015; Shah et al., 2014), and sorafenib (Flaherty et al., 2015; Zhu et al.,

Table 1 Selected Licensed Inhibitors of Tyrosine Kinase Receptors

Inhibitor	Main Receptor Tyrosine Kinase and Other Kinase Targets	Tumors Licensed for Treatment
Dasatinib (Iriyama et al., 2015; Shah et al., 2014)	Bcr-Abl and Src family tyrosine kinase	Ph+ CML and Ph+ ALL
Imatinib (Hochhaus et al., 2009; Joensuu et al., 2012, 2015)	Bcr-Abl, c-kit, and PDGFR	Ph+ CML, Ph+ ALL, GIST, PDGF receptor gene rearrangement myelodysplastic/myeloproliferative diseases, rearrangements, systemic mastocytosis, and hypereosinophilic syndrome +/− chronic eosinophilic leukemia with FIP1L1-PDGFRα fusion kinase deletion
Nilotinib (Blay et al., 2015; Hughes et al., 2014; Kim et al., 2015; Shimoni et al., 2015)	Bcr-Abl, DDR-1 and -2, PDGFR alpha and beta, c-kit, and CSF-1R	Imatinib-resistant Ph+ CML
Sorafenib (Flaherty et al., 2015; Zhu et al., 2015)	Primary kidney cancer (advanced renal cell carcinoma), advanced primary liver cancer (hepatocellular carcinoma), and radioactive iodine resistant advanced thyroid carcinoma	Advanced primary renal cell carcinoma, advanced primary (hepatocellular carcinoma), and advanced thyroid carcinoma resistant to radioactive iodine
Sunitinib (Montemurro et al., 2013)	VEGFRs, PDGFRs, and c-kit	Renal cell carcinoma and imatinib-resistant GIST

ALL, acute lymphoblastic leukemia; CML, chronic myeloid leukemia; CSF-1-R, colony stimulating factor 1 receptor; DDR, discoidin domain deceptor; Ph+, Philadelphia chromosome positive; PDGFR, platelet-derived growth factor receptor; VEGFR, vascular endothelial growth factor receptor.

2015; Table 1)) are now also clinically available or are under assessment for treating tyrosine kinase receptor-driven cancers, with the expectation that using inhibitors with actions against multiple receptor tyrosine kinases (Jain et al., 2013; Manley et al., 2010), e.g., both causing direct inhibition of cancer growth and anti-angiogenesis, may improve their anti-cancer effectiveness (Montemurro et al., 2013).

10.6 Multiple Companion Diagnostics to Guide Selection of Anti-cancer Tyrosine Kinase Receptor Inhibitors to Treat Colorectal Cancer

Several companion diagnostic markers may need to be combined before selecting effective treatments for cancer. This is seen, for example, with the need for combined EGFR-1 and *KRAS* testing for selected cancers.

The transmembrane EGF receptor 1 (HER1) is a target for anti-cancer treatment for a range of cancers, including breast cancer, colorectal cancer, and head and neck cancer (Bemanian et al., 2015; Bennett et al., 2011). The human monoclonal antibody panitumumab inhibits growth of selected EGF receptor 1-dependent tumors by binding to the extracellular domain of the receptor to prevent pathway activation (Bennett et al., 2011). However, KRAS, the protein product of the proto-oncogene *KRAS* with GTPase activity, helps to initiate EGF receptor 1-associated cell signaling and mutations of KRAS make it constitutively active in many cancers. Patients with these constitutively active variants of GTPase encoded by *KRAS* fail to respond to panitumumab (Amado et al., 2008; Peeters et al., 2009).

A *KRAS* Companion Diagnostic was approved by the FDA after licensing of panitumumab to predict response to the antibody in patients with EGF receptor 1-expressing refractory, progressive, and metastatic colorectal cancer (Mack, 2009). Regulators have limited approval of panitumumab for use in patients with metastatic colorectal cancer expressing both EGF receptor 1 and confirmed using a companion *KRAS* diagnostic to have non-mutated suppressible KRAS (Mack, 2009).

The *KRAS* test is also used to predict response to the EGF receptor 1 antagonist cetuximab, a chimeric (mouse/human) monoclonal antibody used to treat patients with metastatic colorectal cancer or head and neck cancer who do not have KRAS-activating mutations. 37% of patients with metastatic colon cancer in the study by Van Cutsem et al. (2011) were eligible for adding cetuximab to their treatment: this prolonged overall survival from 23.5 to 20.0 months (hazard ratio: 0.80). In that study, the *BRAF* mutation was a strong indicator of adverse prognosis (Van Cutsem et al., 2011).

In contrast, for treatment response of locally advanced or metastatic non-small cell lung cancer to gefitinib, a small molecule selective inhibitor of EGF receptor tyrosine kinase (EGF receptor 1-TK), activating mutations of EGF receptor 1-TK need to be present (Carotenuto et al., 2011; Hirsch et al., 2006).

10.7 Immunological Targets in Neuropsychiatry

A key advance in precision approaches to treating brain disease has come from insight into the immunological basis of a wide range of neuropsychiatric disorders (Baumeister, Russell, Pariante, & Mondelli, 2014), from acute psychosis, to depression (Horowitz & Zunszain, 2015) and dementia. A possible association between altered immunity and schizophrenia was suggested over a century ago, with support from both epidemiological and genetic studies for links between infection, inflammation, and schizophrenia (Khandaker et al., 2015).

Evidence acquired over the past decade has led to development of new immunological biomarkers, with the implication that anti-inflammatory treatment may be disease and/or symptom modifying in some psychiatric diseases—from syndromes to well-phenotyped disease subtypes.

10.7.1 Inflammation as a Target in Acute Psychosis

Raised inflammatory markers are typical of many settings associated with psychosis, including psychological stress, the puerperal period (Bergink et al., 2015), and infective syndromes. Several inflammatory markers may be raised in first psychosis. Tumor necrosis factor-alpha (TNF-alpha) was reported (Di Nicola et al., 2013) to be ~threefold higher in adult patients with first psychosis versus controls. For example, C-reactive protein (CRP) has been reported to be raised in patients with first psychosis and even higher in patients with first psychosis who also have a history of psychological trauma in with childhood (Di Nicola et al., 2013; Hepgul et al., 2012). In a meta-analysis of results on associations between childhood stress and inflammation in adults with first psychosis, Baumeister et al. identified 18 studies of 16,870 individuals for the inflammatory marker CRP, 15 studies of 3751 individuals for interleukin-6 (IL-6), and 10 studies of 881 individuals for TNF-alpha (Baumeister, Akhtar, Ciufolini, Pariante, & Mondelli, 2015). All three inflammatory markers were raised in adults with first psychosis, independent of age, gender, and body mass index (Baumeister et al., 2015).

10.7.2 Anti-inflammatory Treatment of Neuropsychiatric Disorders

The above observations raise the question whether anti-inflammatory treatment would be helpful in complementing classical pharmacological treatments in psychiatry. Evidence from a double-blind randomized placebo-controlled trial of the COX-2 inhibitor celecoxib in 46 subjects suggests that it is an effective adjuvant therapy for manic episodes of bipolar mood disorder without psychotic features. Patients with raised cortisol and inflammatory markers have a poorer response to treatment of a first psychosis (Mondelli et al., 2015). Raison et al. conducted a double-blind, placebo-controlled RCT of three infusions of the TNF-alpha antagonist infliximab in 60 outpatients with major depression either on a stable antidepressant regimen ($n=37$) or medication free ($n=23$) for 4 weeks or more and were moderately resistant to treatment (Raison et al., 2013). Infliximab was infused at 5 mg/kg at baseline and after 2 and 6 weeks of a 12-week study. There was no overall effect of the treatment on depression scores. However, baseline concentrations of TNF and its soluble receptors were significantly higher in infliximab-treated clinical responders versus nonresponders. Furthermore, patients with the highest level of high-sensitivity C-reactive protein (hs-CRP) were most likely to response to the anti-TNF-alpha treatment (Raison et al., 2013) and to have significantly greater decreases in hs-CRP from baseline to week 12 versus placebo-treated responders.

10.7.3 New Targets for Treating Dementia

The common forms of dementia are degenerative, typically affecting the elderly. They tend to by insidious in onset, slowly progressive, with little in the way of neurological features, and with the brain on MRI scan becoming atrophic. The two major degenerative dementias are commonly either caused by Alzheimer's disease, characterized by cerebral plaques of beta-amyloid or neurofibrillary tangles or atheromatous vascular disease, with similar cardiovascular risk factors and treatments as for coronary artery and peripheral vascular disease, i.e., management of hypercholesterolemia, hypertension, smoking, diabetes mellitus, and chronic kidney disease.

Autoimmune dementia is now being increasingly recognized as amenable to anti-inflammatory treatment, with good outcomes if intensive immunosuppressive and anti-inflammatory treatment is started early after presentation. In their pioneering paper, Vincent and colleagues reported an association between antibodies against potassium channels and limbic encephalitis (Vincent et al., 2004), which they recognized as a potentially immunotherapy-responsive form of encephalitis. Clues to the diagnosis are now recognized to include onset in the younger or middle aged, a subacute onset, the presence of confusion or delirium, rapid progression (e.g., accelerated forgetting over hours to days and weeks), overt neurological features, focal signal change on brain magnetic resonance imaging, and supportive initial and serial immunological tests.

Historically, brain biopsies were commonly performed to obtain pathological confirmation of the diagnosis. Biopsies have now largely been superseded both because of improvements in brain imaging and algorithm-supported guidance (Schott et al., 2010) for diagnostic options. Suspicion of auto-immune etiology is now usually based on detection of a spectrum of antibodies (Camdessanche et al., 2002; Carvajal-Gonzalez et al., 2014; FDA, 2011; Hart, Maddison, Newsom-Davis, Vincent, & Mills, 2002; Vincent et al., 2004; Wright et al., 2015) complemented by observing recovery on general anti-inflammatory treatment or specific biological treatments targeted against selective inflammatory cytokines. If these patients are untreated, dense amnesia may occur associated with severe brain injury, including major atrophy of the hippocampus.

10.7.4 Systemic Inflammation as a Target for Treating Neurodegeneration

A typical clinical syndrome is seen in the elderly patient who becomes transiently severely confused in association with an episode of infection of the urinary tract or chest. Earlier hypotheses for mechanisms included disorientation when the infection was treated in the hospital setting away from a familiar home environment, toxic effects on brain neurones from systemic inflammatory cytokines generated in response to the infection and direct effects of bacteremia on brain cells.

A new concept is that priming of microglial and endothelial cells in the brain, by a minor, subclinical inflammatory or other insult, may lead to these cells becoming hyperresponsive to inflammatory stimuli during a subsequent infection, microglial cells then themselves having toxic effects on neighboring brain neurones to cause confusion. In experimental studies, mouse microglial cells activated by very low concentration lipopolysaccharide became hyperresponsive to stimulation by live *Salmonella* typhimurium, the effects being much greater in the aged than in control young mouse brains (Lunnon et al., 2011).

Systemic inflammation results in altered blood–brain barrier permeability after 2 months, still present after 6 months (Nagele, Han, Demarshall, Belinka, & Nagele, 2011). Thus, understanding the pathways for microglial priming and subsequent responsiveness may lead to new possible targets for a range of serious psychiatric disorders: in at-risk patients, prophylaxis, against microglial activation between and during events of infection may reduce the incidence and severity of infection-associated confusion. Current work is focused on addressing the microglial mechanisms responsible and a recent safety study suggests that the anti-TNF-alpha antibody etanercept may protect from confusion (Butchart et al., 2015). A larger randomized controlled treatment trial powered to look at efficacy endpoints is currently planned.

10.8 Targeting the Genetic Abnormality in Cystic Fibrosis

Cystic fibrosis is inherited as an autosomal recessive condition. Clinical disease occurs when there are loss-of-function mutations in both copies of the gene for the cystic fibrosis transmembrane conductance regulator (CFTR) protein. Affected homozygous patients have thicker secretions in the lungs, leading to recurrent chest infections and respiratory failure. Thick secretions from the pancreas may also lead to malabsorption due to failure of digestive enzymes to enter the lumen of the small intestine. Public–private partnerships have led to discovery of modulators of selected abnormalities in the CFTR protein. There are multiple molecular defects associated with different mutations in the CFTR protein (Collawn, Fu, Bartoszewski, & Matalon, 2014).

The phe508del CFTR mutation causes the disease by limiting how much CFTR protein reaches the surface of bronchial epithelial cells (Wainwright et al., 2015). Lumacaftor increases trafficking of phe508del CFTR to the cell surface and ivacaftor enhances chloride transport of CFTR on the cell surface (Wainwright et al., 2015). In combination, these drugs resulting in modest improvements in lung function and ~30% reduction in the number of respiratory infections (Wainwright et al., 2015).

10.9 Drug Selection Based on Pharmacogenetic Variants in Drug-Metabolizing Enzymes

Tests that are mandatory, recommended, or under consideration because of within individual variability in drug-metabolizing enzymes include thiopurine methyl transferase testing (McLeod & Siva, 2002) for thiopurine drugs (azathioprine, 6-mercaptopurine) to minimize risk of serious bone marrow suppression, UDP-1A1 to reduce risk of irinotecan-induced diarrhea (McLeod & Siva, 2002), CYP2C19 testing for risk of resistance to the antiplatelet prodrug clopidogrel (Mega et al., 2010), CYP2C9 and vitamin K epoxyreductase testing for anticoagulant response to warfarin (Avery et al., 2011; Maitland-van der Zee et al., 2014), and markers of poor metabolizer CYP2D6 activity for risk of resistance to tamoxifen (Welzen et al., 2015) and blunted pain relief from the prodrug codeine (Crews et al., 2014).

Testing for UDP-glucoronosyltransferase (UGT) gene variations illustrates how to avoid an important preventable adverse effect of the anti-cancer drug irinotecan, which prevents DNA from uncoiling by inhibiting the enzyme topoisomerase 1, and so reduces cell growth in selected colorectal and lung cancers (Lévesque et al., 2013). The main dose-limiting toxicity of irinotecan treatment is diarrhea, secondary to enteric injury following biliary excretion of SN-38 (Lévesque et al., 2013). The active metabolite of irinotecan SN-38 is glucoronidated by hepatic UGTs and those with low activity are at increased risk of serious diarrhea (Lévesque et al., 2013). Patients homozygous for *UGT1A1*28* are also at greater risk of neutropoenia from irinotecan than those with one or two copies of the wild-type allele *UGT1A1*1* (Lévesque et al., 2013).

10.10 Histocompatibility Antigen Companion Diagnostics

Histocompatibility antigen (HLA) companion diagnostics are useful for reducing risk of the serious drug-induced skin ADRs: Stevens–Johnson Syndrome and toxic epidermal necrolysis. Serious skin hypersensitivity reactions (Stevens–Johnson Syndrome) due to the antiretroviral agent abacavir were shown be associated with the gene *HLA-B*5701* (Pavlos, Mallal, & Phillips, 2012). Prospective screening significantly reduced the incidence of clinical suspected hypersensitivity from 8% to 3% (Pavlos et al., 2012). Approval for clinical use of abacavir should now be coupled to a lower risk result from the *HLA-B*5701* diagnostic test (Pavlos et al., 2012).

Carbamazepine is used to treat epilepsy and neuropathic pain, bipolar disorder, and acute manic and mixed episodes associated with bipolar disorder. Stevens–Johnson Syndrome and toxic epidermal necrolysis are associated with the presence of the *HLA-B*1502* allele (FDA, 2007). FDA advice is that "Patients with ancestry from areas in which *HLA-B*1502* is present should be screened for the *HLA-B*1502* allele before starting treatment with carbamazepine" (FDA, 2007). The patient populations at the greatest risk are from Asia, including South Asia (FDA, 2007). Innovative approaches are needed to make these HLA tests available cost effectively on a national scale.

11. NETWORK PHARMACOLOGY

It is being increasingly recognized that drugs in clinical use may have several potential beneficial or adverse effects. Finding these pleiotropic benefit and risk targets allows multiple indications for new drugs, the opportunity to reprofile use of existing drugs for which safety profiles in other contexts are already established, and to predict potential adverse effects of treatments. The concept of network pharmacology (Hopkins, 2008) arises from the observation that many drugs share the same type of ADR-causing mechanism and that many drugs share similar therapeutic effects (Janga & Tzakos, 2009; Pujol, Mosca, Farres, & Aloy, 2010). These beneficial or adverse effects may be through actions on known or hidden targets, which may be unmasked through considering networks of known disease mechanisms and known drug targets (Fig. 3), complemented by discovery work *in silico* and using a wide range of "omic," chemical biology (Taylor et al., 2013), pharmacology, and clinical methods (Watkins, Marsh, Taylor, & Singer, 2010).

Emerging tools offer a platform to identify and explore molecular complexity of a particular disease. These tools are helping both in identifying targets for new therapeutics and in supporting companion diagnostics by, identification of new disease genes and pathways, molecular relationships among apparently distinct disease phenotypes, the biological significance of disease-associated mutations, new biomarkers for complex diseases, and thus new drug targets (Barabasi, Gulbahce, & Loscalzo, 2011).

Network pharmacology approaches provide important potential tools for accelerating development of precision medicines both in strategy for reprofiling existing drugs by helping to identify previous unrecognized treatment targets and for improving benefit-risk profiles when developing new therapeutic agents (Watkins et al., 2010).

An example of this approach is the Cancer Cell Map Initiative (CCMI), a new venture aimed at "systematically detailing these complex interactions among cancer genes and how they differ between diseased and healthy states" (Krogan, Lippman, Agard, Ashworth, & Ideker, 2015). The initiative aims to create cancer cell maps for different tumor types as a tool for targeting disrupted molecular networks, and as a way to accelerate the development of precision medicines against cancers. Developing further targeted medicines for cancer chemotherapy needs to involve detailed understanding of the molecular networks which allow both oncogenesis—initial formation of a cancer, allow cancer cells to develop resistance to treatment, and recognition that, within individual patients, there may be heterogeneity within cancer tissue of mechanisms for sustained cancer growth and resistance to endogenous tumor immunity and cancer chemotherapy.

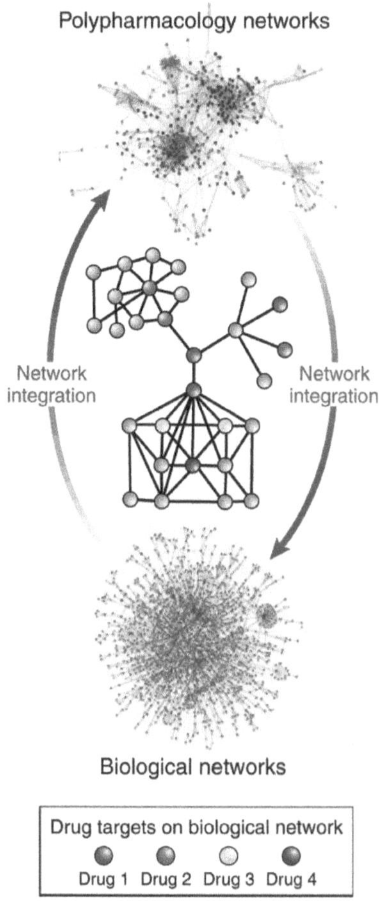

Figure 3 A network-centric view of drug action maps drug-target (polypharmacology) networks to biological networks. In center part of the biological network, nodes (proteins) targeted by same drug are represented in the same color. Drug efficacy and toxicity are understood by actions at specific nodes and hubs. *Reproduced with permission from Hopkins (2007).*

12. FUTURE DEVELOPMENTS

President Obama announced a Precision Medicine initiative in his State of the Union Address in 2015 (Riley, 2015). This provides a strong signal that the health policy community is being successful in persuading government leaders and funders that precision will be better both for the health of patients and for the financial health of the economy. For the life science community, major new investment is expected to support chemists, biochemists, geneticists, data analysts, and a wide range of health professionals to develop public–private partnerships. Members should include experts from academia, health systems, pharmaceutical and smaller biotechnology companies, and patient charities, to accelerate development and implementation in clinical practice of new precision medicines. This will need corresponding radical changes by drug regulators to ensure that a streamlined system can process in a timely way drug submission data based on new methodologies focusing on precision medicines for small subgroups of patients (Riley, 2015).

ACKNOWLEDGMENTS

D.S. holds a patent for siRNA technology. He has worked with but has no financial interest in the Magic Tag® chemical genomics technology. D.S. is a member of the Healthcare Professionals Working Party of the European Medicines Agency.

REFERENCES

ADAPT SMART: An enabling platform for the coordination of Medicines Adaptive Pathways to Patients (MAPPs) activities. (2015). http://www.adaptsmart.eu. Accessed on 5th November, 2015).

Amado, R. G., Wolf, M., Peeters, M., Van Cutsem, E., Siena, S., Freeman, D. J., et al. (2008). Wild-type KRAS is required for panitumumab efficacy in patients with metastatic colorectal cancer. *Journal of Clinical Oncology*, *26*(10), 1626–1634. http://dx.doi.org/10.1200/jco.2007.14.7116.

Avery, P. J., Jorgensen, A., Hamberg, A. K., Wadelius, M., Pirmohamed, M., & Kamali, F. (2011). A proposal for an individualized pharmacogenetics-based warfarin initiation dose regimen for patients commencing anticoagulation therapy. *Clinical Pharmacology and Therapeutics*, *90*(5), 701–706. http://dx.doi.org/10.1038/clpt.2011.186.

Balduzzi, S., Mantarro, S., Guarneri, V., Tagliabue, L., Pistotti, V., Moja, L., et al. (2014). Trastuzumab-containing regimens for metastatic breast cancer. *The Cochrane Database of Systematic Reviews*, *6*, Cd006242. http://dx.doi.org/10.1002/14651858.CD006242.pub2.

Banerjee, S., Dowsett, M., Ashworth, A., & Martin, L. A. (2007). Mechanisms of disease: Angiogenesis and the management of breast cancer. *Nature Clinical Practice. Oncology*, *4*(9), 536–550. http://dx.doi.org/10.1038/ncponc0905.

Barabasi, A. L., Gulbahce, N., & Loscalzo, J. (2011). Network medicine: A network-based approach to human disease. *Nature Reviews. Genetics*, *12*(1), 56–68. http://dx.doi.org/10.1038/nrg2918.

Baselga, J., Cortes, J., Kim, S. B., Im, S. A., Hegg, R., Im, Y. H., et al. (2012). Pertuzumab plus trastuzumab plus docetaxel for metastatic breast cancer. *The New England Journal of Medicine*, *366*(2), 109–119. http://dx.doi.org/10.1056/NEJMoa1113216.

Baumeister, D., Akhtar, R., Ciufolini, S., Pariante, C. M., & Mondelli, V. (2015). Childhood trauma and adulthood inflammation: A meta-analysis of peripheral C-reactive protein, interleukin-6 and tumour necrosis factor-alpha. *Molecular Psychiatry*. http://dx.doi.org/10.1038/mp.2015.67.

Baumeister, D., Russell, A., Pariante, C. M., & Mondelli, V. (2014). Inflammatory biomarker profiles of mental disorders and their relation to clinical, social and lifestyle factors. *Social Psychiatry and Psychiatric Epidemiology*, *49*(6), 841–849. http://dx.doi.org/10.1007/s00127-014-0887-z.

Bemanian, V., Sauer, T., Touma, J., Lindstedt, B. A., Chen, Y., Odegard, H. P., et al. (2015). The epidermal growth factor receptor (EGFR / HER-1) gatekeeper mutation T790M is present in European patients with early breast cancer. *PLoS One*, *10*(8). e0134398. http://dx.doi.org/10.1371/journal.pone.0134398.

Bennett, L., Zhao, Z., Barber, B., Zhou, X., Peeters, M., Zhang, J., et al. (2011). Health-related quality of life in patients with metastatic colorectal cancer treated with panitumumab in first- or second-line treatment. *British Journal of Cancer*, *105*(10), 1495–1502. http://dx.doi.org/10.1038/bjc.2011.409.

Bergink, V., Laursen, T. M., Johannsen, B. M., Kushner, S. A., Meltzer-Brody, S., & Munk-Olsen, T. (2015). Pre-eclampsia and first-onset postpartum psychiatric episodes: A Danish population-based cohort study. *Psychological Medicine*, *45*(16), 3481–3489. http://dx.doi.org/10.1017/S0033291715001385.

Blay, J. Y., Shen, L., Kang, Y. K., Rutkowski, P., Qin, S., Nosov, D., et al. (2015). Nilotinib versus imatinib as first-line therapy for patients with unresectable or metastatic gastrointestinal stromal tumours (ENESTg1): A randomised phase 3 trial. *The Lancet Oncology*, *16*(5), 550–560. http://dx.doi.org/10.1016/s1470-2045(15)70105-1.

Broad. (2015). *The RNAi consortium at the broad institute*. https://www.broadinstitute.org/rnai/trc. Accessed on 5th November, 2015.

Butchart, J., Brook, L., Hopkins, V., Teeling, J., Püntener, U., Culliford, D., et al. (2015). Etanercept in Alzheimer disease: A randomized, placebo-controlled, double-blind, phase 2 trial. *Neurology*, *84*, 2161–2168.

Buxhofer-Ausch, V., Secky, L., Wlcek, K., Svoboda, M., Kounnis, V., Briasoulis, E., et al. (2013). Tumor-specific expression of organic anion-transporting polypeptides: Transporters as novel targets for cancer therapy. *Journal of Drug Delivery*, *2013*, 863539. http://dx.doi.org/10.1155/2013/863539.

Camdessanche, J. P., Antoine, J. C., Honnorat, J., Vial, C., Petiot, P., Convers, P., et al. (2002). Paraneoplastic peripheral neuropathy associated with anti-Hu antibodies. A clinical and electrophysiological study of 20 patients. *Brain*, *125*(Pt. 1), 166–175.

Carotenuto, P., Roma, C., Rachiglio, A. M., Pasquale, R., Franco, R., Antinolfi, G., et al. (2011). Optimizing response to gefitinib in the treatment of non-small-cell lung cancer. *Pharmacogenomics and Personalized Medicine*, *4*, 1–9. http://dx.doi.org/10.2147/PGPM.S6626.

Carvajal-Gonzalez, A., Leite, M. I., Waters, P., Woodhall, M., Coutinho, E., Balint, B., et al. (2014). Glycine receptor antibodies in PERM and related syndromes: Characteristics, clinical features and outcomes. *Brain*, *137*(Pt 8), 2178–2192. http://dx.doi.org/10.1093/brain/awu142.

Chapman, P. B., Hauschild, A., Robert, C., Haanen, J. B., Ascierto, P., Larkin, J., et al. (2011). Improved survival with vemurafenib in melanoma with BRAF V600E mutation. *The New England Journal of Medicine*, *364*(26), 2507–2516. http://dx.doi.org/10.1056/NEJMoa1103782.

Charvat, H., Sasazuki, S., Inoue, M., Iwasaki, M., Sawada, N., Shimazu, T., et al. (2015). Prediction of the 10-year probability of gastric cancer occurrence in the Japanese population: The JPHC study cohort II. *International Journal of Cancer*. http://dx.doi.org/10.1002/ijc.29705.

Chen, C. Genetic-testing startup 23andMe seeks to raise $150 million. *Bloomberg Business*. July 7, 2015. http://www.bloomberg.com/news/articles/2015-07-06/genetic-testing-startup-23andme-seeks-to-raise-150-million.

Collawn, J. F., Fu, L., Bartoszewski, R., & Matalon, S. (2014). Rescuing DeltaF508 CFTR with trimethylangelicin, a dual-acting corrector and potentiator. *American Journal of Physiology. Lung Cellular and Molecular Physiology, 307*(6), L431–434. http://dx.doi.org/10.1152/ajplung.00177.2014.

Collins, F. S. (2015). Exceptional opportunities in medical science: A view from the National Institutes of Health. *Journal of the American Medical Association, 313*(2), 131–132. http://dx.doi.org/10.1001/jama.2014.16736.

Connolly, S. J., Ezekowitz, M. D., Yusuf, S., Eikelboom, J., Oldgren, J., Parekh, A., et al. (2009). Dabigatran versus warfarin in patients with atrial fibrillation. *The New England Journal of Medicine, 361*(12), 1139–1151. http://dx.doi.org/10.1056/NEJMoa0905561.

Crabb, N., Marlow, M., Bell, H., & Newland, A. (2012). The NICE Diagnostics Assessment Programme. *Health Policy and Technology, 1*, 5–7.

Crews, K. R., Gaedigk, A., Dunnenberger, H. M., Leeder, J. S., Klein, T. E., Caudle, K. E., et al. (2014). Clinical Pharmacogenetics Implementation Consortium guidelines for cytochrome P450 2D6 genotype and codeine therapy: 2014 update. *Clinical Pharmacology and Therapeutics, 95*(4), 376–382. http://dx.doi.org/10.1038/clpt.2013.254.

Davis, J. C., Furstenthal, L., & Desai, A. A. (2009). The microeconomics of personalized medicine: Today's challenge and tomorrow's promise. *Nature Reviews. Drug Discovery, 8*, 279–286.

Di Nicola, M., Cattaneo, A., Hepgul, N., Di Forti, M., Aitchison, K. J., Janiri, L., et al. (2013). Serum and gene expression profile of cytokines in first-episode psychosis. *Brain, Behavior, and Immunity, 31*, 90–95. http://dx.doi.org/10.1016/j.bbi.2012.06.010.

Dilly, S. J., Bell, M. J., Clark, A. J., Marsh, A., Napier, R. M., Sergeant, M. J., et al. (2007). A photoimmobilisation strategy that maximises exploration of chemical space in small molecule affinity selection and target discovery. *Chemical Communications (Cambridge, England), 27*, 2808–2810. http://dx.doi.org/10.1039/b704271j.

FDA. (2007). *Information for healthcare professionals: Dangerous or even fatal skin reactions—Carbamazepine (marketed as Carbatrol, Equetro, Tegretol, and generics)*. http://www.fda.gov/DrugsDrugSafety/PostmarketDrugSafetyInformationforPatientsandProviders/ucm124718.htm. Accessed on 5th November, 2015.

FDA. (2011). *Food and Drug Administration Amendments Act (FDAAA) of 2007*. http://www.fda.gov/RegulatoryInformation/Legislation/SignificantAmendmentstotheFDCAct/FoodandDrugAdministrationAmendmentsActof2007/default.htm. Accessed on 5th November, 2015.

FDA. (2013). *Sprue-like enteropathy—FDA Drug Safety Communication*. Food and Drug Administration. http://www.fda.gov/Drugs/DrugSafety/ucm359477.htm. Accessed on 5th November, 2015.

FDA. (2015a). *List of cleared or approved companion diagnostic devices (in vitro and imaging tools)*. US FDA. http://www.fda.gov/MedicalDevices/ProductsandMedicalProcedures/InVitroDiagnostics/ucm301431.htm. Accessed on 21st October, 2015.

FDA. (2015b). *US FDA: Current good manufacturing practice for combination products (draft guidance)*. http://www.fda.gov/RegulatoryInformation/Guidances/ucm126198.htm. Accessed on 5th November, 2015.

Flaherty, K. T., Manola, J. B., Pins, M., McDermott, D. F., Atkins, M. B., Dutcher, J. J., et al. (2015). BEST: A Randomized Phase II Study of vascular endothelial growth factor, RAF kinase, and mammalian target of rapamycin combination targeted therapy with bevacizumab, sorafenib, and temsirolimus in advanced renal cell carcinoma—A trial of the ECOG-ACRIN Cancer Research Group (E2804). *Journal of Clinical Oncology, 33*(21), 2384–2391. http://dx.doi.org/10.1200/jco.2015.60.9727.

Franke, R. M., Gardner, E. R., & Sparreboom, A. (2010). Pharmacogenetics of drug transporters. *Current Pharmaceutical Design, 16*(2), 220–230.

Freemantle, N., Marston, L., Walters, K., Wood, J., Reynolds, M. R., & Petersen, I. (2013). Making inferences on treatment effects from real world data: Propensity scores, confounding by indication, and other perils for the unwary in observational research. *BMJ*, *347*, f6409. http://dx.doi.org/10.1136/bmj.f6409.

Fujita, K., Sugiura, T., Okumura, H., Umeda, S., Nakamichi, N., Watanabe, Y., et al. (2014). Direct inhibition and down-regulation by uremic plasma components of hepatic uptake transporter for SN-38, an active metabolite of irinotecan, in humans. *Pharmaceutical Research*, *31*(1), 204–215. http://dx.doi.org/10.1007/s11095-013-1153-x.

Ghatak, A., Faheem, O., & Thompson, P. D. (2010). The genetics of statin-induced myopathy. *Atherosclerosis*, *210*(2), 337–343. http://dx.doi.org/10.1016/j.atherosclerosis.2009.11.033.

Gijsen, M., King, P., Perera, T., Parker, P. J., Harris, A. L., Larijani, B., et al. (2010). HER2 phosphorylation is maintained by a PKB negative feedback loop in response to anti-HER2 herceptin in breast cancer. *PLoS Biology*, *8*(12). e1000563. http://dx.doi.org/10.1371/journal.pbio.1000563.

Goldstein, J. L., & Brown, M. S. (1990). Regulation of the mevalonate pathway. *Nature*, *343*(6257), 425–430. http://dx.doi.org/10.1038/343425a0.

Gomez-Martin, C., Lopez-Rios, F., Aparicio, J., Barriuso, J., Garcia-Carbonero, R., Pazo, R., et al. (2014). A critical review of HER2-positive gastric cancer evaluation and treatment: From trastuzumab, and beyond. *Cancer Letters*, *351*(1), 30–40. http://dx.doi.org/10.1016/j.canlet.2014.05.019.

Gong, I. Y., & Kim, R. B. (2013). Impact of genetic variation in OATP transporters to drug disposition and response. *Drug Metabolism and Pharmacokinetics*, *28*(1), 4–18.

Hagenbuch, B., & Stieger, B. (2013). The SLCO (former SLC21) superfamily of transporters. *Molecular Aspects of Medicine*, *34*(2–3), 396–412. http://dx.doi.org/10.1016/j.mam.2012.10.009.

Hart, I. K., Maddison, P., Newsom-Davis, J., Vincent, A., & Mills, K. R. (2002). Phenotypic variants of autoimmune peripheral nerve hyperexcitability. *Brain*, *125*(Pt. 8), 1887–1895.

Hauck, L., Harms, C., Grothe, D., An, J., Gertz, K., Kronenberg, G., et al. (2007). Critical role for FoxO3a-dependent regulation of p21CIP1/WAF1 in response to statin signaling in cardiac myocytes. *Circulation Research*, *100*(1), 50–60. http://dx.doi.org/10.1161/01.RES.0000254704.92532.b9.

Henricks, L. M., Schellens, J. H., Huitema, A. D., & Beijnen, J. H. (2015). The use of combinations of monoclonal antibodies in clinical oncology. *Cancer Treatment Reviews*, *41*(10), 859–867. http://dx.doi.org/10.1016/j.ctrv.2015.10.008.

Hepgul, N., Pariante, C. M., Dipasquale, S., DiForti, M., Taylor, H., Marques, T. R., et al. (2012). Childhood maltreatment is associated with increased body mass index and increased C-reactive protein levels in first-episode psychosis patients. *Psychological Medicine*, *42*(9), 1893–1901. http://dx.doi.org/10.1017/S0033291711002947.

Hermida, N., Markl, A., Hamelet, J., Van Assche, T., Vanderper, A., Herijgers, P., et al. (2013). HMGCoA reductase inhibition reverses myocardial fibrosis and diastolic dysfunction through AMP-activated protein kinase activation in a mouse model of metabolic syndrome. *Cardiovascular Research*, *99*(1), 44–54. http://dx.doi.org/10.1093/cvr/cvt070.

Hirsch, F. R., Varella-Garcia, M., Bunn, P. A., Jr., Franklin, W. A., Dziadziuszko, R., Thatcher, N., et al. (2006). Molecular predictors of outcome with gefitinib in a phase III placebo-controlled study in advanced non-small-cell lung cancer. *Journal of Clinical Oncology*, *24*(31), 5034–5042. http://dx.doi.org/10.1200/JCO.2006.06.3958.

Hochhaus, A., O'Brien, S. G., Guilhot, F., Druker, B. J., Branford, S., Foroni, L., et al. (2009). Six-year follow-up of patients receiving imatinib for the first-line treatment

of chronic myeloid leukemia. *Leukemia*, *23*(6), 1054–1061. http://dx.doi.org/10.1038/leu.2009.38.
Holland, I. B. (2011). ABC transporters, mechanisms and biology: An overview. *Essays in Biochemistry*, *50*(1), 1–17. http://dx.doi.org/10.1042/bse0500001.
Hopkins, A. L. (2007). Network pharmacology. *Nature Biotechnology*, *25*(10), 1110–1111. http://dx.doi.org/10.1038/nbt1007-1110.
Hopkins, A. L. (2008). Network pharmacology: The next paradigm in drug discovery. *Nature Chemical Biology*, *4*(11), 682–690. http://dx.doi.org/10.1038/nchembio.118.
Horowitz, M. A., & Zunszain, P. A. (2015). Neuroimmune and neuroendocrine abnormalities in depression: Two sides of the same coin. *Annals of the New York Academy of Sciences*, *1351*(1), 68–79. http://dx.doi.org/10.1111/nyas.12781.
Huang, L., Zhang, T., Xie, C., Liao, X., Yu, Q., Feng, J., et al. (2013). SLCO1B1 and SLC19A1 gene variants and irinotecan-induced rapid response and survival: A prospective multicenter pharmacogenetics study of metastatic colorectal cancer. *PLoS One*, *8*(10). e77223. http://dx.doi.org/10.1371/journal.pone.0077223.
Hughes, T. P., Hochhaus, A., Branford, S., Muller, M. C., Kaeda, J. S., Foroni, L., et al. (2010). Long-term prognostic significance of early molecular response to imatinib in newly diagnosed chronic myeloid leukemia: An analysis from the International Randomized Study of Interferon and STI571 (IRIS). *Blood*, *116*(19), 3758–3765. http://dx.doi.org/10.1182/blood-2010-03-273979.
Hughes, T. P., Saglio, G., Kantarjian, H. M., Guilhot, F., Niederwieser, D., Rosti, G., et al. (2014). Early molecular response predicts outcomes in patients with chronic myeloid leukemia in chronic phase treated with frontline nilotinib or imatinib. *Blood*, *123*(9), 1353–1360. http://dx.doi.org/10.1182/blood-2013-06-510396.
Hurvitz, S. A., Betting, D. J., Stern, H. M., Quinaux, E., Stinson, J., Seshagiri, S., et al. (2012). Analysis of Fcgamma receptor IIIa and IIa polymorphisms: Lack of correlation with outcome in trastuzumab-treated breast cancer patients. *Clinical Cancer Research*, *18*(12), 3478–3486. http://dx.doi.org/10.1158/1078-0432.ccr-11-2294.
Investigators, T. C. A. S. T. C. (1989). Preliminary report: Effect of encainide and flecainide on mortality in a randomized trial of arrhythmia suppression after myocardial infarction. *The New England Journal of Medicine*, *321*(6), 406–412. http://dx.doi.org/10.1056/nejm198908103210629.
Iriyama, N., Fujisawa, S., Yoshida, C., Wakita, H., Chiba, S., Okamoto, S., et al. (2015). Early cytotoxic lymphocyte expansion contributes to a deep molecular response to dasatinib in patients with newly diagnosed chronic myeloid leukemia in the chronic phase: Results of the D-first study. *American Journal of Hematology*, *90*(9), 819–824. http://dx.doi.org/10.1002/ajh.24096.
Jain, P., Kantarjian, H., Nazha, A., O'Brien, S., Jabbour, E., Romo, C. G., et al. (2013). Early responses predict better outcomes in patients with newly diagnosed chronic myeloid leukemia: Results with four tyrosine kinase inhibitor modalities. *Blood*, *121*(24), 4867–4874. http://dx.doi.org/10.1182/blood-2013-03-490128.
Janga, S. C., & Tzakos, A. (2009). Structure and organization of drug-target networks: Insights from genomic approaches for drug discovery. *Molecular BioSystems*, *5*(12), 1536–1548. http://dx.doi.org/10.1039/B908147j.
Jenkins, G. H., Grieve, L. A., Yacoub, M. H., & Singer, D. R. (1996). Effect of simvastatin on ejection fraction in cardiac transplant recipients. *The American Journal of Cardiology*, *78*(12), 1453–1456.
Joensuu, H., Eriksson, M., Sundby Hall, K., Hartmann, J. T., Pink, D., Schutte, J., et al. (2012). One vs three years of adjuvant imatinib for operable gastrointestinal stromal tumor: A randomized trial. *JAMA*, *307*(12), 1265–1272. http://dx.doi.org/10.1001/jama.2012.347.
Joensuu, H., Eriksson, M., Sundby Hall, K., Reichardt, A., Hartmann, J. T., Pink, D., et al. (2015). Adjuvant imatinib for high-risk GI stromal tumor: Analysis of a randomized

trial. *Journal of Clinical Oncology*. http://dx.doi.org/10.1200/jco.2015.62.9170. Epub ahead of print.

Kataoka, K., Tokunaga, M., Mizusawa, J., Machida, N., Katayama, H., Shitara, K., et al. (2015). A randomized Phase II trial of systemic chemotherapy with and without trastuzumab followed by surgery in HER2-positive advanced gastric or esophagogastric junction adenocarcinoma with extensive lymph node metastasis: Japan Clinical Oncology Group study JCOG1301 (Trigger Study). *Japanese Journal of Clinical Oncology*, *45*(11), 1082–1086. http://dx.doi.org/10.1093/jjco/hyv134.

Khandaker, G. M., Cousins, L., Deakin, J., Lennox, B. R., Yolken, R., & Jones, P. B. (2015). Inflammation and immunity in schizophrenia: Implications for pathophysiology and treatment. *The Lancet Psychiatry*, *2*(3), 258–270. http://dx.doi.org/10.1016/S2215-0366(14)00122-9.

Kim, D. Y., Joo, Y. D., Lim, S. N., Kim, S. D., Lee, J. H., Lee, J. H., et al. (2015). Nilotinib combined with multiagent chemotherapy for newly diagnosed Philadelphia-positive acute lymphoblastic leukemia. *Blood*, *126*(6), 746–756. http://dx.doi.org/10.1182/blood-2015-03-636548.

Koh, K. K., Sakuma, I., & Quon, M. J. (2011). Differential metabolic effects of distinct statins. *Atherosclerosis*, *215*(1), 1–8. http://dx.doi.org/10.1016/j.atherosclerosis.2010.10.036.

Kou, B., Vatish, M., & Singer, D. R. (2007). Effects of angiotensin II on human endothelial cells survival signalling pathways and its angiogenic response. *Vascular Pharmacology*, *47*(4), 199–208. http://dx.doi.org/10.1016/j.vph.2007.06.011.

Krogan, N. J., Lippman, S., Agard, D. A., Ashworth, A., & Ideker, T. (2015). The cancer cell map initiative: Defining the hallmark networks of cancer. *Molecular Cell*, *58*(4), 690–698. http://dx.doi.org/10.1016/j.molcel.2015.05.008.

Ladwa, S. R., Dilly, S. J., Clark, A. J., Marsh, A., & Taylor, P. C. (2008). Rapid identification of a putative interaction between beta2-adrenoreceptor agonists and ATF4 using a chemical genomics approach. *ChemMedChem*, *3*(5), 742–744. http://dx.doi.org/10.1002/cmdc.200700317.

Le, X. F., Pruefer, F., & Bast, R. C., Jr. (2005). HER2-targeting antibodies modulate the cyclin-dependent kinase inhibitor p27Kip1 via multiple signaling pathways. *Cell Cycle*, *4*(1), 87–95.

Lévesque, E., Bélanger, A. S., Harvey, M., Couture, F., Jonker, D., Innocenti, F., et al. (2013). Refining the UGT1A haplotype associated with irinotecan-induced hematological toxicity in metastatic colorectal cancer patients treated with 5-fluorouracil/irinotecan-based regimens. *The Journal of Pharmacology and Experimental Therapeutics*, *345*, 95–101.

Liao, J. K., & Laufs, U. (2005). Pleiotropic effects of statins. *Annual Review of Pharmacology and Toxicology*, *45*(1), 89–118. http://dx.doi.org/10.1146/annurev.pharmtox.45.120403.095748.

Link, E., Parish, S., Armitage, J., Bowman, L., Heath, S., Matsuda, F., et al. (2008). SLCO1B1 variants and statin-induced myopathy—A genomewide study. *The New England Journal of Medicine*, *359*(8), 789–799. http://dx.doi.org/10.1056/NEJMoa0801936.

Lunnon, K., Teeling, J. L., Tutt, A. L., Cragg, M. S., Glennie, M. J., & Perry, V. H. (2011). Systemic inflammation modulates Fc receptor expression on microglia during chronic neurodegeneration. *Journal of Immunology*, *186*, 7215–7224.

Lv, Y., Yang, Z., Zhao, L., Zhao, S., Han, J., & Zheng, L. (2015). The efficacy and safety of adding bevacizumab to cetuximab- or panitumumab-based therapy in the treatment of patients with metastatic colorectal cancer (mCRC): A meta-analysis from randomized control trials. *International Journal of Clinical and Experimental Medicine*, *8*(1), 334–345.

Ma, P. T., Gil, G., Südhof, T. C., Bilheimer, D. W., Goldstein, J. L., & Brown, M. S. (1986). Mevinolin, an inhibitor of cholesterol synthesis, induces mRNA for low density lipoprotein receptor in livers of hamsters and rabbits. *Proceedings of the National Academy of Sciences*

of the United States of America, 83(21), 8370–8374. http://dx.doi.org/10.1073/pnas.83.21.8370.
Macdougall, I. C., Roger, S. D., de Francisco, A., Goldsmith, D. J., Schellekens, H., Ebbers, H., et al. (2012). Antibody-mediated pure red cell aplasia in chronic kidney disease patients receiving erythropoiesis-stimulating agents: New insights. *Kidney International, 81*(8), 727–732. http://dx.doi.org/10.1038/ki.2011.500.
Mack, G. S. (2009). FDA holds court on post hoc data linking KRAS status to drug response. *Nature Biotechnology, 27*(2), 110–112. http://dx.doi.org/10.1038/nbt0209-110c.
Maitland-van der Zee, A. H., Daly, A. K., Kamali, F., Manolopoulous, V. G., Verhoef, T. I., Wadelius, M., et al. (2014). Patients benefit from genetics-guided coumarin anticoagulant therapy. *Clinical Pharmacology and Therapeutics, 96*(1), 15–17. http://dx.doi.org/10.1038/clpt.2014.44.
Manley, P. W., Drueckes, P., Fendrich, G., Furet, P., Liebetanz, J., Martiny-Baron, G., et al. (2010). Extended kinase profile and properties of the protein kinase inhibitor nilotinib. *Biochimica et Biophysica Acta, 1804*(3), 445–453. http://dx.doi.org/10.1016/j.bbapap.2009.11.008.
Marburger, J. H., III, & Kvamme, E. F. (2008). President's Council of Advisors on Science and Technology. *Priorities for personalized medicine.* Office of Science Technology and Policy: Washington, DC, USA.
McArthur, G. A., Chapman, P. B., Robert, C., Larkin, J., Haanen, J. B., Dummer, R., et al. (2014). Safety and efficacy of vemurafenib in BRAF(V600E) and BRAF(V600K) mutation-positive melanoma (BRIM-3): Extended follow-up of a phase 3, randomised, open-label study. *The Lancet Oncology, 15*(3), 323–332. http://dx.doi.org/10.1016/s1470-2045(14)70012-9.
McLeod, H. L., & Siva, C. (2002). The thiopurine S-methyltransferase gene locus—Implications for clinical pharmacogenomics. *Pharmacogenomics, 3*(1), 89–98. http://dx.doi.org/10.1517/14622416.3.1.89.
Mega, J. L., Simon, T., Collet, J. P., Anderson, J. L., Antman, E. M., Bliden, K., et al. (2010). Reduced-function CYP2C19 genotype and risk of adverse clinical outcomes among patients treated with clopidogrel predominantly for PCI: A meta-analysis. *Journal of the American Medical Association, 304*(16), 1821–1830. http://dx.doi.org/10.1001/jama.2010.1543.
Megaraj, V., Zhao, T., Paumi, C. M., Gerk, P. M., Kim, R. B., & Vore, M. (2011). Functional analysis of nonsynonymous single nucleotide polymorphisms of multidrug resistance-associated protein 2 (ABCC2). *Pharmacogenetics and Genomics, 21*(8), 506–515. http://dx.doi.org/10.1097/FPC.0b013e328348c786.
Millot, F., Guilhot, J., Baruchel, A., Petit, A., Bertrand, Y., Mazingue, F., et al. (2014). Impact of early molecular response in children with chronic myeloid leukemia treated in the French Glivec phase 4 study. *Blood, 124*(15), 2408–2410. http://dx.doi.org/10.1182/blood-2014-05-578567.
Mondelli, V., Ciufolini, S., Belvederi Murri, M., Bonaccorso, S., Di Forti, M., Giordano, A., et al. (2015). Cortisol and inflammatory biomarkers predict poor treatment response in first episode psychosis. *Schizophrenia Bulletin, 41*(5), 1162–1170. http://dx.doi.org/10.1093/schbul/sbv028.
Montemurro, M., Gelderblom, H., Bitz, U., Schutte, J., Blay, J. Y., Joensuu, H., et al. (2013). Sorafenib as third- or fourth-line treatment of advanced gastrointestinal stromal tumour and pretreatment including both imatinib and sunitinib, and nilotinib: A retrospective analysis. *European Journal of Cancer, 49*(5), 1027–1031. http://dx.doi.org/10.1016/j.ejca.2012.10.009.
Nagele, E., Han, M., Demarshall, C., Belinka, B., & Nagele, R. (2011). Diagnosis of Alzheimer's disease based on disease-specific autoantibody profiles in human sera. *PLoS One, 6*(8), e23112.

Niemi, M., Kivisto, K. T., Diczfalusy, U., Bodin, K., Bertilsson, L., Fromm, M. F., et al. (2006). Effect of SLCO1B1 polymorphism on induction of CYP3A4 by rifampicin. *Pharmacogenetics and Genomics*, *16*(8), 565–568. http://dx.doi.org/10.1097/01.fpc.0000215070.52212.0e.

Niemi, M., Pasanen, M. K., & Neuvonen, P. J. (2011). Organic anion transporting polypeptide 1B1: A genetically polymorphic transporter of major importance for hepatic drug uptake. *Pharmacological Reviews*, *63*(1), 157–181. http://dx.doi.org/10.1124/pr.110.002857.

NORD. (2015). *NORD (National Organization for Rare Disorders)*. http://rarediseases.org/. Accessed on 5th November, 2015.

O'Brien, S. G., Guilhot, F., Larson, R. A., Gathmann, I., Baccarani, M., Cervantes, F., et al. (2003). Imatinib compared with interferon and low-dose cytarabine for newly diagnosed chronic-phase chronic myeloid leukemia. *The New England Journal of Medicine*, *348*(11), 994–1004. http://dx.doi.org/10.1056/NEJMoa022457.

Ogbagaber, S. B., Karp, J., & Wahed, A. S. (2015). Design of sequentially randomized trials for testing adaptive treatment strategies. *Statistics in Medicine*. 10.1002/sim.6747. Epub ahead of print.

Ozawa, T., Oda, H., Oda, M., Hosaka, Y., Kashimura, T., Ozaki, K., et al. (2009). Improved cardiac function after sirolimus-eluting stent placement in diabetic patients by pioglitazone: Combination therapy with statin. *Journal of Cardiology*, *53*(3), 402–409. http://dx.doi.org/10.1016/j.jjcc.2009.01.011.

Ozvegy-Laczka, C., Hegedus, T., Varady, G., Ujhelly, O., Schuetz, J. D., Varadi, A., et al. (2004). High-affinity interaction of tyrosine kinase inhibitors with the ABCG2 multidrug transporter. *Molecular Pharmacology*, *65*(6), 1485–1495. http://dx.doi.org/10.1124/mol.65.6.1485.

Pavlos, R., Mallal, S., & Phillips, E. (2012). HLA and pharmacogenetics of drug hypersensitivity. *Pharmacogenomics*, *13*(11), 1285–1306. http://dx.doi.org/10.2217/pgs.12.108.

Peeters, M., Siena, S., Van Cutsem, E., Sobrero, A., Hendlisz, A., Cascinu, S., et al. (2009). Association of progression-free survival, overall survival, and patient-reported outcomes by skin toxicity and KRAS status in patients receiving panitumumab monotherapy. *Cancer*, *115*(7), 1544–1554. http://dx.doi.org/10.1002/cncr.24088.

Pichler, M., & Calin, G. A. (2015). MicroRNAs in cancer: From developmental genes in worms to their clinical application in patients. *British Journal of Cancer*, *113*(4), 569–573. http://dx.doi.org/10.1038/bjc.2015.253.

Pujol, A., Mosca, R., Farres, J., & Aloy, P. (2010). Unveiling the role of network and systems biology in drug discovery. *Trends in Pharmacological Sciences*, *31*(3), 115–123. http://dx.doi.org/10.1016/j.tips.2009.11.006.

Quintas-Cardama, A., & Cortes, J. (2009). Molecular biology of bcr-abl1-positive chronic myeloid leukemia. *Blood*, *113*(8), 1619–1630. http://dx.doi.org/10.1182/blood-2008-03-144790.

Raison, C. L., Rutherford, R. E., Woolwine, B. J., Shuo, C., Schettler, P., Drake, D. F., et al. (2013). A randomized controlled trial of the tumor necrosis factor antagonist infliximab for treatment-resistant depression: The role of baseline inflammatory biomarkers. *JAMA Psychiatry*, *70*(1), 31–41. http://dx.doi.org/10.1001/2013.jamapsychiatry.4.

Rajtik, T., Carnicka, S., Szobi, A., Mesarosova, L., Matus, M., Svec, P., et al. (2012). Pleiotropic effects of simvastatin are associated with mitigation of apoptotic component of cell death upon lethal myocardial reperfusion-induced injury. *Physiological Research*, *61*(Suppl. 2), S33–41.

Ramanan, S., & Grampp, G. (2014). Drift, evolution, and divergence in biologics and biosimilars manufacturing. *BioDrugs*, *28*(4), 363–372. http://dx.doi.org/10.1007/s40259-014-0088-z.

Rha, B., Tate, J. E., Weintraub, E., Haber, P., Yen, C., Patel, M., et al. (2014). Intussusception following rotavirus vaccination: An updated review of the available evidence. *Expert Review of Vaccines*, *13*(11), 1339–1348. http://dx.doi.org/10.1586/14760584.2014.942223.

Riley, M. F. (2015). An unfulfilled promise: Changes needed to the drug approval process to make personalized medicine a reality. *Food and Drug Law Journal*, *70*(2), 289–314. ii–iii.

Roche, P. C., & Ingle, J. N. (1999). Increased HER2 with U.S. Food and Drug Administration-approved antibody. *Journal of Clinical Oncology*, *17*(1), 434.

Sattar, N., Preiss, D., Murray, H. M., Welsh, P., Buckley, B. M., de Craen, A. J., et al. (2010). Statins and risk of incident diabetes: A collaborative meta-analysis of randomised statin trials. *Lancet*, *375*(9716), 735–742. http://dx.doi.org/10.1016/S0140-6736(09)61965-6.

Schmidt, A. F., Klungel, O. H., & Groenwold, R. H. (2016). GetReal Consortium. Adjusting for confounding in early postlaunch settings: Going beyond logistic regression models. *Epidemiology*, **27**(1), 133–142. http://dx.doi.org/10.1097/EDE.0000000000000388.

Schott, J. M., Reiniger, L., Thom, M., Holton, J. L., Grieve, J., Brandner, S., et al. (2010). Brain biopsy in dementia: Clinical indications and diagnostic approach. *Acta Neuropathologica*, *120*(3), 327–341. http://dx.doi.org/10.1007/s00401-010-0721-y.

Schulthess, D., Chlebus, M., Bergstrm, R., & Baelen, K. V. (2014). Medicine adaptive pathways to patients (MAPPs): Using regulatory innovation to defeat Eroom's law. *Chinese Clinical Oncology*, *3*(2), 21. http://dx.doi.org/10.3978/j.issn.2304-3865.2014.05.07.

Sebat, J., Lakshmi, B., Troge, J., Alexander, J., Young, J., Lundin, P., et al. (2004). Large-scale copy number polymorphism in the human genome. *Science*, *305*(5683), 525–528. http://dx.doi.org/10.1126/science.1098918.

Shah, N. P., Guilhot, F., Cortes, J. E., Schiffer, C. A., le Coutre, P., Brummendorf, T. H., et al. (2014). Long-term outcome with dasatinib after imatinib failure in chronic-phase chronic myeloid leukemia: Follow-up of a phase 3 study. *Blood*, *123*(15), 2317–2324. http://dx.doi.org/10.1182/blood-2013-10-532341.

Shan, G., Wilding, G. E., Hutson, A. D., & Gerstenberger, S. (2015). Optimal adaptive two-stage designs for early phase II clinical trials. *Statistics in Medicine*. http://dx.doi.org/10.1002/sim.6794.

Shi, Z., Peng, X. X., Kim, I. W., Shukla, S., Si, Q. S., Robey, R. W., et al. (2007). Erlotinib (Tarceva, OSI-774) antagonizes ATP-binding cassette subfamily B member 1 and ATP-binding cassette subfamily G member 2-mediated drug resistance. *Cancer Research*, *67*(22), 11012–11020. http://dx.doi.org/10.1158/0008-5472.CAN-07-2686.

Shimoni, A., Volchek, Y., Koren-Michowitz, M., Varda-Bloom, N., Somech, R., Shem-Tov, N., et al. (2015). Phase 1/2 study of nilotinib prophylaxis after allogeneic stem cell transplantation in patients with advanced chronic myeloid leukemia or Philadelphia chromosome-positive acute lymphoblastic leukemia. *Cancer*, *121*(6), 863–871. http://dx.doi.org/10.1002/cncr.29141.

Singer, D. R. J., & Watkins, J. (2012). Using companion and coupled diagnostics within strategy to personalize targeted medicines. *Personalized Medicine*, *9*, 751–761.

Slamon, D., Eiermann, W., Robert, N., Pienkowski, T., Martin, M., Press, M., et al. (2011). Adjuvant trastuzumab in HER2-positive breast cancer. *The New England Journal of Medicine*, *365*(14), 1273–1283. http://dx.doi.org/10.1056/NEJMoa0910383.

Southworth, M. R., Reichman, M. E., & Unger, E. F. (2013). Dabigatran and postmarketing reports of bleeding. *The New England Journal of Medicine*, *368*(14), 1272–1274. http://dx.doi.org/10.1056/NEJMp1302834.

Soverini, S., Hochhaus, A., Nicolini, F. E., Gruber, F., Lange, T., Saglio, G., et al. (2011). BCR-ABL kinase domain mutation analysis in chronic myeloid leukemia patients treated with tyrosine kinase inhibitors: Recommendations from an expert panel on behalf of European LeukemiaNet. *Blood*, *118*(5), 1208–1215. http://dx.doi.org/10.1182/blood-2010-12-326405.

Stallard, N., Hamborg, T., Parsons, N., & Friede, T. (2014). Adaptive designs for confirmatory clinical trials with subgroup selection. *Journal of Biopharmaceutical Statistics*, *24*(1), 168–187. http://dx.doi.org/10.1080/10543406.2013.857238.
Stern, H. M., Gardner, H., Burzykowski, T., Elatre, W., O'Brien, C., Lackner, M. R., et al. (2015). PTEN loss is associated with worse outcome in HER2-amplified breast cancer patients but is not associated with trastuzumab resistance. *Clinical Cancer Research*, *21*(9), 2065–2074. http://dx.doi.org/10.1158/1078-0432.ccr-14-2993.
Sugimoto, Y., Tsukahara, S., Ishikawa, E., & Mitsuhashi, J. (2005). Breast cancer resistance protein: Molecular target for anticancer drug resistance and pharmacokinetics/pharmacodynamics. *Cancer Science*, *96*(8), 457–465. http://dx.doi.org/10.1111/j.1349-7006.2005.00081.x.
Sun, H., Depraetere, K., De Roo, J., Mels, G., De Vloed, B., Twagirumukiza, M., et al. (2015). Semantic processing of EHR Data for clinical research. *Journal of Biomedical Informatics*. http://dx.doi.org/10.1016/j.jbi.2015.10.009.
Svoboda, M., Wlcek, K., Taferner, B., Hering, S., Stieger, B., Tong, D., et al. (2011). Expression of organic anion-transporting polypeptides 1B1 and 1B3 in ovarian cancer cells: Relevance for paclitaxel transport. *Biomedicine & Pharmacotherapy*, *65*(6), 417–426. http://dx.doi.org/10.1016/j.biopha.2011.04.031.
Taylor, P. C., Clark, A. J., Marsh, A., Singer, D. R., & Dilly, S. J. (2013). A chemical genomics approach to identification of interactions between bioactive molecules and alternative reading frame proteins. *Chemical Communications (Cambridge, England)*, *49*(83), 9588–9590. http://dx.doi.org/10.1039/c3cc44647f.
Thuc, L. C., Teshima, Y., Takahashi, N., Nagano-Torigoe, Y., Ezaki, K., Yufu, K., et al. (2010). Mitochondrial K(ATP) channels-derived reactive oxygen species activate pro-survival pathway in pravastatin-induced cardioprotection. *Apoptosis*, *15*(6), 669–678. http://dx.doi.org/10.1007/s10495-010-0473-0.
Van Cutsem, E., Kohne, C. H., Lang, I., Folprecht, G., Nowacki, M. P., Cascinu, S., et al. (2011). Cetuximab plus irinotecan, fluorouracil, and leucovorin as first-line treatment for metastatic colorectal cancer: Updated analysis of overall survival according to tumor KRAS and BRAF mutation status. *Journal of Clinical Oncology*, *29*(15), 2011–2019. http://dx.doi.org/10.1200/jco.2010.33.5091.
Vincent, A., Buckley, C., Schott, J. M., Baker, I., Dewar, B. K., Detert, N., et al. (2004). Potassium channel antibody-associated encephalopathy: A potentially immunotherapy-responsive form of limbic encephalitis. *Brain*, *127*(Pt. 3), 701–712. http://dx.doi.org/10.1093/brain/awh077.
Wainwright, C. E., Elborn, J. S., Ramsey, B. W., Marigowda, G., Huang, X., Cipolli, M., et al. (2015). Lumacaftor-ivacaftor in patients with cystic fibrosis homozygous for Phe508del CFTR. *The New England Journal of Medicine*, *373*(3), 220–231. http://dx.doi.org/10.1056/NEJMoa1409547.
Watkins, J., Marsh, A., Taylor, P. C., & Singer, D. R. (2010). Personalized medicine: The impact on chemistry. *Therapeutic Delivery*, *1*(5), 651–665.
Welzen, M. E., Dezentje, V. O., van Schaik, R. H., Colbers, A. P., Guchelaar, H. J., van Erp, N. P., et al. (2015). The effect of tamoxifen dose increment in patients with impaired CYP2D6 activity. *Therapeutic Drug Monitoring*, *37*(4), 501–507. http://dx.doi.org/10.1097/ftd.0000000000000195.
Wen, J., Luo, J., Huang, W., Tang, J., Zhou, H., & Zhang, W. (2015). The pharmacological and physiological role of multidrug-resistant protein 4. *The Journal of Pharmacology and Experimental Therapeutics*, *354*(3), 358–375. http://dx.doi.org/10.1124/jpet.115.225656.
Williamson, B., Soars, A. C., Owen, A., White, P., Riley, R. J., & Soars, M. G. (2013). Dissecting the relative contribution of OATP1B1-mediated uptake of xenobiotics into human hepatocytes using siRNA. *Xenobiotica*, *43*(10), 920–931. http://dx.doi.org/10.3109/00498254.2013.776194.

Wright, S., Hashemi, K., Stasiak, L., Bartram, J., Lang, B., Vincent, A., et al. (2015). Epileptogenic effects of NMDAR antibodies in a passive transfer mouse model. *Brain, 138*(Pt. 11), 3159–3167. http://dx.doi.org/10.1093/brain/awv257.

Yang, H., Higgins, B., Kolinsky, K., Packman, K., Bradley, W. D., Lee, R. J., et al. (2012). Antitumor activity of BRAF inhibitor vemurafenib in preclinical models of BRAF-mutant colorectal cancer. *Cancer Research, 72*(3), 779–789. http://dx.doi.org/10.1158/0008-5472.can-11-2941.

Zair, Z. M., Eloranta, J. J., Stieger, B., & Kullak-Ublick, G. A. (2008). Pharmacogenetics of OATP (SLC21/SLCO), OAT and OCT (SLC22) and PEPT (SLC15) transporters in the intestine, liver and kidney. *Pharmacogenomics, 9*(5), 597–624. http://dx.doi.org/10.2217/14622416.9.5.597.

Zheng, X., & Hu, S. J. (2005). Effects of simvastatin on cardiac performance and expression of sarcoplasmic reticular calcium regulatory proteins in rat heart. *Acta Pharmacologica Sinica, 26*(6), 696–704. http://dx.doi.org/10.1111/j.1745-7254.2005.00105.x.

Zhu, A. X., Park, J. O., Ryoo, B. Y., Yen, C. J., Poon, R., Pastorelli, D., et al. (2015). Ramucirumab versus placebo as second-line treatment in patients with advanced hepatocellular carcinoma following first-line therapy with sorafenib (REACH): A randomised, double-blind, multicentre, phase 3 trial. *The Lancet Oncology, 16*(7), 859–870. http://dx.doi.org/10.1016/s1470-2045(15)00050-9.

CHAPTER FIVE

Personalized Medicine in Respiratory Disease: Role of Proteomics

V.S. Priyadharshini, Luis M. Teran[1]
Instituto Nacional de Enfemedades Respiratorias, Delegación Tlalpan, Mexico
[1]Corresponding author: e-mail addresses: teranjlm@gmail.com; lmteran@iner.gob.mx

Contents

Abstract

Respiratory diseases affect humanity globally, with chronic lung diseases (e.g., asthma, chronic obstructive pulmonary disease, idiopathic pulmonary fibrosis, among others) and lung cancer causing extensive morbidity and mortality. These conditions are highly heterogeneous and require an early diagnosis. However, initial symptoms are non-specific, and the clinical diagnosis is made late frequently. Over the last few years, personalized medicine has emerged as a medical care approach that uses novel technology aiming to personalize treatments according to the particular patient's medical needs. This review highlights the contributions of proteomics toward the understanding of personalized medicine in respiratory disease and its potential applications in the clinic.

Advances in Protein Chemistry and Structural Biology, Volume 102
ISSN 1876-1623
http://dx.doi.org/10.1016/bs.apcsb.2015.11.008

1. INTRODUCTION

Respiratory diseases are common cause of deaths in the world: it is estimated that asthma affects 300 million people in the world, and chronic obstructive pulmonary disease (COPD) ranks 4th as a major cause of deaths around the globe. Similarly, idiopathic pulmonary fibrosis (IPF) is a fatal fibrotic lung disease characterized by progressive interstitial fibrosis preventing gaseous exchange leading to a rapid decline in the quality of life; the average survival for patients suffering this disease is between 3 and 5 years, once the diagnosis is made. The ultimate cause of this situation is the lack of efficacy in diagnostics and treatment, it is so because, these diseases involve complex molecular interactions and diverge with several molecular pathways making it difficult for scientists to understand the exact disease pathogenesis. In respiratory diseases, most patients do not show early symptoms, leading to late clinical diagnosis. Current clinical diagnostic tests, such as chest X-ray, lung function tests, and histopathological techniques are useful; however, they are not specific. For example, tuberculosis (TB) and community-acquired pneumonia symptoms represent similar radiological findings which could lead to false diagnosis (Krysl, Korzeniewska-Kosela, Muller, & FitzGerald, 1994; Woodring et al., 1986). So, clinicians usually prescribe antibiotics to treat pneumonia before TB is correctly diagnosed (Grossman, Hsueh, Gillespie, & Blasi, 2014; Wang et al., 2011). Similarly, spirometry measures pulmonary function for the diagnosis of obstructive airway diseases such as COPD, restrictive pulmonary diseases, asthma and it is invaluable to test respiratory health. However, recent ancestry-based genetic studies have challenged this traditional view by showing that ancestry and genetic factors play a role in determining lung function. Indeed, Kumar et al. evaluated three independent African-American general population cohorts and found that global estimates of percentage African ancestry were inversely associated with baseline lung function measures: an African-American man with 25% African ancestry would have a predicted FEV1 of 407 mL higher compared to an African-American man with 75% African ancestry of similar age and height. In addition, it was shown that asthma severity, based on percentage-predicted FEV1, was misclassified in up to 5% of asthmatics when using race-based lung function equations as opposed to ancestry-based equations (Kumar et al., 2010). GWAS studies have identified candidate genes for lung function as examples of genetic variation which could constitute personalized genetic profiles predictive of lung function and risk for impairment in disease states such as asthma and COPD (Hancock et al., 2010; Repapi et al., 2010; Wilk et al., 2009).

2. PERSONALIZED MEDICINE

Over the last few years, personalized medicine has emerged as a medical care approach that uses novel technology aiming to personalize treatments according to the particular patient medical needs. Early investigations in personalized medicine were made looking at genetic mutations that have eventually led to significant discoveries. For example, a rare SNP in the exon region of β2-adrenergic receptor (*ADBR2*) gene Thr164Ile has been associated with life-threatening events in asthmatics. Ortega et al. estimated that 150 non-Hispanic whites would need to be genotyped to identify three subjects with this rare Ile164 allele, while 100 African Americans genotyped to identify two subjects with this SNP in order to prevent a single hospitalization during LABA treatment over a 1-year period (Ortega et al., 2014). Although genomics has proved useful, post-translational modifications such as methylation, acetylation, glycosylation and protein folding alter the protein product or phenotype. However, a single gene can also generate different proteins through alternative splicing and post-translational modifications producing thousands of protein types in each cell. Proteins are the major components of the human body and they are the main structural component of all cells. Thus, unraveling the respiratory proteome, it is expected to play a significant role in the development of new systems supporting health care. Proteomics allows to analyze global protein expression rather than using single biomarker approach, and developments on new technologies have begun to unravel the complexity of the proteome, indeed, while the DNA carries the genetic information. The recent mapping of the proteome in 2014 is providing an unprecedented opportunity: a couple of human proteome drafts were created using mass spectrometry (MS) including the Human Proteome Map (Kim et al., 2014; Reardon, 2015) and the proteomics database (Wilhelm et al., 2014), and the tissue human protein map was built from tissues that allowed discovering at least one major isoform of 85% of the translated human genome (Uhlén et al., 2015). This last approach allowed providing spatial information on expression pattern in both some cells and subcellular structures in human tissues. Interestingly, The President Barack Obama launched the precision medicine initiative (Collins & Varmus, 2015), last January 2015 raising hopes for precision medicine and future innovative treatments.

Personalized medicine is also known as "P4 medicine" (Personalized, Predictive, Preventive, and Participatory) (Hood & Friend, 2011). This approach extends beyond genomic medicine because genes never act alone, instead proteins control biological functions both in health and disease.

3. PROTEOMICS

MS-based proteomics is crucial in medical research because it allows to analyze the whole proteome in health and disease. Expression proteomics involving either two-dimensional electrophoresis (2DE) or liquid chromatography–MS (LC–MS) analysis is the most common method used in investigating respiratory illnesses (Fig. 1 illustrates protein fractionation methods). In 2DE, proteins are separated by isoelectric focusing strips according to their net charge (first dimension), followed by molecular mass separation in a polyacrylamide gel electrophoresis (PAGE) (second dimension) (Carrette, Burkhard, Sanchez, & Hochstrasser, 2006; Holcăpek, Jiraśko, & Lísa, 2012). MS-peptide mass fingerprinting along with sequencing of selected peptides using tandem MS (also known as MS/MS) is the ideal protein identification method (Fig. 2). Any two samples can be qualitatively and quantitatively compared, and the presence or absence of spots provides information about differentially expressed proteins including isoforms. Primary limitations, it involves laborious laboratory procedure, and more than one protein can be frequently resolved in the same spot, making quantification difficult. While the use of 2D PAGE is still standard, the majority of research work undertaken in the field nowadays utilizes LC–MS.

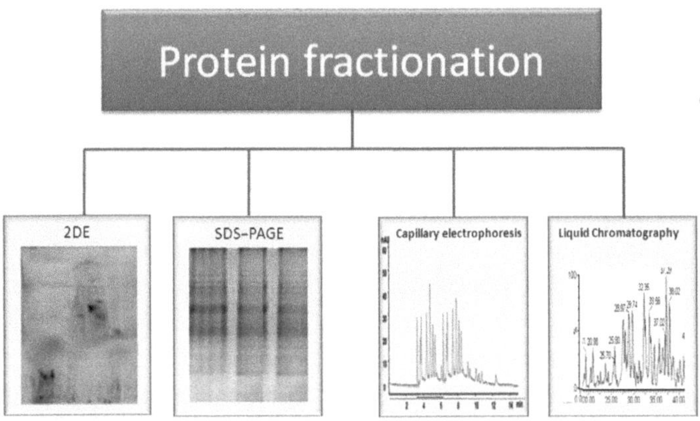

Figure 1 Protein fractionation methods.

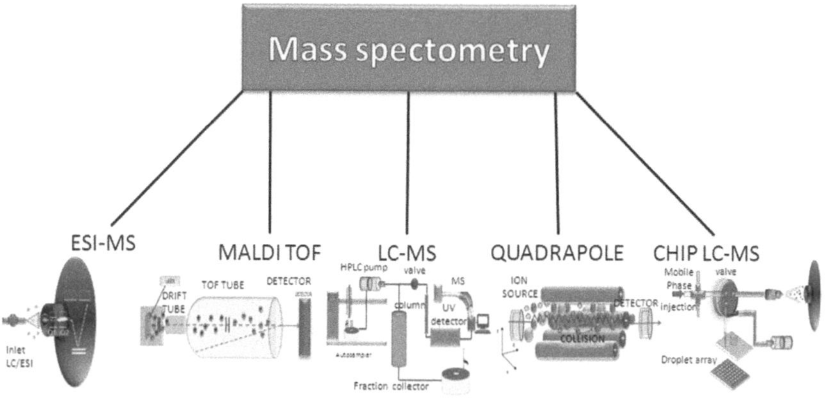

Figure 2 Mass spectrometry (MS)-based proteomics methods including ESI–MS, MALDI-TOF, LC–MS, Quadrapole MS, and ChIP LC–MS. *Figure adaptation: ESI–MS (http://www.pharmacelsus.de/esi/), MALDI-TOF (http://www.mpipz.mpg.de/44542/MALDI-TOF-TOF_MS_MS), LC–MS (http://home.iprimus.com.au/pakholt/lcms/details.html), and ChIP LC–MS (Wang, Zhu, & Fang, 2014).*

LC–MS is a powerful technique that it is oriented toward the identification of proteins in complex mixtures (Horvatovich, Hoekman, Govorukhina, & Bischoff, 2010), Fig. 2 describes various MS-based methods. In this strategy, proteins are enzymatically digested before LC. Then, the mass spectra generated by MS analysis are compared with the theoretical MS/MS spectra from the databases (also known as shotgun proteomics). Some limitations to this technology are (a) most proteins are identified based on few peptides, (b) protein isoforms and post-translational modifications are often missed, (c) there is a lack of sensitivity to perform quantitative proteomics, and (d) low-abundant and small proteins often will be lost. To date, the current shotgun proteomics approach has become a routine to study proteins in biological samples. In the search for disease biomarkers, quantitative proteomics has, therefore, arisen. Quantitative proteomics can be categorized into relative and absolute quantitation. Relative quantitative methods like isotope-coded affinity tag, isobaric tags for relative and absolute quantitation (iTRAQ), and stable isotope labeling by amino acids in cell culture (SILAC) are used to compare proteins (on the level of peptide differences): several peptides of a protein must be found to get a reliable relative quantitation between samples. Among these methods, SILAC is the most used technique (Gygi et al., 1999; Horvatovich et al., 2010; Mann, 2006; Ross et al., 2004). For the detection of low-abundant proteins, targeted proteomics technologies have been developed including the antibody-based enrichment method named stable isotope standards and capture by anti-peptide antibodies (SISCAPA) immobilized on novel nanoaffinity columns are used to concentrate and enrich specific peptides along with stable isotope-labeled internal standard of the same sequences. On elution from the anti-peptide antibody affinity column, electrospray ionization-mass spectrometry (ESI–MS) is then used to determine the quantities of peptides by comparison of the signals from the peptides with those of the corresponding stable isotope-labeled internal standards. Selected reaction monitoring (SRM), is a procedure used in tandem MS also called multiple reaction monitoring (MRM), has emerged as a powerful tool in biomarker discovery, which can detect proteins expressed at concentrations above 50–100 copies per cell (Elschenbroich & Kislinger, 2011; Picotti & Aebersold, 2012). Triple-quadrupole MS (TQ-MS) is integrated into the SRM technology; the first and the third quadrupoles select predefined *m/z* values of the targeted peptide ion or particular fragment, whereas the second quadrupole serves as a collision cell. Key features of SRM assays include the ability to target specific peptide sequences and quantify protein isoforms or protein post-translational modifications and the capacity for multiplexing that allows analysis of hundreds of peptides. SRM can be made more quantitative by spiking a known concentration of a peptide isotopically labeled, which is used as an internal standard, corresponding to a specific target protein. The ratio between the endogenous and the synthetic peptide is used to calculate the absolute amount of protein after LC–ESI–MS. This strategy is known as absolute quantification (termed AQUA) of proteins.

4. RESPIRATORY PROTEOMICS

The respiratory tract extends from the nose and mouth to the alveoli, and its primary function is to deliver oxygen into the body and carbon dioxide out it. Under normal conditions, many cell types produce proteins throughout the respiratory system to maintain the proteome biological functions. However, harmful agents may alter it leading to disease. Lindskog et al. have recently mapped the normal human lung proteome by using a combination of transcriptomics and antibody-based proteomics, and demonstrated, 344 proteins expressed in the lung (70% of all human proteins; 166 of these genes were found to be increased in the lung compared to other tissue types). For example, the surfactants A, B, C, and D were highly expressed in type II pneumocytes, while EMP2, a membrane protein, was localized to type I pneumocytes. MSR1, MARCO, MRC1, and C19ORF59 were expressed in macrophages, and three genes encoding the dynein proteins (DNAH5, DNAI1, and DNAI2) along with proteins of unknown function (CCDC17 and FAM92B) were expressed in ciliated cells. To study respiratory disease, proteomics has used several biological samples including serum, blood cells, bronchoalveolar lavage fluid (BAL), nasal lavage fluid (NLF), sputum, exhaled breath condensate, lung tissue, among others. While analysis of these samples has provided valuable data, proteomics findings may vary depending on the material type to be investigated. The primary objective of using proteomics in medical research is to detect proteins associated with disease for therapeutic intervention and to identify novel biomarkers (Mikami, Aoki, & Kimura, 2012; Tambor et al., 2010). A biomarker is a measurable molecule that reflects the disease process that can be found in body fluids or tissues. Thus, establishing new biomarkers has a range of applications in respiratory medicine including early detection of disease, patients' stratification risk, disease monitoring, and development of effective targeted therapies that may tailor patients' needs. To date, the impact of proteomics on personalized respiratory medicine has been limited. However, novel molecular biomarkers are beginning to emerge. In this chapter, a summary of the most recent findings on potential biomarkers in chronic lung diseases and lung cancer is made. Nasal proteomic studies have recently been reviewed (Teran, Montes-Vizuet, Li, & Franz, 2015).

5. ASTHMA

The occurrence and prevalence of asthma are increasing globally and pose an economic burden on society. In children and adolescent, this disease is mostly of allergic origin. Exposure of these patients to specific aeroallergens leads to a series of immunological changes culminating in airway inflammation. Moreover, environmental irritants such as diesel fumes, environmental tobacco smoke, endotoxin, and air pollution can exacerbate the disease and trigger asthma symptoms. Studies using proteomics technology allowed detecting a broad range of proteins both in BAL fluid and bronchial biopsies (Cederfur et al., 2012; O'Neil et al., 2011; Wu et al., 2005). In 2005, Wu et al. analyzed BAL fluid after segmental allergen challenge using SDS–PAGE, followed by LC–MS/MS, which resulted in the identification of 160 differentially expressed proteins representing a wide range of biological categories including metabolic enzymes, serum proteins, cytokine/

chemokines, calcium-binding proteins, matrix metalloproteinases, signaling, and proteins involved in lung remodeling (Wu et al., 2005). Matrix metalloproteinases 9 (MMP 9) was validated as a potential marker. However, only four asthmatic patients were included in the study. In the search of galectin proteins, Cederfur et al. applied galectin affinity chromatography to analyze BAL fluid and identified a specific glycoform haptoglobin, among 175 proteins (Cederfur et al., 2012). In 2012, O'Neil et al. also investigated the asthma proteome profile in bronchial biopsies and reported seven proteins differentially expressed (O'Neil et al., 2011) including those involved in cellular movement and immune cell trafficking with specific roles on collagen fibrillogenesis, protein elongation, and chemotaxis. In this study, treating patients with budesonide-induced activation of prothrombin pathway followed by actin motility, and signaling with a decreased expression of alpha-2-macroglobulin (α-2-M) and vientin. The authors proposed that the downregulation of these two proteins could have been resulted from the budesonide-induced suppression of inflammatory response. Because the former studies analyzed two different sample types (BALF vs. bronchial biopsies), it is not surprising that they reported different proteomics profiles. It is well established that the asthmatic response to allergen challenge is characterized by an early phase followed by a late stage, which last several hours (this is usually more severe). Singh et al. analyzed plasma samples derived from asthma subjects who were exposed to allergen challenge to identify potential markers specific to any of these two asthmatic responses. Interestingly, increased fibronectin and low levels of inter-alpha-inhibitor H4 (ITIH4) were found to characterize dual asthmatic responders (patients who develop both early and late response). However, there was not a specific marker that could distinguish early from late response (Singh et al., 2012; Wu et al., 2005). Fibronectin deposition in the subepithelium is a known feature of asthma. Sputum proteomics has been proposed as a suitable means to investigate the airways in view that sample can be safely obtained. Indeed, a shotgun proteomics (LC–MS/MS) of sputum in 10 asthmatic patients showed 17 expressed proteins among which SERPINA1 (serpin peptidase inhibitor), SCGB1A1 (secretoblin), and SMR3B (submaxillary gland androgen-regulated protein 3B) was proposed as potential markers: SERPINA1 was upregulated, while SCGB1A1 and SMR3B were downregulated. In this study, complement 3a was found to be associated with the exercise-induced asthma phenotype, which is known to occur following vigorous exercise (Gharib et al., 2011). Neutrophilic asthma is a severe asthma phenotype that is associated with lung neutrophilia, lower lung function, air trapping, and thicker airway walls. Lee et al. conducted a study in five subjects with neutrophilic uncontrolled asthma and 10 subjects with neutrophilic controlled asthma using 2DE (pooled sputum) followed by matrix-assisted laser desorption/ionization-time of light-mass spectrometry (MALDI-TOF–MS). Among 14 differentially expressed proteins, they identified S100A9 (calgranulin B) as a specific marker for neutrophilic uncontrolled asthma (Lee et al., 2013). CD4+ T lymphocytes derived from uncontrolled asthma patients have also been investigated and found to produce 13 differentially expressed proteins 47 including vimentin, which was described in O'Neil et al.'s report (O'Neil et al., 2011).

While the studies described above helped to understand the pathogenesis of asthma, targeted therapies neutralizing Th2-type cytokines (IL-4, IL-5, and IL-13) are emerging (Dunn & Wechsler, 2015) as novel drugs for personalizing treatment. IL-4 and IL-13 regulate IgE synthesis, while IL-5 regulates differentiation and survival of eosinophils. Clinical studies targeting IL-5 with the humanized monoclonal antibody mepolizumab, in late onset eosinophil asthma, have shown to reduce asthma exacerbations as well as decreasing the use of oral corticosteroids and improved asthma symptoms (Bel et al., 2014, Pavord et al., 2012). Similarly, the monoclonal antibody to IL-13 (lebrikizumab) results in improved FEV1 in moderate to severe asthmatics exhibiting a high-Th2 profile (Corren et al., 2011). Additional therapies include those targeting IL-4, IL-9, IL-17, and TSLP, and they may hold the promise of personalized medicine treatment in asthma (Table 1).

Table 1 Targeted Therapies for Asthma

Cytokine	Biologics	Mode of Action
IL-4	Dupilumab	Mab to IL-4 receptor (IL-4 α)
	Pitrakinra	Antagonist α subunit chain receptor of IL-4 and IL-13
IL-5	Mepolizumab	Mab to IL-5
	Reslizumab	Mab to IL-5
	Benralizumab	Mab to IL-5
IL-9	MEDI-528	Mab to IL-9
IL-13	Lebrikizumab	Mab antibody to IL-13
IL-17	Brodalumab	Mab to IL-17 (α chain)
	Secukinumab	Mab to IL-17
TSLP	AMG 157	Mab to TSLP

Mab, monoclonal antibody.

6. CHRONIC OBSTRUCTIVE PULMONARY DISEASE

COPD is characterized by a gradual decline in pulmonary function and increasing airway obstruction, which leads to emphysema. Smoking is the most usual cause of COPD, although infections and air pollution may exacerbate the disease. Many potential biomarkers using proteomics have been associated with COPD including neutrophil defensins (NAP) 1 and 2, S100A8, and S100A9, TIMP1, CTSD, the Rho-GDP dissociation inhibitor protein, and IL-8, among others (Stockley, 2014). Using nano-LC–MS, Tu et al. identified 76 differentially expressed proteins in BAL fluid from COPD patients that belong to several processes including alcohol metabolism, gluconeogenesis, and glycolysis (Tu et al., 2014). Alcohol exhaled through the lung is routinely used to estimate blood alcohol concentration (BAC). Interestingly, COPD patients show a significant underestimation of BAC and authors speculated that the alcohol-metabolizing proteins ADH1B and ALDH2 could account for the BAC changes detected in these patients. In this study, however, subjects suffering other chronic lung diseases were not included to cross-validate findings. To investigate more specific markers, Pastor et al. studied COPD patients with and without lung cancer (LC), which allowed the detection of four upregulated proteins common to both COPD and LC/COPD patients (Pastor et al., 2013). However, none of them was unique to COPD. Instead, both Cathepsin D (CTSD) preprotein and Ezrin (EZR) distinguished LC patients from the other two groups. As discussed above, analysis of BAL fluid has provided valuable information on the proteins present in the airway epithelial lining fluid (ELF). However, BAL fluid represents a fraction of the ELF as this is diluted by the BAL (proteins in ELF are released either by epithelial and inflammatory cells or from plasma exudation). In order to obtain higher concentrations of molecules, Franciosi et al. collected ELF directly from the airways and identified four proteins as candidate biomarkers of COPD including lactotransferrin, cofilin-1, high-mobility group box 1 (HMGB1), and alpha 1-antichymotrypsin (serpin A3) (Franciosi et al., 2013). In the serum, several inflammatory mediators have also been associated with COPD including C-reactive protein (CRP), serum amyloid A (SAA), IL-6, fibrinogen, von Willebrand factor, surfactant protein-A (SP-A), and receptor for advanced glycation end products (RAGE), (Bozinovski et al., 2008; Mazur et al., 2011). Fibrinogen is one of the most promising biomarkers in COPD. Bandow et al. identified fibrinogen in serum collected from 24 patients with stage 2a COPD using an improved image analysis workflow for 2DE gel followed by LC–MS/MS (Bandow et al., 2008). Two clinical studies conducted in thousands of patients showed that high fibrinogen levels in serum predict mortality and COPD-related hospitalization (Thomsen et al., 2013; Valvi, Mannino, Müllerova, & Tal-Singer, 2012). Fibrinogen is synthesized primarily in the liver and converted by thrombin into fibrin during blood coagulation. Under physiological conditions, fibrin protein is formed and degraded in a counterbalanced manner. However, chronic inflammation as that seen in COPD may result in profound clinical consequences. *In vivo* evidence for the role of fibrinogen in inflammation derives from fibrinogen-deficient mice in collagen-induced arthritis showing fewer affected joints and a significant overall amelioration of disease severity. The role of fibrinogen in COPD has been reviewed (Duvoix et al., 2013). Additional biomarkers in serum include inter-R-trypsin inhibitor heavy chain H3 (ITIHC3) and vitamin D-binding protein (VDBP) as early biomarkers in healthy smokers (Bortner et al., 2011) and glucose-regulated protein-78 (GRP78) and soluble cluster of differentiation 163 (CD163) as inflammatory proteins potentially in involved lung remodeling (Merali et al., 2014). Lung tissue has also been explored using proteomics technology: Ohlmeier et al. analyzed tissue samples from stage IV COPD and detected high expression of surfactant protein-A (SP-A), and RAGE in two separate studies (Ohlmeier et al., 2010, 2008). In the first study, SP-A was found to be associated with COPD, but not to fibrotic lung or normals (these results were further demonstrated by immunohistochemistry and Western blotting). In the second report, three different RAGE variants were identified (FL-RAGE, cRAGE, and esRAGE) using 2-DE, MS, and Western blot in lung tissues: FL-RAGE (full-length RAGE) and cRAGE (C-terminal processed full-length RAGE), but not esRAGE (spliced endogenous secretory RAGE), were declined in COPD lungs. cRAGE expression was found to correlate with COPD progression (when compare with normals cRAGE was 2.1-fold lower and 3.4-fold lower in mild and very severe COPD, respectively). In contrast, the three RAGE variants were decreased in IPF (esRAGE expression changed in IPF, but not in COPD). Thus, cRAGE and esRAGE could be used as a marker of COPD progression and IPF, respectively. In another study, both MMP-13 and thioredoxin-like 2 (TXL2) were increased in COPD lung tissue obtained from COPD patients who underwent surgery for LC (Lee et al., 2009). TXNL2 has been found overexpressed in mice with LC

(Lee et al., 2009). This finding suggests that TXNL2 in humans could derive from the LC region in COPD patients.

With the aim of improving therapy, the Global Initiative for Chronic Obstructive Lung Disease (GOLD) created a four-quadrant system assessment to phenotype patients based on symptoms and risk of exacerbations that resulted in a better COPD treatment (Vestbo et al., 2013). Thus, combining clinical useful information with well-characterized markers may allowed novel COPD personalized treatments to be made.

7. IDIOPATHIC PULMONARY FIBROSIS

IPF is distinguished by progressive lung fibrosis, which causes extracellular matrix deposition and distortion of the alveolar architecture. Some of the early proposed markers for pulmonary fibrosis include Apo A1 and RAGE (Govender, Baugh, Pennington, Dunn, & Donnelly, 2007; Gucek, 2012). In 2011, Ishikawa et al. showed low levels of hemoglobin (Hb) α and β in IPF patients as compared with COPD and normal subjects (Ishikawa et al., 2010). Hbα and Hbβ were expressed as monomers and complexes in the lung tissues, with the Hbα complex entirely missing in the IPF lung. MS analysis revealed Hbα modification at cysteine 105, preventing the formation of the Hbα complexes. By immunohistochemistry, Hbα and -β were localized to alveolar cells in both controls and COPD patients, but feeble staining was seen in IPF. Kim et al. also reported reduced Hbβ along with Apo A1 among 16 differentially expressed proteins in BAL fluid derived from IPF using LC–MS/MS (Kim et al., 2010). In this last study, however, Apo A1 was proposed to play the most significant role in the pathogenesis of IPF given its anti-inflammatory activity. By treating sensitized mice with Apo A-I, authors showed substantial inhibition of the bleomycin-induced lung fibrosis, which places Apo A-I as a candidate for IPF treatment. In two additional studies, Korfei et al. analyzed lung tissue (Korfei et al., 2011, 2013). In 2011, they compared sporadic IPF with healthy controls using 2-DE, MALDI-TOF–MS and showed 89 differentially expressed proteins in IPF (Korfei et al., 2011). The predominant markers included unfolded protein response (UPR), heat-shock proteins, and DNA damage stress markers: by immunohistochemistry, UPR was found to localize type II alveolar epithelial cells. In 2013, they compared the proteomics profile between idiopathic interstitial pneumonia (IIP) and nonspecific interstitial pneumonia (NSIP) (Korfei et al., 2013)—these idiopathic pneumonias show similar clinical pattern; however, they exhibit different prognosis (IPF patients survive longer). Results from this last study indicated that both diseases have a similar proteomics profile. However, peroxiredoxin (PRDX)1, PRDX6, and proteasome activator complex subunit 1 (PSME1) were upregulated in fibrotic nonspecific interstitial pneumonia (fNSIP) but downregulated in IPF. These are involved in protection mechanisms against oxidative and ER stress, which may illustrate improved outcome and survival in patients with fNSIP. In a separate study, the BAL fluid proteomics profile of IPF patients was compared with that of other interstitial lung diseases (ILDs) including Sarcoidosis (Sar), IPF, fibrosis associated with systemic sclerosis (SSc), and pulmonary Langerhans cell histiocytosis (PLCH) (Landi et al., 2013). Principal component analysis clustered all ILDs into six distinct groups with IPF showing a differential protein pattern from the other conditions. Authors hypothesized that the position of the IPF cluster probably reflected the severity of fibrotic involvement in this disease. Validation by Western blot showed that 14-3-3ε was weakly expressed in IPF as compared with both PLCH and SSc. In contrast, ANXA3 was upregulated in IPF, SSc, PLCH with respect to Sar, while peroxiredoxin 1 and GSTP1 predominated in Sar, IPF, and SSc. Wang et al. have quoted that frequently detected proteins such as annexins and peroxiredoxins are very closely related molecules that exert overlapping functions, making it difficult to draw a conclusion as to why they are differentially expressed (Wang, Bouwman, & Mariman, 2009). They proposed that cellular stress can be the universal reason for these proteins to be produced. Multiplex bead-based immunoassay has been used to investigate biomarkers in plasma. Richards et al. studied 95 proteins in 241 IPF patients and identified five proteins including MMP-7, ICAM-1, IL-8, VCAM-1, and S100A12, which predicted poor outcome (Richards et al., 2012). In this study, the shortest median survival was associated with high S100A12 concentrations (above 26.5 ng/mL), while the longest median survival (4.6 years) was observed with low levels of MMP-7 (below 4.3 ng/mL). Experimental evidence of an *in vivo* role of this metalloproteinase derives from an MMP7 knockout mice study showing that these mice are substantially protected from pulmonary fibrosis in response to intratracheal bleomycin (Zuo et al., 2002). Using genome-wide transcriptomic analysis of lung biopsy, DePianto reported upregulation of two clusters of coregulated genes in IPF including a cluster of genes related to the bronchiolar epithelium, and another containing T- and B-cell markers (DePianto et al., 2015). A relevant finding in this study was the demonstration that the genes encoding MMP3 and CXCL13 were upregulated and correlate negatively with survival over a 3-year follow-up period. Thus, as the novel biomarkers continue to emerge, the development of targeted therapies in stratified IPF patients may pave the way to personalized interventions to be made.

8. ASPIRIN-EXACERBATED RESPIRATORY DISEASE

Aspirin-exacerbated respiratory disease (AERD) affects both upper and lower airways: in the initial phase, AERD is manifested as nasal congestion, which progresses to chronic rhinosinusitis (CRS), nasal polyposis, asthma, and intolerance to aspirin or another nonsteroidal anti-inflammatory drugs (Tonge et al., 2001). This disease usually appears in 30- to 40-year old patients who refer an acute viral-like nasal episode hat never resolved completely afterward (rhinovirus has been linked to AERD development). Overproduction of cysteinyl leukotrienes by LTC4 synthase has been postulated as a central component of AERD pathology. To identify the proteins involved in the airway inflammation, Choi et al. analyzed NLF derived from 18 AERD patients exposed to lysine-aspirin (LASA) challenge (Shevchenko, Wilm, Vorm, & Mann, 1996): using 2DE and MALDI-TOF/TOF, they found 18 protein peptides differentially expressed, and upon validation studies, three potential biomarkers were identified including apolipoprotein A1 (ApoA1), α2 microglobulin (α2M), and ceruloplasmin. The kinetics of APOA1 was investigated: levels of this protein in NLF peaked 1 h following LASA challenge (Beavis & Chait, 1992). ApoA1 is a major component of high-density lipoprotein in the plasma and plays roles in cholesterol transport while it exerts antimicrobial activity in the nasal epithelial lying fluid. *Staphylococcus aureus* is a common organism in both CRS and NPs, and its biofilms appear to be associated with severe disease. Thus, ApoA1 may not only protect the sinonasal mucosa from bacterial infection but also prevent the inflammatory process associated with the release of enterotoxins released from *S. aureus*, which act as superantigens. The authors speculated that some of these proteins were secreted into NLF as a result of increased vascular permeability following aspirin exposure. Proteomic technology has also been applied to the analysis of serum derived from patients suffering AERD. Lee et al. described high concentrations of C3a in serum derived from this group of patients both at baseline and following aspirin challenge (Lee et al., 2006). In a separate study, fatty acid-binding protein 1 (FABP1) was found to be highly upregulated in AERD (sixfold) in NPs, while the heat-shock protein 70 (HSP70) was downregulated twofold. (Beavis & Chait, 1992). FABPs are cytoplasmic proteins that bind eicosanoids that have the ability to cause cysteinyl leukotriene overproduction, and they may enhance the inflammatory response in AERD patients. About HSP70, Zander et al. contend with the former study by detecting upregulation of this protein along with beta-adaptin in the nasal polyps derived from aspirin sensitive patients using a protein microarray (Vorm, Roepstorff, & Mann, 1994). These studies can be differentiated by the two distinct proteomics technologies used (2DE–LC–MS vs. protein microarrays). Interestingly, treating patients suffering NPs with glucocorticoids induced upregulation of both HSP70 and HSP90 (FarajzadehDeroee et al., 2009). The HSP family of proteins has cytoprotective effects by preventing aggregation and promotion of correct folding and assembly of proteins: when this process is ineffective or downregulated, the misfolded proteins accumulate and trigger apoptosis. While HSPs proteins may reverse the pathological mucosal changes in NP by decreasing misfolding of proteins, further studies must define the role of these proteins in AERD patients with nasal polyps.

9. CYSTIC FIBROSIS

Cystic fibrosis (CF) is an autosomal recessive disorder, caused by a mutation in the CF transmembrane conductance regulator (CFTR) gene. Under normal conditions, the CFTR gene results in expression of the CFTR protein: a transmembrane channel situated in the apical membrane of epithelial cells responsible for the transport of chloride and bicarbonate ions into and out of the cell. Δ F508 CFTR is the most common mutation (seen in 70% of patients). It is a three base deletion that induces a deletion of a phenylalanine at position 508, which results in a misfolded protein that is prematurely targeted for degradation (99% of the protein is degraded before it reaches the plasma membrane) (Lukacs et al., 1994; Yang, Janich, Cohn, & Wilson, 1993; Zhang, Kartner, & Lukacs, 1998). Approximately, 2000 mutations have been identified so far (http://www.genet.sickkids.on.ca/app; http://www.cftr2.org). To date, CFTR mutations have been classified into six distinct classes as shown in Table 2 and depicted in Fig. 3. Clinical symptoms are characterized by recurrent lower respiratory infections, exocrine pancreatic insufficiency, and increased electrolyte concentration in sweat. Less common symptoms include meconium ileus, liver disease, and pancreatitis (Hartl & Amaral, 2015).

Proteomic studies have contributed to better understand proteins involved in CF. Using 2DE, Pollard et al. examined protein expression in

Table 2 CFTR Mutations

Mutation Class	CFTR Defect	Functional Consequence
I	Protein synthesis	Reduced CFTR protein expression
II	Protein processing	Misfolded CFTR and lack of transport to cell surface
III	channel gating	Reduced/lack of CFTR channel opening
IV	Channel conductance	Altered CFTR pore restricts Cl-movement
V	Reduced CFTR protein production	Low levels of CFTR protein
VI	High CFTR protein turnover at cell surface	Functional but nonstable CFTR protein on cell surface

CF cells compared with wild-type CFTR-corrected matched controls and identified 20 differentially expressed proteins (Pollard et al., 2006). Interestingly, by radiolabeling newly synthesized proteins in CF cells, authors in this study identified 31 additional proteins including those associated with NF-κB signaling and molecules known to be involved in proteasomal degradation. Proteolysis of CFTR is a feature of CF disease. In another *in vitro* study, Gomes-Alves et al. identified 71 proteins differentially expressed in baby hamster kidneys overexpressing Δ F508 CFTRm of which, 35% were known to interact with CFTR or have involvement in the cellular stress UPR (Gomes-Alves, Couto, Pesquita, Coelho, & Penque, 2010). *In vivo* protein analysis has also applied using proteomics: Carvalho-Oliveira et al. studied protein expression in Δ F508 CFTR mice whole lung extracts and noticed reduced concentrations of lung carbonyl reductase (CBR2) (Carvalho-Oliveira et al., 2009). Authors hypothesized that the use of whole lung homogenates with cell types lacking CFTR expression could have prevented the detection of more proteins, including candidates for CFTR proteolysis. Srivistava et al. employed an antibody capture microarray to analyze 507 proteins in CF sera and detected 46 proteins elevated in CF versus controls (Srivastava et al., 2006): 16 of them had been implicated in CF disease in earlier studies. Sputum and sera samples are rich in secreted proteins and inflammatory molecules are not enriched in CFTR-expressing cells, hence, are not best for identifying proteins involved in CFTR processing. By using multidimensional protein identification technology (MudPIT), a shotgun tandem MS (MS/MS) approach, Koulov et al. investigated the global interactome of CFTR to elucidate the Aha regulation of

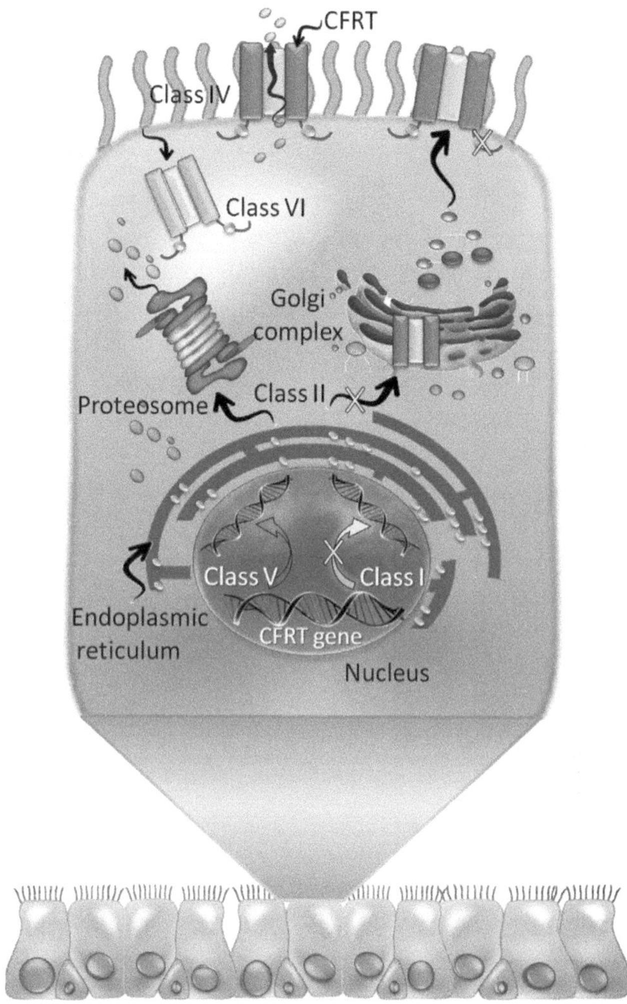

Figure 3 Cystic fibrosis transmembrane conductance regulator (CFTR) mutations: mutation classes I, II, V, and VI lead to under production of CFTR protein at the cell membrane, whereas mutation classes III and IV influence the function or activity of CFTR at the cell membrane. Class I mutations are associated with the greatest disruption to CFTR-mediated chloride transport.

HSP90: this study identified approximately 170 CFTR binding partners in the "CFTR proteome," including three HSP90 cochaperones; Aha1, p23, and FKBP8. Aha1 protein was demonstrated to enhance CFTR degradation through the regulation of ATP-dependent folding steps (Koulov et al., 2010).

Defective secretion of digestive enzymes with fat malabsorption in the gastrointestinal system is a feature of the pancreatic insufficiency that occurs in CF. The gut environment undergoes a series of changes including alteration of the gut microbiota, intestinal inflammation, steatorrhea, and failure to thrive (Raia et al., 2000). Debyser et al. analyzed fecal protein extracts from patients with CF and pancreatic insufficiency and their unaffected siblings using shotgun proteomics: they identified 1676 proteins, 495 proteins were unique to patients, and 793 were unique to siblings (Debyser et al., 2015). Fecal protein extracts from patients contained a higher number of host proteins involved in inflammation and mucus formation (*Enterobacteriaceae*, *Ruminococcus gnavus*, and *Clostridia* species) and reduction of butyrate reducers such as *Faecalibacterium prausnitzii*. Also, potential biomarkers to monitor the gut health in patients with CF showed significantly higher levels of lipocalin-2 (NGAL), zinc-alpha2 glycoprotein (ZAG), and orosomucoid1 (ORM). To gain insight into the biochemical mechanisms of CF pulmonary exacerbation, Laguna et al. used a metabolomics platform to compare the plasma metabolite profiles of patients during an exacerbation with a steady-state CF. Five metabolites were found to be altered during CF exacerbations including perturbations in hypoxanthine, N^4-acetylcytidine, *N*-acetylmethionine, mannose, and cortisol metabolism. While these approaches provide further insight into the pathogenesis of this disease, analysis of CFTR is among the primary challenges due to its low expression in the plasma membrane. Significant advances in both MS-based quantitative proteomics and tandem enrichment using a sequential use of cell surface biotinylation and gel electrophoretic enrichment allowed measuring the absolute amount of apical plasma membrane expression of CFTR in CFBE 41o-cells. The former workflow resulted in 40 fmol of apical PM CFTR (23,000 molecules per cell) per million cells, and the vCFBE 41o-cells expressing wild-type CFTR (wtCFTR) were found to contain 2.9 ng of the protein in the apical PM 106 cells. The absolute quantitation of apical PM CFTR in CFBE 41 was carried out using LC–MRM/MS. This improvement in proteomics technologies will make possible to develop new disease-modifying drugs. In this regard, specific therapies targeting CFTR have been developed.

Ivacaftor (VX-770), a potentiator of CFTR function, was introduced in 2012 as a successful targeted treatment for patients with CF and represents a powerful example of precision medicine (Riordan, 2005). This compound was initially tested in the CFTR Δ Gly551Asp (p.G551D) mutation, which is the most frequent class III mutation, occurring in 2% (heterozygous) of patients with CF worldwide (Van Goor et al., 2009). Major clinical benefits have been demonstrated both for respiratory status (10% of increase for the percent-predicted FEV1 with a decrease in the number of exacerbations) and for the general condition (weight gain and quality of life). A significant decline in the sweat chloride levels was also demonstrated (Davies et al., 2013). VX-770 was indeed able to potentiate the mutated CFTR protein, to increase the open probability of the chloride channel, to decrease the excessive absorption of sodium, and to restore the ciliary motility.

Lumacaftor (VX-809), a corrector of CFTR, was designed to improve the intracellular processing and delivery of Δ F508 mutant. By treating bronchial epithelial cells expressing the Δ F508 homozygous mutation with VX-809, it was demonstrated that this drug increase the chloride transport (Davies et al., 2013 ; Ramsey et al., 2011; Van Goor et al., 2009, 2011; Yu et al., 2012).

However, a clinical trial phase II showed a limited therapeutic effect on the sweat chloride levels in patients homozygous for the F508del mutation. Thus, it was hypothesized that inducing CFTR expression was not sufficient and that CFTR function should be promoted. Interestingly, a combination of ivacaftor and lumafactor, in a phase II trial improved lung function and decreased the sweat chloride levels in adults homozygous or heterozygous for the p.F508del mutation (Boyle et al., 2014).

The US Food and Drug Administration (FDA) approved ivacaftor and lumafactor, for use in CF patients with most common CFTR mutation, Phe508del in July 2015 (Du & Lukacs, 2009; Pitonzo, Yang, Matsumura, Johnson, & Skach, 2009). Another promising compound similar to lumacaftor, VX-661, has also been developed, and clinical trials are underway. Interim results of a phase II, double-blind placebo-controlled study of VX-661 (various daily doses) alone or in combination with ivacaftor have been reported for p.F508del homozygous patients (Donaldson, Pilewski, Griese, Dong, & Lee, 2013). The treatment led to an improvement in the percent-predicted FEV1 (maximum 9% of gain with a combination VX-661 and ivacaftor) and a decreased sweat chloride level.

Ataluren is an agent that has been used in class I CFTR mutations, characterized by which produce a shortened, and nonfunctional CFTR protein as a result of premature termination codon (PTC) (Thursfield & Davies, 2012; Welch et al., 2007). Ataluren enables ribosomes to bypass this PTC leading to the formation of functional protein. A phase II prospective clinical trial of ataluren was performed resulted in improved electrophysiological profile in the majority of patients and chloride transport entered the normal range (Kerem et al., 2008). However, a phase III RCT did not show a significant improvement in neither percentage-predicted FEV1 and the number of pulmonary exacerbations (Kerem et al., 2014). A further analysis of this study revealed that subset of patients is irregularly taking inhaled tobramycin had a 5% increase in FEV1 with ataluren treatment compared with placebo as well as a reduction pulmonary exacerbations (Kerem et al., 2014). The authors suggested that the tobramycin treatment may have interfered with the ataluren therapeutic effect. A proteomic study presented at the 2015 human proteomic organization (HuPO) conference, but not yet published, (Yates, HuPO Vancouver 2015) identified 48 novel interactors that critically influence the Δ F508 CFTR maturation which contribute to the failure of the mutant proteins (four novel phosphosites were identified, and phosphorylation by CSNK2A1 was found to be essential for rescue of F508 CFTR). In this work, the mechanism by which lumacaftor interact with Δ F508 CFTR was also established. Thus, proteomic studies of targeted therapies may provide further insights into CF pathophysiology.

10. LUNG CANCER

Lung cancer arises from the interplay between an individual's genetic susceptibilities and his personal exposure history. The vast majority results from exposure to tobacco smoke, although other environmental risk factors can also be involved including exposure to radiation, asbestos, heavy metals, and polycyclic aromatic hydrocarbons. Detection of LC in early phases is vital because it is a leading cause of death worldwide. LC is usually classified as small-cell lung cancer (SCLC) and non-small-cell lung cancer (NSCLC). NSCLC includes squamous cell carcinoma (SQLC), adenocarcinoma (ADC), and large cell carcinoma (LCC), which accounts for 80% of LCs. LC proteomics studies will be mentioned briefly in view that this subject has recently been reviewed extensively (Indovina et al., 2013). A number of potential markers have been described in sera including haptaglobin α chain (Kang et al., 2011) AA, (Sung et al., 2011) aberrantly glycosylated markers (i.e., fucosylated forms of complement component 9) (Li, Gabrielson, Askin, Chan, & Zhang, 2013). With an aptamer-based proteomics technology, Ostroff et al. analyzed serum samples from 1326 subjects diagnosed with clinical stages I–III NSCLC and identified 12 proteins including CD30 ligand, leucine-rich repeats, and immunoglobulin-like domains 3 (LRIG3), macrophage inflammatory protein-4 (MIP-4), protein kinase C (PRKCI), repulsive guidance molecule C (RGMC), stem cell factor soluble receptor (SCF-sR), sL-selectin, and YES (Ostroff et al., 2010). These proteins efficiently discriminated NSCLC from controls. Similarly, many lung tissue markers with potential clinical applications have been reported; for example, phosphopeptide enrichment followed by MS/MS led to the identification of Cavin-1 and MIF proteins as potential biomarkers in NSCLC (Gamez-Pozo et al., 2012). By using MALDI-TOF–MS, Rahman et al. showed that MIF (macrophage migration inhibitory factor) was present in preinvasive NSCLC lesion along with other five potential biomarkers (ubiquitin, thymosin b4, cytochrome *c*, acyl-coA-binding protein, and cystatin A (CSTA)). The search for prognosis biomarkers has also become of paramount importance because only 10–15% of the patients with LC will ultimately be cured: annexin A2 and annexin A3 (ANXA2 and ANXA3) were discovered as prognostic markers in lung SQLC and ADC (Indovina et al., 2013; Yao et al., 2009), respectively. Upregulation of ANXA2 and HSP27 showed more advanced clinical stage and more poor differentiation, while ANXA3 was found to be associated with increased relapse rate and decreased survival in ADC patients. The prognosis biomarkers application in cancer metastasis may help to determine the severity of cancer and the probability of responding to treatment functions both in health and disease. An example of personalized medicine is the use of Herceptin, which targets HER2-positive breast cancer cells, blocking their ability to receive chemical signals and precluding the cells from growing.

Lung cancer is one of the diseases where proteomic-based approaches have made substantial progress in identifying biomarkers that have the potential to be used in respiratory personalized medicine. Lately, Taguchi et al. researched plasma protein profiles in four mouse models of LC based on quantitative proteomics that allowed the identification Titf1/Nkx2-1 as serum protein signature in lung ADC, while an epidermal growth factor receptor (EGFR) signature (Adam10, Cdh1, and Cd44) was found in plasma of the NSCLC EGFR mutant mice model (Tanaka et al., 2009). Mutations in the EGFR have been targeted with the inhibitors of the EGFR tyrosine kinase gefitinib and erlotinib, which leads to a significant reduction in the tumor size. Most of these mutations are in exon 19 or a missense mutation, leading to a substitution of arginine for leucine in exon 21 (Pallis & Syrigos,

2013) In 2007, an eight-peak profile signature was established in a subset of NSCLC patients that classify patients as good or poor by comparing the mass spectra intensity of the eight peaks with the outcome from EGFR-tyrosine kinase inhibitor (TKI) treatment, as assessed by progression and overall survival (Taguchi et al., 2007). This proteomics profile has evolved into a prognostic test (VeriStrat) (Kuiper et al., 2012) which has been used in several studies to investigate patients' outcome when treated with EGFR-TKI (Amann et al., 2010; Carbone et al., 2012; Lazzari et al., 2012; Stinchcombe et al., 2013). In a recent biomarker-stratified randomized phase 3 trial, Gregorc et al. evaluated the predictive value of VeriStrat in patients with NSCLC treated with either erlotinib or chemotherapy (Gregorc et al., 2014), which revealed that patients classified as poor were better treated with chemotherapy in comparison with erlotinib in terms of patients survival. In contrast, there was no significant difference in overall survival when the proteomics test was classified as good. It was suggested that the biomarker-stratified subgroups could not have been sufficiently powered for comparison of treatments. Interestingly, serum amyloid A protein 1 (SAA1), together with its two truncated forms, was found to be the protein that generates four out of the eight mass signals that integrate VeriStrat (Milan et al., 2012). Salmon et al. reported another proteomics profile based on 11 peaks that predicted overall survival and progression-free survival outcome of patients treated with erlotinib (Salmon et al., 2009). Two metabolomic studies have also been able to predict the response to TKIs treatment in NSCLC patients (Fan et al., 2009; Hori et al., 2011). Pharmacoproteomics studies using imaging MS have had a tremendous impact because it allows both localizing the distribution of the drug and determining the deposition of its metabolites in lung tissues, opening a new opportunity for personalized pharmaceutical applications. By applying this technology, Marko-Varga et al. were able to localize the EGFR-TKIs erlotinib and gefitinib to different tissue compartments in three different LC phenotypes (Marko-Varga et al., 2011). The goal of personalized medicine is not only to individualize treatment to increase clinical benefits of therapy but also to reduce possible adverse reactions in patients. Proteomics has helped us to gain some insight into the proteins associated with ILD, which has been found as an adverse reaction to EGF-TKIs (Atagi et al., 2013; Nyberg et al., 2011). Nyberg et al. analyzed 43 gefitinib-treated NSCLC patients who developed ILD and identified 29 proteins, of which 17 belonged to the acute phase response pathway (Nyberg et al., 2011). The authors proposed that these 17 proteins could be used as potential biomarkers for the increased risk of developing ILD. Atagi et al. have sought to identify proteomics biomarkers in a phase IV erlotinib study where patients developed ILD within 120 days of erlotinib treatment (Atagi et al., 2013). In this study, three proteins were differentially expressed between the ILD patients and controls including C3, C4A/C4B, and APOA1: susceptibility to ILD developed was more frequent when C3 levels were higher than the median.

Afatinib (Gilotrif) is a TKI that was approved in the United States in July 2013 for first-line treatment of patients with metastatic NSCLC whose tumors have EGFR exon 19 deletions or exon 21 (L858R) substitution mutations as detected by the diagnostic companion test, therascreen EGFR RGQ PCR Kit. Approval was based on data from the LUX-Lung 3 trial, comparing afatinib to chemotherapy with pemetrexed/cisplatin. Data interpreted from the study, afatinib group's progression-free survival (PFS), were 11.1 months compared with 6.9 months pemetrexed/cisplatin-treated subjects. Additionally, patients with tumors expressing the two most common EGFR mutations (Del19 or L858R) taking afatinib lived longer than a year free from tumor progression (PFS of 13.6 months) in comparison with 6.9 months for those in the comparator arm (Sequist et al., 2013).

Another successful story in LC-personalized medicine is the use of the kinase inhibitor crizotinib to treat NSCLC patients who exhibit a chromosomal transformation that activates the anaplastic lymphoma kinase (ALK gene) (O'Bryant, Wenger, Kim, & Thompson, 2013). Phase I and II trials assessing crizotinib in ALK-positive NSCLC patients showed that 60.8% of patients showed a good response, and median progression-free survival was 9.7 months. A second ALK TKI, ceritinib (Zykadia)i was FDA approved in 2014 for or ALK-positive, metastatic NSCL patients who have progressed on or are intolerant to crizotinib. Indeed, early-stage results found showed that ceritinib was highly active in patients with advanced, ALK-rearranged NSCLC, even those who experienced disease progression during crizotinib treatment, despite the presence of resistance mutations in ALK (Shaw et al., 2014). Ceritinib was shown to overcome crizotinib resistance mutations in a preclinical trial. *In vitro* and *in vivo* models of acquired resistance to crizotinib were evaluated using cell lines produced from crizotinib-resistant NSCLC patients biopsies. Results revealed that ceritinib overcame crizotinib resistance mutations (ALK-harboring L1196M, I1171T, G1269A, and S1206Y) (Friboulet et al., 2014). These observations altogether support the view that proteomics has begun to be translated from the bench to the patients' bedside, to deliver personalized care in respiratory medicine.

11. CONCLUSIONS

It is now well recognized that respiratory diseases are a group heterogeneous and complex diseases that required sophisticated tools to optimize diagnostics, prognostic, and therapeutics. The proteomics technology has developed at a fast pace over the past decade allowing to characterize thousands of proteins in a single experiment. Recently, human proteome has been mapped including the lung proteome that create an unprecedented opportunity for health. In respiratory medicine, proteomics has made substantial progress in identifying therapeutic targets that have the potential to be used in personalized lung oncology. Similarly, strategies targeting type 2 inflammation in asthma are evolving in novel therapies. Moreover, MED-ALL (Mechanisms of the Development of Allergy), an ongoing project, has been designed to understand the allergy mechanisms from childhood to young adulthood (Antó et al., 2012). Targeted therapies in CF, has been highly successful for patients suffering this disease, showing the potential of personalized medicine. Indeed, the use of small-molecule drugs targeting CFTR, improve lung function and reduce sweat chloride levels as demonstrated in randomized controlled trials. However, the path to personalized medicine in respiratory disease is still facing many challenges: for example, the cost of ivacaftor is very expensive (around 311,000 US $/year). It is essential that healthcare managers and industry policymakers cooperate with each other to make a substantial impact. It is expected that the initiative launched by President Obama will lead precision medicine to every day clinical practice in the future.

ACKNOWLEDGMENTS

We are grateful to the Alexander von Humboldt Foundation for the financial support and to Julio Guerrero for helping with figures design.

REFERENCES

Amann, J. M., Lee, J. W., Roder, H., Brahmer, J., Gonzalez, A., Schiller, J. H., et al. (2010). Genetic and proteomic features associated with survival after treatment with erlotinib in first-line therapy of non-small cell lung cancer in Eastern Cooperative Oncology Group 3503. *Journal of Thoracic Oncology*, *5*, 169–178.

Antó, J. M., Pinart, M., Akdis, M., Auffray, C., Bachert, C., Basagaña, X., et al. (2012). WHO collaborating centre on asthma and rhinitis (Montpellier). Understanding the complexity of IgE-related phenotypes from childhood to young adulthood: A mechanisms of the development of allergy (MeDALL) seminar. *Journal of Allergy and Clinical Immunology*, *129*, 943–954.

Atagi, S., Katakami, N., Yoshioka, H., Fukuoka, M., Kudoh, S., Ogiwara, A., et al. (2013). Nested case control study of proteomic biomarkers for interstitial lung disease in Japanese patients with non-small-cell lung cancer treated with erlotinib: A multicenter phase IV study (JO21661). *Clinical Lung Cancer*, *14*, 407–417.

Bandow, J. E., Baker, J. D., Berth, M., Painter, C., Sepulveda, O. J., Clark, K. A., et al. (2008). Improved image analysis workflow for 2-D gels enables large-scale 2-D gel-based proteomics studies—COPD biomarker discovery study. *Proteomics*, *8*, 3030–3041.

Beavis, R. C., & Chait, B. T. (1992). Cyano-4-hydroxycinnamic acid as a matrix for matrix assisted laser desorption mass spectrometry. *The International Journal of Mass Spectrometry and Gas Phase Ion Chemistry*, 7, 156–158.

Bel, E. H., Wenzel, S. E., Thompson, P. J., Prazma, C. M., Keene, O. N., Yancey, S. W., et al. (2014). Oral glucocorticoid-sparing effect of mepolizumab in eosinophilic asthma. *New England Journal of Medicine*, *371*, 1189–1197.

Bortner, J. D., Jr., Richie, J. P., Jr., Das, A., Liao, J., Umstead, T. M., Stanley, A., et al. (2011). Proteomic profiling of human plasma by Itraq reveals down-regulation of ITIHC3 and VDBP by cigarette smoking. *Journal of Proteome Research*, *10*, 1151–1159.

Boyle, M. P., Bell, S. C., Konstan, M. W., McColley, S. A., Rowe, S. M., Rietschel, E., et al. (2014). A CFTR corrector (lumacaftor) and a CFTR potentiator (ivacaftor) for treatment of patients with cystic fibrosis who have a phe508del CFTR mutation: A phase 2 randomised controlled trial. *The Lancet Respiratory Medicine*, *2*, 527–538.

Bozinovski, S., Hutchinson, A., Thompson, M., Macgregor, L., Black, J., Giannakis, E., et al. (2008). Serum amyloid is a biomarker of acute exacerbations of chronic obstructive pulmonary disease. *American Journal of Respiratory and Critical Care Medicine*, *177*, 269–278.

Carbone, D. P., Ding, K., Roder, H., Grigorieva, J., Roder, J., Tsao, M. S., et al. (2012). Prognostic and predictive role of the VeriStrat plasma test in patients with advanced non-small-cell lung cancer treated with erlotinib or placebo in the NCIC Clinical Trials Group BR.21 trial. *Journal of Thoracic Oncology*, 7, 1653–1660.

Carrette, O., Burkhard, P. R., Sanchez, J. C., & Hochstrasser, D. F. (2006). State-of-the-art two-dimensional gel electrophoresis: A key tool of proteomics research. *Nature Protocols*, *1*, 812–823.

Carvalho-Oliveira, I. M1., Charro, N., Aarbiou, J., Buijs-Offerman, R. M., Wilke, M., Schettgen, T., et al. (2009). Proteomic analysis of naphthalene-induced airway epithelial injury and repair in a cystic fibrosis mouse model. *Journal of Proteome Research*, *8*(7), 3606–3616.

Cederfur, C., Malmström, J., Nihlberg, K., Block, M., Breimer, M. E., Bjermer, L., et al. (2012). Glyco-proteomic identification of galectin-3 and -8 ligands in bronchoalveolar lavage of mild asthmatics and healthy subjects. *Biochimica et Biophysica Acta*, *1820*, 1429–1436.

Collins, F. S., & Varmus, H. (2015). A new initiative on precision medicine. *New England Journal of Medicine*, *372*, 793–795.

Corren, J., Lemanske, R. F., Hanania, N. A., Korenblat, P. E., Parsey, M. V., Arron, J. R., et al. (2011). Lebrikizumab treatment in adults with asthma. *New England Journal of Medicine*, *365*, 1088–1098.

Davies, J. C., Wainwright, C. E., Canny, G. J., Chilvers, M. A., Howenstine, M. S., Munck, A., et al. (2013). Efficacy and safety of ivacaftor in patients aged 6 to 11 years with cystic fibrosis with a G551D mutation. *American Journal of Respiratory and Critical Care Medicine*, *187*, 1219–1225.

Debyser, G., Mesuere, B., Clement, L., Van de Weygaert, J., Van Hecke, P., Duytschaever, G., et al. (2015). Faecal proteomics: A tool to investigate dysbiosis and inflammation in patients with cystic fibrosis. *Journal of Cystic Fibrosis*. http://dx.doi.org/10.1016/j.jcf.2015.08.003. [Epub ahead of print] PubMed PMID: 26330184. pii: S1569-1993(15)00177-0.

DePianto, D. J., Chandriani, S., Abbas, A. R., Jia, G., N'Diaye, E. N., Caplazi, P., et al. (2015). Heterogeneous gene expression signatures correspond to distinct lung pathologies and biomarkers of disease severity in idiopathic pulmonary fibrosis. *Thorax, 70,* 48–56.

Donaldson, S., Pilewski, J., Griese, M., Dong, Q., & Lee, P. S. (2013). WS7.3 VX-661, an investigational CFTR corrector, in combination with ivacaftor, a CFTR potentiator, in patients with CF and homozygous for the F508Del-CFTR mutation: Interim analysis. *Journal of Cystic Fibrosis, 12*(Suppl. 1), S14.

Du, K., & Lukacs, G. L. (2009). Cooperative assembly and misfolding of CFTR domains *in vivo*. *Molecular Biology of the Cell, 20*(7), 1903–1915.

Dunn, R. M., & Wechsler, M. E. (2015). Anti-interleukin therapy in asthma. *Clinical Pharmacology & Therapeutics, 97*(1), 55–65.

Duvoix, A., Dickens, J., Haq, I., Mannino, D., Miller, B., TalSinger, R., et al. (2013). Blood fibrinogen as a biomarker of chronic obstructive pulmonary disease. *Thorax, 68,* 670–676.

Elschenbroich, S., & Kislinger, T. (2011). Targeted proteomics by selected reaction monitoring mass spectrometry: Applications to systems biology and biomarker discovery. *Molecular BioSystems, 7,* 292–303.

Fan, T. W., Lane, A. N., Higashi, R. M., Bousamra, M., 2nd., Kloecker, G., & Miller, D. M. (2009). Metabolic profiling identifies lung tumor responsiveness to erlotinib. *Experimental and Molecular Pathology, 87,* 83–86.

FarajzadehDeroee, A., Oweinah, J., Naraghi, M., Hosemann, W., Athari, B., Völker, U., et al. (2009). Regression of polypoid nasal mucosa after systemic corticosteroid therapy: A proteomics study. *American Journal of Rhinology & Allergy, 23,* 480–485.

Franciosi, L., Govorukhina, N., Fusetti, F., Poolman, B., Lodewijk, M. E., Timens, W., et al. (2013). Proteomic analysis of human epithelial lining fluid by microfluidics based nano LC-MS/MS: A feasibility study. *Electrophoresis, 34,* 2683–2694.

Friboulet, L., Li, N., Katayama, R., Lee, C. C., Gainor, J. F., Crystal, A. S., et al. (2014). The ALK inhibitor ceritinib overcomes crizotinib resistance in non-small cell lung cancer. *Cancer Discovery, 4*(6), 662–673.

Gamez-Pozo, A., Sánchez-Navarro, I., Calvo, E., Agulló-Ortuño, M. T., Lopez-Vacas, R., Díaz, E., et al. (2012). PTRF/cavin-1 and MIF proteins are identified as non-small cell lung cancer biomarkers by label-free proteomics. *PLoS One, 7*(3), e33752.

Gharib, S. A., Nguyen, E. V., Lai, Y., Plampin, J. D., Goodlett, D. R., & Hallstrand, T. S. (2011). Induced sputum proteome in healthy subjects and asthmatic patients. *Journal of Allergy and Clinical Immunology, 128,* 1176–1184.

Gomes-Alves, P., Couto, F., Pesquita, C., Coelho, A. V., & Penque, D. (2010). Rescue of F508del-CFTR by RXR motif inactivation triggers proteome modulation associated with the unfolded protein response. *Biochimica et Biophysica Acta, 1804*(4), 856–865.

Govender, P., Baugh, J. A., Pennington, S. R., Dunn, M. J., & Donnelly, S. C. (2007). Role of proteomics in the investigation of pulmonary fibrosis. *Expert Review of Proteomics, 4,* 379–388.

Gregorc, V., Novello, S., Lazzari, C., Barni, S., Aieta, M., Mencoboni, M., et al. (2014). Predictive value of a proteomic signature in patients with non-small-cell lung cancer treated with second-line erlotinib or chemotherapy (PROSE): A biomarker-stratified, randomised phase 3 trial. *The Lancet Oncology, 15,* 713–721.

Grossman, R. F., Hsueh, P. R., Gillespie, S. H., & Blasi, F. (2014). Community-acquired pneumonia and tuberculosis: Differential diagnosis and the use of fluoroquinolones. *International Journal of Infectious Diseases, 18,* 14–21.

Gucek, M. (2012). Proteomics approaches to fibrotic disorders. *Fibrogenesis & Tissue Repair, 5*(Suppl. 1), S10.

Gygi, S. P., Rist, B., Gerber, S. A., Turecek, F., Gelb, M. H., & Aebersold, R. (1999). Quantitative analysis of complex protein mixtures using isotope coded affinity tags. *Nature Biotechnology, 199*(17), 994–999.

Hancock, D. B., Eijgelsheim, M., Wilk, J. B., Gharib, S. A., Loehr, L. R., Marciante, K. D., et al. (2010). Meta-analyses of genome-wide association studies identify multiple loci associated with pulmonary function. *Nature Genetics, 42*(1), 45–52.

Hartl, D., & Amaral, M. (2015). Cystic fibrosis—From basic science to clinical benefit: A review series. *Journal of Cystic Fibrosis, 14*(4), 415–416.

Holcăpek, M., Jiraško, R., & Lísa, M. (2012). Recent developments in liquid chromatography-mass spectrometry and related techniques. *Journal of Chromatography A, 1259*, 3–15.

Hood, L., & Friend, S. H. (2011). Predictive, personalized, preventive, participatory (P4) cancer medicine, Nature Reviews. *Clinical Oncology, 8*(3), 184–187. http://dx.doi.org/10.1038/nrclinonc.2010.227. PubMed PMID: 21364692.

Hori, S., Nishiumi, S., Kobayashi, K., Shinohara, M., Hatakeyama, Y., Kotani, Y., et al. (2011). A metabolomic approach to lung cancer. *Lung Cancer, 74*, 284–292.

Horvatovich, P., Hoekman, B., Govorukhina, N., & Bischoff, R. (2010). Multidimensional chromatography coupled to mass spectrometry in analyzing complex proteomics samples. *Journal of Separation Science, 33*, 1421–1437.

Indovina, P., Marcelli, E., Pentimalli, F., Tanganelli, P., Tarro, G., & Giordano, A. (2013). Mass spectrometry-based proteomics: The road to lung cancer biomarker discovery. *Mass Spectrometry Reviews, 32*, 129–142.

Ishikawa, N., Ohlmeier, S., Salmenkivi, K., Myllarniemi, M., Rahman, I., Mazur, W., et al. (2010). Hemoglobin α and β are ubiquitous in the human lung, decline in idiopathic pulmonary fibrosis but not in COPD. *Respiratory Research, 11*, 123.

Kang, S. M., Sung, H. J., Ahn, J. M., Park, J. Y., Lee, S. Y., Park, C. S., et al. (2011). The haptoglobin β chain as a supportive biomarker for human lung cancers. *Molecular BioSystems*, 7, 1167–1175.

Kerem, E., Hirawat, S., Armoni, S., Yaakov, Y., Shoseyov, D., Cohen, M., et al. (2008). Effectiveness of PTC124 treatment of cystic fibrosis caused by nonsense mutations: A prospective phase II trial. *The Lancet, 372*, 719–727.

Kerem, E., Konstan, M. W., De Boeck, K., Accurso, F. J., Sermet-Gaudelus, I., Wilschanski, M., et al. (2014). Ataluren for the treatment of nonsense-mutation cystic fibrosis: A randomised, double-blind, placebo-controlled phase 3 trial. *The Lancet. Respiratory Medicine, 2*, 539–547.

Kim, T. H., Lee, Y. H., Kim, K. H., Lee, S. H., Cha, J. Y., Shin, E. K., et al. (2010). Role of lung apolipoprotein A-I in idiopathic pulmonary fibrosis: Antiinflammatory and antifibrotic effect on experimental lung injury and fibrosis. *American Journal of Respiratory and Critical Care Medicine, 182*, 633–642.

Kim, M. S., Pinto, S. M., Getnet, D., Nirujogi, R. S., Manda, S. S., Chaerkady, R., et al. (2014). A draft map of the human proteome. *Nature, 509*, 575–581.

Korfei, M., Schmitt, S., Ruppert, C., Henneke, I., Markart, P., Loeh, B., et al. (2011). Comparative proteomic analysis of lung tissue from patients with idiopathic pulmonary fibrosis (IPF) and lung transplant donor lungs. *Journal of Proteome Research, 10*, 2185–2205.

Korfei, M., von der Beck, D., Henneke, I., Markart, P., Ruppert, C., Mahavadi, P., et al. (2013). Comparative proteome analysis of lung tissue from patients with idiopathic pulmonary fibrosis (IPF), non-specific interstitial pneumonia (NSIP) and organ donors. *Journal of Proteomics, 85*, 109–128.

Koulov, A. V., LaPointe, P., Lu, B., Razvi, A., Coppinger, J., Dong, M. Q., et al. (2010). Biological and structural basis for Aha1 regulation of Hsp90 ATPase activity in maintaining proteostasis in the human disease cystic fibrosis. *Molecular Biology of the Cell, 21*(6), 871–884. http://dx.doi.org/10.1091/mbc.E09-12-1017. Epub 2010 Jan 20. PubMed PMID: 20089831; PubMed Central PMCID: PMC2836968.

Krysl, J., Korzeniewska-Kosela, M., Muller, N. L., & FitzGerald, J. M. (1994). Radiologic features of pulmonary tuberculosis: An assessment of 188 cases. *Canadian Association of Radiologists Journal, 45*, 101–107.

Kuiper, J. L., Lind, J. S., Groen, H. J., Roder, J., Grigorieva, J., Roder, H., et al. (2012). VeriStrat(®) has prognostic value in advanced stage NSCLC patients treated with erlotinib and sorafenib. *British Journal of Cancer, 107*, 1820–1825.

Kumar, R., Seibold, M. A., Aldrich, M. C., Williams, L. K., Reiner, A. P., Colangelo, L., et al. (2010). Genetic ancestry in lung-function predictions. *New England Journal of Medicine, 363*(4), 321–330.

Landi, C., Bargagli, E., Bianchi, L., Gagliardi, A., Carleo, A., Bennett, D., et al. (2013). Towards a functional proteomics approach to the comprehension of idiopathic pulmonary fibrosis, sarcoidosis, systemic sclerosis and pulmonary Langerhans cell histiocytosis. *Journal of Proteomics, 83*, 60–75.

Lazzari, C., Spreafico, A., Bachi, A., Roder, H., Floriani, I., Garavaglia, D., et al. (2012). Changes in plasma mass spectral profile in course of treatment of nonsmall cell lung cancer patients with epidermal growth factor receptor tyrosine kinase inhibitors. *Journal of Thoracic Oncology*, 7, 40–48.

Lee, E. J., In, K. H., Kim, J. H., Lee, S. Y., Shin, C., Shim, J. J., et al. (2009). Proteomic analysis in lung tissue of smokers and COPD patients. *Chest, 135*(2), 344–352. http://dx.doi.org/10.1378/chest.08-1583. Epub 2008 Aug 27. PubMed PMID: 18753468.

Lee, T. H., Jang, A. S., Park, J. S., Kim, T. H., Choi, Y. S., Shin, H. R., et al. (2013). Elevation of S100 calcium binding protein A9 in sputum of neutrophilic inflammation in severe uncontrolled asthma. *Annals of Allergy, Asthma & Immunology, 111*, 268–275.

Lee, S. H., Rhim, T., Choi, Y. S., Min, J. W., Kim, S. H., Cho, S. Y., et al. (2006). Complement C3a and C4a increased in plasma of patients with aspirin-induced asthma. *American Journal of Respiratory and Critical Care Medicine, 173*(4), 370–378. Epub 2005 November 17.

Li, Q. K., Gabrielson, E., Askin, F., Chan, D. W., & Zhang, H. (2013). Glycoproteomics using fluid-based specimens in the discovery of lung cancer protein biomarkers: Promise and challenge. *Proteomics. Clinical Applications*, 7, 55–69.

Lukacs, G. L., Mohamed, A., Kartner, N., Chang, X. B., Riordan, J. R., & Grinstein, S. (1994). Conformational maturation of CFTR but not its mutant counterpart (ΔF508) occurs in the endoplasmic reticulum and requires ATP. *The EMBO Journal, 13*(24), 6076–6086.

Mann, M. (2006). Functional and quantitative proteomics using SILAC. *Nature Reviews. Molecular Cell Biology*, 7, 952–958.

Marko-Varga, G., Fehniger, T. E., Rezeli, M., Döme, B., Laurell, T., & Vegvári, A. (2011). Drug localization in different lung cancer phenotypes by MALDI mass spectrometry imaging. *Journal of Proteomics, 74*, 982–992.

Mazur, W., Toljamo, T., Ohlmeier, S., Vuopala, K., Nieminen, P., Kobayashi, H., et al. (2011). Elevation of surfactant protein A in plasma and sputum in cigarette smokers. *The European Respiratory Journal, 38*(2), 277–284. http://dx.doi.org/10.1183/09031936.00110510. Epub 2011 Jan 27. PubMed PMID: 21273386.

Merali, S., Barrero, C. A., Bowler, R. P., Chen, D. E., Criner, G., Braverman, A., et al. (2014). Analysis of the plasma proteome in COPD: Novel low abundance proteins reflect the severity of lung remodeling. *International Journal of Chronic Obstructive Pulmonary Disease, 11*, 177–178.

Mikami, T., Aoki, M., & Kimura, T. (2012). The application of mass spectrometry to proteomics and metabolomics in biomarker discovery and drug development. *Molecular Pharmacology, 5*, 301–316.

Milan, E., Lazzari, C., Anand, S., Floriani, I., Torri, V., Sorlini, C., et al. (2012). SAA1 is over-expressed in plasma of non small cell lung cancer patients with poor outcome after treatment with epidermal growth factor receptor tyrosine-kinase inhibitors. *Journal of Proteomics*, *76*(Spec. No.), 91–101.

Nyberg, F., Ogiwara, A., Harbron, C. G., Kawakami, T., Nagasaka, K., Takami, S., et al. (2011). Proteomic biomarkers for acute interstitial lung disease in gefitinib-treated Japanese lung cancer patients. *PLoS One*, *6*(7), e22062.

O'Bryant, C. L., Wenger, S. D., Kim, M., & Thompson, L. A. (2013). Crizotinib: A new treatment option for ALK-positive non-small cell lung cancer. *Annals of Pharmacotherapy*, *47*, 189–197.

O'Neil, S. E., Sitkauskiene, B., Babusyte, A., Krisiukeniene, A., Stravinskaite-Bieksiene, K., Sakalauskas, R., et al. (2011). Network analysis of quantitative proteomics on asthmatic bronchi: Effects of inhaled glucocorticoid treatment. *Respiratory Research*, *12*, 124.

Ohlmeier, S., Mazur, W., Salmenkivi, K., Myllarniemi, M., Bergmann, U., & Kinnula, V. L. (2010). Proteomic studies on receptor for advanced glycation end product variants in idiopathic pulmonary fibrosis and chronic obstructive pulmonary disease. *Proteomics. Clinical Applications*, *4*, 97–105.

Ohlmeier, S., Vuolanto, M., Toljamo, T., Vuopala, K., Salmenkivi, K., Myllarniemi, M., et al. (2008). Proteomics of human lung tissue identifies surfactant protein A as a marker of chronic obstructive pulmonary disease. *Journal of Proteome Research*, *7*, 5125–5132.

Ortega, V. E., Hawkins, G. A., Moore, W. C., Hastie, A. T., Ampleford, E. J., Busse, W. W., et al. (2014). Effect of rare variants in ADRB2 on risk of severe exacerbations and symptom control during long-acting beta-agonist treatment in a multiethnic asthma population: A genetic study. *The Lancet. Respiratory Medicine*, *2*(3), 204–213.

Ostroff, R. M., Bigbee, W. L., Franklin, W., Gold, L., Mehan, M., Miller, Y. E., et al. (2010). Unlocking biomarker discovery: Large scale application of aptamer proteomic technology for early detection of lung cancer. *PLoS One*, *5*(12), e15003.

Pallis, A. G., & Syrigos, K. N. (2013). Epidermal growth factor receptor tyrosine kinase inhibitors in the treatment of NSCLC. *Lung Cancer*, *80*, 120–130.

Pastor, M. D., Nogal, A., Molina-Pinelo, S., Meléndez, R., Salinas, A., Gonzalez De la Peña, M., et al. (2013). Identification of proteomic signatures associated with lung cancer and COPD. *Journal of Proteomics*, *89*, 227–237.

Pavord, I. D., Korn, S., Howarth, P., Bleecker, E. R., Buhl, R., Keene, O. N., et al. (2012). Mepolizumab for severe eosinophilic asthma (DREAM): A multicentre, double-blind, placebo-controlled trial. *The Lancet*, *380*(9842), 651–659.

Picotti, P., & Aebersold, R. (2012). Selected reaction monitoring-based proteomics: Workflows, potential, pitfalls and future directions. *Nature Methods*, *9*, 555–566. 916.

Pitonzo, D., Yang, Z., Matsumura, Y., Johnson, A. E., & Skach, W. R. (2009). Sequence-specific retention and regulated integration of a nascent membrane protein by the endoplasmic reticulum Sec61 translocon. *Molecular Biology of the Cell*, *20*(2), 685–698.

Pollard, H. B1., Eidelman, O., Jozwik, C., Huang, W., Srivastava, M., Ji, X. D., et al. (2006). *De novo* biosynthetic profiling of high abundance proteins in cystic fibrosis lung epithelial cells. *Molecular & Cellular Proteomics*, *5*(9), 1628–1637.

Raia, V., Maiuri, L., de Ritis, G., de Vizia, B., Vacca, L., Conte, R., et al. (2000). Evidence of chronic inflammation in morphologically normal small intestine of cystic fibrosis patients. *Pediatric Research*, *47*, 344–350.

Ramsey, B. W., Davies, J., McElvaney, N. G., Tullis, E., Bell, S. C., Dřevínek, P., et al. (2011). A CFTR potentiator in patients with cystic fibrosis and the G551D mutation. *New England Journal of Medicine*, *365*, 1663–1672.

Reardon, S. (2015). Precision-medicine plan raises hopes. *Nature*, *517*(7536), 540.

Repapi, E., Sayers, I., Wain, L. V., Burton, P. R., Johnson, T., Obeidat, M., et al. (2010). Genome-wide association study identifies five loci associated with lung function. *Nature Genetics*, *42*(1), 36–44.

Richards, T. J., Kaminski, N., Baribaud, F., Flavin, S., Brodmerkel, C., Horowitz, D., et al. (2012). Peripheral blood proteins predict mortality in idiopathic pulmonary fibrosis. *American Journal of Respiratory and Critical Care Medicine*, *185*, 67–76.

Riordan, J. R. (2005). Assembly of functional CFTR chloride channels. *Annual Review of Physiology*, *67*, 701–718.

Ross, P. L., Huang, Y. N., Marchese, J. N., Williamson, B., Parker, K., Hattan, S., et al. (2004). Multiplexed protein quantitation in *Saccharomyces cerevisiae* using amine-reactive isobaric tagging reagents. *Molecular & Cellular Proteomics*, *3*, 1154–1169.

Salmon, S., Chen, H., Chen, S., Herbst, R., Tsao, A., Tran, H., et al. (2009). Classification by mass spectrometry can accurately and reliably predict outcome in patients with non-small cell lung cancer treated with erlotinib-containing regimen. *Journal of Thoracic Oncology*, *4*, 689–696.

Sequist, L. V., Yang, J. C., Yamamoto, N., O'Byrne, K., Hirsh, V., Mok, T., et al. (2013). Phase III study of afatinib or cisplatin plus pemetrexed in patients with metastatic lung adenocarcinoma with EGFR mutations. *Journal of Clinical Oncology*, *31*(27), 3327–3334. http://dx.doi.org/10.1200/JCO.2012.44.2806. Epub 2013 July 1.

Shaw, A. T., Kim, D. W., Mehra, R., Tan, D. S., Felip, E., Chow, L. Q., et al. (2014). Ceritinib in ALK-rearranged non-small-cell lung cancer. *New England Journal of Medicine*, *370*(13), 1189–1197.

Shevchenko, A., Wilm, M., Vorm, O., & Mann, M. (1996). Mass spectrometric sequencing of proteins silver-stained polyacrylamide gels. *Analytical Chemistry*, *68*(5), 850–858. PubMed PMID: 8779443.

Singh, A., Cohen Freue, G. V., Oosthuizen, J. L., Kam, S. H., Ruan, J., Takhar, M. K., et al. (2012). Plasma proteomics can discriminate isolated early from dual responses in asthmatic individuals undergoing an allergen inhalation challenge. *Proteomics. Clinical Applications*, *6*, 476–485.

Srivastava, M1., Eidelman, O., Jozwik, C., Paweletz, C., Huang, W., Zeitlin, P. L., et al. (2006). Serum proteomic signature for cystic fibrosis using an antibody microarray platform. *Molecular Genetics and Metabolism*, *87*(4), 303–310.

Stinchcombe, T. E., Roder, J., Peterman, A. H., Grigorieva, J., Lee, C. B., Moore, D. T., et al. (2013). A retrospective analysis of VeriStrat status on outcome of a randomized phase II trial of first-line therapy with gemcitabine, erlotinib, or the combination in elderly patients (age 70 years or older) with stage IIIB/IV non-small-cell lung cancer. *Journal of Thoracic Oncology*, *8*, 443–451.

Stockley, R. A. (2014). Biomarkers in chronic obstructive pulmonary disease: Confusing or useful? *International Journal of Chronic Obstructive Pulmonary Disease*, *9*, 163–177.

Sung, H. J., Ahn, J. M., Yoon, Y. H., Rhim, T. Y., Park, C. S., Park, J. Y., et al. (2011). Identification and validation of SAA as a potential lung cancer biomarker and its involvement in metastatic pathogenesis of lung cancer. *Journal of Proteome Research*, *10*, 1383–1395.

Taguchi, F., Solomon, B., Gregorc, V., Roder, H., Gray, R., Kasahara, K., et al. (2007). Mass spectrometry to classify non-small-cell lung cancer patients for clinical outcome after treatment with epidermal growth factor receptor tyrosine kinase inhibitors: A multicohort cross-institutional study. *Journal of the National Cancer Institute*, *99*, 838–846.

Tambor, V., Fucíkova, A., Lenco, J., Kacerovsky, M., Rehaćek, V., Stulík, J., et al. (2010). Application of proteomics in biomarker discovery: A primer for the clinician. *Physiological Research*, *59*, 471–497.

Tanaka, Y., Nishida, N., Sugiyama, M., Kurosaki, M., Matsuura, K., Sakamoto, N., et al. (2009). Genome-wide association of IL28B with response to pegylated interferon-alpha and ribavirin therapy for chronic hepatitis C. *Nature Genetics*, *41*, 1105–1109.

Teran, L. M., Montes-Vizuet, R., Li, X., & Franz, T. (2015). Respiratory proteomics: From descriptive studies to personalized medicine. *Journal of Proteome Research*, *14*(1), 38–50.

Thomsen, M., Ingebrigtsen, T. S., Marott, J. L., Dahl, M., Lange, P., Vestbo, J., et al. (2013). Inflammatory biomarkers and exacerbations in chronic obstructive pulmonary disease. *JAMA*, *309*(22), 2353–2361. http://dx.doi.org/10.1001/jama.2013.5732. PubMed PMID: 23757083.

Thursfield, R. M., & Davies, J. C. (2012). Cystic fibrosis: Therapies targeting specific gene defects. *Paediatric Respiratory Reviews*, *13*, 215–219.

Tonge, R., Shaw, J., Middleton, B., Rowlinson, R., Rayner, S., Young, J., et al. (2001). Validation and development of fluorescence two-dimensional differential gel electrophoresis proteomics technology. *Proteomics*, *1*, 377–396.

Tu, C., Mammen, M. J., Li, J., Shen, X., Jiang, X., Hu, Q., et al. (2014). Large-scale, ion-current-based proteomics investigation of bronchoalveolar lavage fluid in chronic obstructive pulmonary disease patients. *Journal of Proteome Research*, *13*, 627–639.

Uhlén, M., Fagerberg, L., Hallström, B. M., Lindskog, C., Oksvold, P., Mardinoglu, A., et al. (2015). Proteomics. Tissue-based map of the human proteome. *Science*, *347*, 126041.

Valvi, D., Mannino, D. M., Müllerova, H., & Tal-Singer, R. (2012). Fibrinogen, chronic obstructive pulmonary disease (COPD) and outcomes in two United States cohorts. *International Journal of Chronic Obstructive Pulmonary Disease*, 7, 173–182.

Van Goor, F., Hadida, S., Grootenhuis, P. D., Burton, B., Cao, D., Neuberger, T., et al. (2009). Rescue of CF airway epithelial cell function in vitro by a CFTR potentiator, VX-770. *Proceedings of the National Academy of Sciences of the United States of America*, *106*, 18825–18830.

Van Goor, F1., Hadida, S., Grootenhuis, P. D., Burton, B., Stack, J. H., Straley, K. S., et al. (2011). Correction of the F508del-CFTR protein processing defect in vitro by the investigational drug VX-809. *Proceedings of the National Academy of Sciences of the United States of America*, *108*, 18843–18848.

Vestbo, J., Hurd, S. S., Agustí, A. G., Jones, P. W., Vogelmeier, C., Anzueto, A., et al. (2013). Global strategy for the diagnosis, management and prevention of chronic obstructive pulmonary disease, GOLD executive summary. *American Journal of Respiratory and Critical Care Medicine*, *187*, 347–365.

Vorm, O., Roepstorff, P., & Mann, M. (1994). Improved resolution and very high sensitivity in MALDI TOF of matrix surfaces made by fast evaporation. *Analytical Chemistry*, *66*, 3281–3287.

Wang, P., Bouwman, F. G., & Mariman, E. C. (2009). Generally detected proteins in comparative proteomics—A matter of cellular stress response? *Proteomics*, *9*, 2955–2966.

Wang, M., Fitzgerald, J. M., Richardson, K., Marra, C. A., Cook, V. J., Hajek, J., et al. (2011). Is the delay in diagnosis of pulmonary tuberculosis related to exposure to fluoroquinolones or any antibiotic? *The International Journal of Tuberculosis and Lung Disease*, *15*, 1062–1068.

Wang, X. L., Zhu, Y., & Fang, Q. (2014). Coupling liquid chromatography/mass spectrometry detection with microfluidic droplet array for label-free enzyme inhibition assay. *Analyst*, *139*(1), 191–197. http://dx.doi.org/10.1039/c3an01917a. Epub 2013 Nov 6. PubMed PMID: 24196165.

Welch, E. M., Barton, E. R., Zhuo, J., Tomizawa, Y., Friesen, W. J., Trifillis, P., et al. (2007). PTC124 targets genetic disorders caused by nonsense mutations. *Nature*, *447*, 87–91.

Wilhelm, M., Schlegl, J., Hahne, H., Moghaddas Gholami, A., Lieberenz, M., Savitski, M. M., et al. (2014). Mass-spectrometry-based draft of the human proteome. *Nature*, *509*, 582–587.

Wilk, J. B., Chen, T. H., Gottlieb, D. J., Walter, R. E., Nagle, M. W., Brandler, B. J., et al. (2009). A genome-wide association study of pulmonary function measures in the Framingham heart study. *PLoS Genetics*, *5*(3), e1000429.

Woodring, J. H., Vandiviere, H. M., Fried, A. M., Dillon, M. L., Williams, T. D., & Melvin, I. G. (1986). Update: The radiographic features of pulmonary tuberculosis. *American Journal of Roentgenology*, *146*, 497–506.

Wu, J., Kobayashi, M., Sousa, E. A., Liu, W., Cai, J., Goldman, S. J., et al. (2005). Differential proteomic analysis of bronchoalveolar lavage fluid in asthmatics following segmental antigen challenge. *Molecular & Cellular Proteomics*, *4*, 1251–1264.

Yang, Y., Janich, S., Cohn, J. A., & Wilson, J. M. (1993). The common variant of cystic fibrosis transmembrane conductance regulator is recognized by hsp70 and degraded in a pre-Golgi non-lysosomal compartment. *Proceedings of the National Academy of Sciences of the United States of America*, *90*(20), 9480–9484.

Yao, H., Zhang, Z., Xiao, Z., Chen, Y., Li, C., Zhang, P., et al. (2009). Identification of metastasis associated proteins in human lung squamous carcinoma using two dimensional difference gel electrophoresis and laser capture microdissection. *Lung Cancer*, *65*, 41–48.

Yu, H., Burton, B., Huang, C. J., Worley, J., Cao, D., Johnson, J. P., Jr., et al. (2012). Ivacaftor potentiation of multiple CFTR channels with gating mutations. *Journal of Cystic Fibrosis*, *11*, 237–245.

Zhang, F., Kartner, N., & Lukacs, G. L. (1998). Limited proteolysis as a probe for arrested conformational maturation of Δ F508 CFTR. *Nature Structural & Molecular Biology*, *5*(3), 180–183.

Zuo, F., Kaminski, N., Eugui, E., Allard, J., Yakhini, Z., Ben-Dor, A., et al. (2002). Gene expression analysis reveals matrilysin as a key regulator of pulmonary fibrosis in mice and humans. *Proceedings of the National Academy of Sciences of the United States of America*, *99*, 6292–6297.

CHAPTER SIX

Computational Approaches to Accelerating Novel Medicine and Better Patient Care from Bedside to Benchtop

Theodore Sakellaropoulos*[,1], Junguk Hur[†], Ioannis N. Melas[‡], Ellen Y. Guo[§], Leonidas Alexopoulos[¶,1], Mohammad Bohlooly[||], Jane P.F. Bai*[,2]

*Office of Clinical Pharmacology, Office of Translational Science, Center for Drug Evaluation and Research, U.S. Food and Drug Administration, Silver Spring, Maryland, USA
[†]Department of Biomedical Sciences, School of Medicine & Health Sciences, University of North Dakota, Grand Forks, North Dakota, USA
[‡]Astrazeneca, Discovery Sciences, Group of Quantitative Biology, Molndal, Sweden
[§]Molecular and Cellular Biology, Liberal Arts and Sciences, University of Illinois at Urbana–Champaign, Urbana–Champaign, Illinois, USA
[¶]School of Mechanical Engineering, National Technical University of Athens, Athens, Greece. ProtAtonce Ltd., Athens, Greece
[||]Astrazeneca, Discovery Sciences, Reagents and Assay Development, Transgenics Group, Molndal, Sweden
[2]Corresponding author: e-mail address: jane.bai@fda.hhs.gov

Contents

[1] Present address: School of Mechanical Engineering, National Technical University of Athens, Athens, Greece.

Advances in Protein Chemistry and Structural Biology, Volume 102
ISSN 1876-1623
http://dx.doi.org/10.1016/bs.apcsb.2015.09.005

Abstract

Some successes have been achieved in the war on cancer over the past 30 years with recent efforts on protein kinase inhibitors. Nonetheless, we are still facing challenges due to cancer evolution. Cancers are complex and heterogeneous due to primary and secondary mutations, with phenotypic and molecular heterogeneity manifested among patients of a cancer, and within an individual patient throughout the disease course. Our understanding of cancer genomes has been facilitated by advances in omics and in bioinformatics technologies; major areas in cancer research are advancing in parallel on many fronts. Computational methods have been developed to decipher the molecular complexity of cancer and to identify driver mutations in cancers. Utilizing the identified driver mutations to develop effective therapy would require biological linkages from cellular context to clinical implication; for this purpose, computational mining of biomedical literature facilitates utilization of a huge volume of biomedical research data and knowledge. In addition, frontier technologies, such as genome editing technologies, are facilitating investigation of cancer mutations, and opening the door for developing novel treatments to treat diseases. We will review and highlight the challenges of treating cancers, which behave like moving targets due to mutation and evolution, and the current state-of-the-art research in the areas mentioned above.

1. INTRODUCTION

Battling cancers is an ongoing public health effort across all stakeholders. The 5-year relative survival rates between 2004 and 2010 for all stages of breast cancer, colorectal cancer (CRC), lung and bronchus cancer, and prostate cancer were 89%, 65%, 17%, and 99%%, respectively (American Cancer Society, 2015). In 2011, the total direct health care costs for cancer were $88.7 billion. It is estimated that there would be 0.59 million cancer deaths and 1.66 million new cancer cases in 2015 (American Cancer Society, 2015). Over the years, cancer therapy has advanced from cytotoxic mechanisms, such as radiation and chemotherapy, to more targeted therapies. Advances in molecular biology and omics technologies have led to the discovery of protein kinases as druggable targets for treating human diseases; receptor tyrosine kinases (RTKs) play critical roles in cellular processes of metabolism, differentiation, migration, cell-cycle control, and survival (Lemmon & Schlessinger, 2010). RTKs have been the targets of cancer research; as a result, there are approximately 30 small-molecule drugs and biological products approved for treating various cancers (Drugs@FDA, 2015).

Cancers are, however, complex diseases. We have come to appreciate the complexity of cancer biology through our clinical experiences in treating patients at bedside. For example, imatinib is the first small molecular drug approved for treating Philadelphia chromosome-positive (Ph+) chronic myeloid leukemia; Kit (CD117)-positive unresectable and/or metastatic malignant gastrointestinal stromal tumors (GISTs). Imatinib-resistant tumors were subsequently discovered in relapsed patients and harbored mutations that could be primary or secondary (acquired), with clonal or polyclonal evolutions (Antonescu et al., 2005; Lim, Huang, et al., 2008). Even in the era of targeted therapy, metastasis and relapse could eventually occur due to mutation and cancer evolution (Kroeger et al., 2014).

Genomic instability processes, as tumor evolves, contribute to spatial and temporal diversity of cancer and pose a great challenge to treating cancer (de Bruin et al., 2014). As of now, we know there are a huge number of points of mutation across the whole cancer genome (Wellcome Trust Sanger Institute, 2015). Fortunately, not all mutations bear the same level of clinical significance. Driver mutations are nonsynonymous mutations causally linked with cancerogenesis (Stratton, Campbell, & Futreal, 2009), while passenger mutations are carried along through cell division and have been suggested to be involved in cancer progression (McFarland, Korolev, Kryukov, Sunyaev, & Mirny, 2013). Therefore, identifying driver mutations precisely is a critical step toward effective treatment. Bioinformatics approaches are often used to identify novel targets using bedside cancer tissues; a recent study, for example, identified novel mutations in EPHB4 RTK in lung cancer (Ferguson et al., 2015). Furthermore, recent advances in genome editing technologies facilitate *in vivo* validation of the point of mutation identified by computational algorithms, thus bridging the gap between computational results and preclinical research for development of effective therapy.

From the publications in PubMed, great advances on many fronts in the war on cancer have been made with efforts devoted to many relevant areas, as evidenced from the trend lines shown in Fig. 1. Over the last decade, new and affordable high-throughput technologies have allowed the molecular profiling of individual patients, and together with the increasing use of electronic medical records (EMRs), have renewed interest in precision medicine. It is hopeful that the 2015's precision medicine initiative (Collins & Varmus, 2015) will enable our advances in understanding cancer biology and in developing cancer therapy. However, there are obstacles to be surmounted on our way to the ultimate goal of curing cancer. In this chapter, we will review the currently approved products for treating cancers via

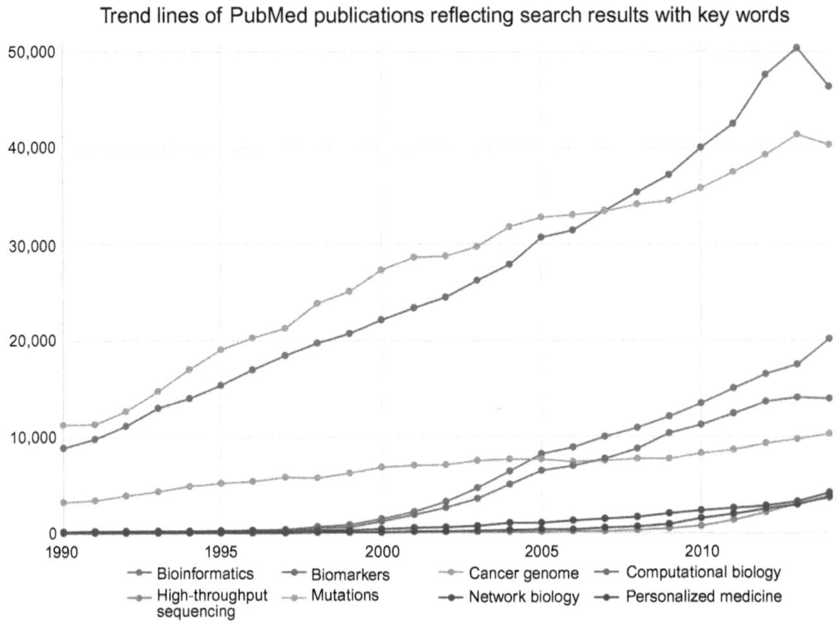

Figure 1 The numbers of PubMed publications found from 1990 to 2014 with keyword search. The keywords included "biomarkers," "bioinformatics," "cancer genome," "computational biology," "high-throughput sequencing," "network biology," and "personalized medicine." "Biomarkers" and "mutations" showed steeper increases starting from 1990 compared to other keywords; however, their numbers decreased from 2013 to 2014.

targeting protein kinases, tumor heterogeneity contributing to metastasis and relapse, and bioinformatics approaches to identifying driver genes as well as mining biomedical literature to uncover hidden knowledge in cancer biology. Last, we will highlight genomic editing, which enables preclinical research of points of mutation for discovering effective therapy.

2. CURRENTLY APPROVED PRODUCTS TARGETING PROTEIN KINASES IN ONCOLOGY

Targeting abnormal protein kinases in tumor tissues has been a strategic focus for treating cancers since the 1990s (Andrejauskas-Buchdunger & Regenass, 1992; Baselga, Norton, Albanell, Kim, & Mendelsohn, 1998; Buchdunger et al., 1996). Following the approval of trastuzumab in 1998 for treating breast cancer with overexpression of epidermal growth factor receptor 2, there have been 29 small-molecule drugs and biologics approved for treating cancers under the premise of targeted therapy (Tables 1 and 2).

Table 1 The List of Approved Small Molecular Kinase Inhibitors, Their Target Diseases, and Associated Markers

Drug	Marker	Indications
Bosutinib	Ph+ (Philadelphia chromosome positive) (chromosome 22 is called Philadelphia chromosome when it contains the abnormal *BCR-ABL* gene)	Ph+ chronic myelogenous leukemia (CML)
Vandetanib	None	Medullary thyroid cancer
Cabozantinib	None	Medullary thyroid cancer
Afatinib	Epidermal growth factor receptor (EGFR) exon 19 deletions or exon 21 (*L858R*) substitution mutations	Metastatic non-small cell lung cancer (NSCLC)
Imatinib	Ph+ (Philadelphia chromosome positive); Kit positive in GISTs where the *KIT* gene is v-kit Hardy-Zuckerman 4 feline sarcoma viral oncogene homolog	Philadelphia chromosome-positive chronic myeloid leukemia; Kit (CD117) positive unresectable and/or metastatic malignant gastrointestinal stromal tumors (GISTs)
Ponatinib	*T315I*+; *T315I*+ Ph+ (Philadelphia chromosome positive)	*T315I*-positive CML; *T315I*-positive Philadelphia chromosome-positive acute lymphoblastic leukemia (ALL)
Ibrutinib	Chronic lymphocytic leukemia with 17p deletion (loss of genetic materials in chromosome 17)	Mantle cell lymphoma; Chronic lymphocytic leukemia; Chronic lymphocytic leukemia with 17p deletion
Axitinib	None	Renal cell carcinoma (RCC)
Gefitinib	None	NSCLC
Trametinib	BRAF *V600E* or *V600K* mutation	Metastatic melanoma
Sorafenib	None	Hepatocellular carcinoma (HC); advanced RCC; thyroid carcinoma
Dasatinib	Ph+ (Philadelphia chromosome positive)	Ph+ CML; Ph+ ALL
Regorafenib	None	Metastatic colorectal cancer (CRC); GIST
Sunitinib	None	Advanced RCC, GIST
Dabrafenib	*BRAF V600E* or *V600K* mutation	Unresectable or metastatic melanoma
Erlotinib	Epidermal growth factor receptor (EGFR) exon 19 deletions or exon 21 (*L858R*) substitution mutations	NSCLC
Nilotinib	Ph+ (Philadelphia chromosome positive)	Ph+ CML
Lapatinib	HER2 overexpression	Breast cancer
Pazopanib	None	Advanced RCC; advanced soft tissue sarcoma
Crizotinib	ALK+ (anaplastic lymphoma linase)	ALK+ NSCLC
Vemurafenib	*BRAF V600E* mutation	Unresectable or metastatic melanoma
Ceritinib	ALK+ (anaplastic lymphoma kinase)	ALK+ NSCLC
Lenvatinib	None	Thyroid cancer

Table 2 The List of Approved Monoclonal Antibodies That Target Protein Kinases, Their Target Diseases, and Associated Markers

Biological Products	Mechanism	Disease and Associated Marker
Bevacizumab	It binds VEGF and prevents the interaction of VEGF to its receptors (Flt-1 and KDR) on the surface of endothelial cells. Reduction of microvascular growth and inhibition of metastatic disease progression	Colorectal cancer , non-squamous non-small cell lung cancer, glioblastoma, ovarian, renal cell carcinoma, and cervical cancer
Ramucirumab	A vascular endothelial growth factor receptor 2 antagonist that specifically binds VEGF receptor 2 and blocks binding of VEGFR ligands, VEGF-A, VEGF-C, and VEGF-D	Cancers (gastric and gastroesophageal junction, colorectal) and non-small cell lung cancer with EGFR or ALK genomic tumor aberrations
Cetuximab	Binding to the EGFR blocks phosphorylation and activation of receptor-associated kinases, resulting in inhibition of cell growth, induction of apoptosis, and decreased matrix metalloproteinase and vascular endothelial growth factor production	KRAS wild type and EGFR-expressing colorectal cancer
Trastuzumab	Trastuzumab, humanized anti-HER2 IgG1, is a mediator of antibody-dependent cellular cytotoxicity (ADCC). Herceptin-mediated ADCC has been shown to be preferentially exerted on HER2-overexpressing cancer cells	HER2-overexpressing breast cancer, gastric, or gastroesophageal junction cancer
Ado-Trastuzumab emtansine	A HER2-targeted antibody–drug conjugate. The antibody is the humanized anti-HER2 IgG1, trastuzumab. The small-molecule cytotoxin, DM1, is a microtubule inhibitor	HER2-overexpressing breast cancer
Pertuzumab	Pertuzumab targets the extracellular dimerization domain (subdomain II) of the human epidermal growth factor receptor 2 protein (HER2) and, thereby, blocks ligand-dependent heterodimerization of HER2 with other HER family members, including EGFR, HER3, and HER4. As a result, pertuzumab inhibits ligand-initiated intracellular major signaling pathways, mitogen-activated protein (MAP) kinase and phosphoinositide 3-kinase (PI3K). Inhibition of these signaling pathways result in cell growth arrest and apoptosis, respectively	HER2-positive breast cancer
Panitumumab	Binds specifically to EGFR on both normal and tumor cells and competitively inhibits the binding of ligands for EGFR	Wild-type KRAS (exon 2) metastatic colorectal cancer

Note: From approved labels at Drugs@FDA.

Based on their mechanisms of action (Drugs@FDA, 2015), these approved products share some common targets of protein kinases. A lack of selectivity seems to be characteristic of protein kinase inhibitors developed so far (Karaman et al., 2008). To determine the pharmacological networks of these products, we manually retrieved all of the targets for each product from Section Mechanism of Action, of its approved label (Drugs@FDA), and then supplemented them with additional ones from DrugBank 4.3 (Law et al., 2014). All together, we identified 73 drug targets (see Supplementary Material on http://dx.doi.org/10.1016/bs.apcsb.2015.09.005). Targets belonging to the same family were clustered together, and the resulting pharmacological network is shown in Fig. 2.

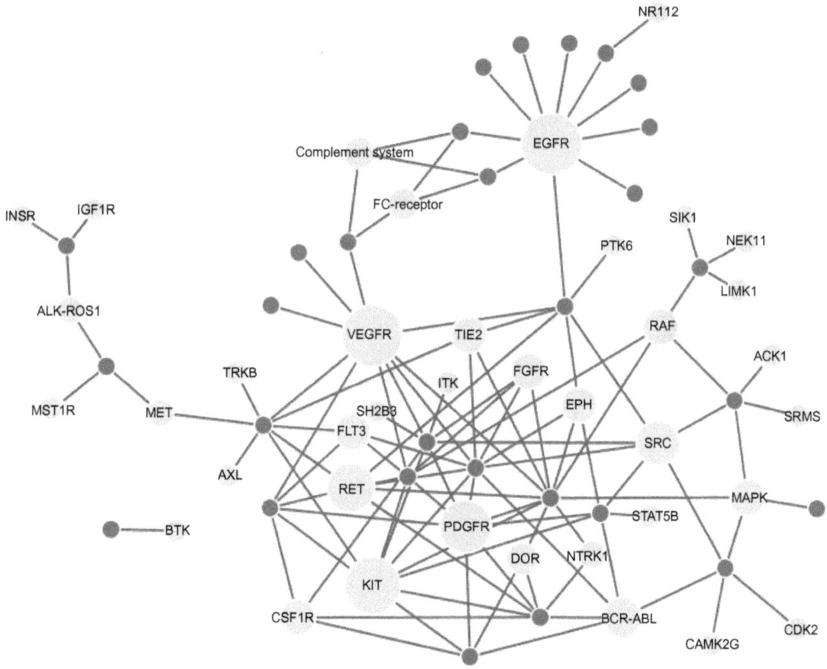

Figure 2 The network of the pharmacological targets of small molecular drugs (blue; gray in the print version) and antibodies (green; light gray in the print version) circles. Yellow (gray in the print version) nodes represent the drug targets, and the size of them is proportional to the number of drugs targeting those receptor tyrosine kinases. A list of 29 drugs was compiled from Drugs@FDA. For each drug in the list, we retrieved all the targets mentioned in its FDA label (Section Mechanism of Action) and also those mentioned in DrugBank 4.3 (Law et al., 2014). Overall, we identified 73 drug targets (see Supplementary Material on http://dx.doi.org/10.1016/bs.apcsb.2015.09.005). Targets belonging to the same family were clustered together, and the resulting pharmacological network is shown above.

As listed in Tables 1 and 2, several products are approved for treating cancers with specific markers. At bedside, treatment regimens are tailored to individual patients' tumor characteristics (Evaluation of Genomic Applications in Practice and Prevention, EGAPP Working Group, 2013). Unfortunately, relapse is often a fate that many a cancer patient would have to face following cycles of treatment combining surgery, radiation, and targeted therapy (plus chemotherapy). For example, a recent study of 1210 metastatic renal cell carcinoma (RCC) patients revealed that 74% and 26% of patients relapsed within and after 5 years, respectively, following surgical removal of localized tumors and targeted therapy (Kroeger et al., 2014). Abnormal complex intracellular biological networks and pathways enable cancer cells to survive through mutations (Cancer Genome Atlas Network, 2012), constituting the major obstacle to curing cancer. From how individual patients with tumors of different molecular characteristics responded to treatment, we are learning the extent of molecular complexity and the magnitude of mutational diversity in various types of cancer cells. Tumor heterogeneity is a challenge to implementation of effective treatment regimens as well as discovery of new effective agents. The types of heterogeneity in relation to clinical consequences will be highlighted in the following section.

3. TUMOR HETEROGENEITY

3.1 Disease and Molecular Heterogeneity

Since the approval of imatinib in 2001 for treating Ph+ chronic myelogenous leukemia (CML) and *kit*-positive GIST, our knowledge of tumor heterogeneity continues to grow from clinical experiences. Cancer heterogeneity manifests at the patient level as well as at the molecular level, with the amount of evidences expanding exponentially due to advances in omics technologies. An illustrative example is chronic eosinophilic leukemia (CEL). Imatinib is known to target Bcr-Abl tyrosine kinase and causes apoptosis in Bcr-Abl-positive cells (Table 1) (Drugs@FDA, 2015), and also an inhibitor of the RTKs for platelet-derived growth factor (PDGF). Interestingly, some patients with idiopathic hypereosinophilic syndrome (HES)/CEL responded to imatinib and had complete remission with low-dose imatinib, even though their tumor did not express BCR-ABL, a primary target of imatinib (Gotlib et al., 2004). Furthermore, some responding patients carried the fusion protein of FIP1L1-PDGFRA, while other responding patients did not (Gotlib et al., 2004). In this fusion protein, the function of PDGFRA (the PDGF receptor alpha) is known but FIP1L1 (factor interacting with PAP-like-1) has not been characterized with respect to its function in humans. Mutations in the coding gene, *FIP1L1–PDGFRA*, are associated with resistance to imatinib. The *FIP1L1–PDGFRA S601P* mutation, for example, rendering destabilization of the kinase domain to which imatinib and sorafenib bind, was found in HES/CEL patients resistant to both imatinib and sorafenib (Salemi et al., 2009). Though sorafenib was effective in treating HES/CEL patients with the *FIP1L1–PDGFRA T674I* mutation (Lierman et al., 2006), another mutation of *D842V* emergent from sorafenib treatment was devastating and rendered HES/CEL patients resistant to imatinib, sorafenib, and dasatinib (Lierman et al., 2009). Compared to the older inhibitors, imatinib, sorafenib, and dasatinib, the most recently approved ponatinib appeared to be more effective in treating HES/CEL patients with the mutations noted above (Sadovnik et al., 2014). This is an example of adapting our strategic target(s) to address mutations seen in clinical failures.

Resistance to imatinib is not limited to *FIP1L1–PDGFRA* mutations. For patients with GISTs, their responses to imatinib depend on the mutational status of *c-kit* and *PDGFRA* (American Cancer Society, 2015). Mutations in imatinib-resistant tumors could be primary or secondary (acquired), with clonal or polyclonal evolutions (Antonescu et al., 2005; Lim, Huang, et al., 2008). Fortunately, sunitinib was reportedly effective in treating GISTs after imatinib failure due to primary and secondary mutations (Heinrich et al., 2008). Another example is non-small cell lung cancer (NSCLC), and erlotinib is indicated for treating advanced NSCLC with mutations in epidermal growth factor receptor (EGFR) (Drugs@FDA, 2015). A case of NSCLC with a point of mutation in exon 21 of *EGFR* plus negative *ALK* rearrangement, a combination known to be clinically sensitive to erlotinib, did not respond to erlotinib and unfortunately expired (Giuliani, Martelli, Remo, & Bonetti, 2015). Failed cases such as this remind us of the importance of elucidating the landscape of cancer genomes.

KRAS mutation is another area of intense research in cancer heterogeneity. KRAS, encoded by the Kirsten rat sarcoma viral oncogene homolog gene, is an important protein targeted for treating CRC; there are two biologics (cetuximab and panitumumab) approved for CRC with wild-type *KRAS* (Table 2). We now know that there are a few thousand mutations detected in *KRAS*, across codons 12, 13, 61, and 146 (Edkins et al., 2006; Wellcome Trust Sanger Institute, 2015). Genetic mutation could lead to gain or loss of function, as well as switch of function (Reva, Antipin, & Sander, 2011) at the cellular level, and to fatal consequence at the patient level. A devastating consequence of a *KRAS* mutation was observed in a case of lung metastasis of CRC (Cejas et al., 2009; Pereira et al., 2015). These individual case reports of relapse and clinical setbacks continue to provide us glimpses of tumor heterogeneity. We have yet to gain a comprehensive understanding of the survival biology of cancer cells through which mutations occur and accumulate as cancer evolves.

3.2 Temporal and Spatial Heterogeneity

Molecular complexity and diversity not only exist among patients of a specific cancer and also in a patient's tumorous tissues. Spatial and temporal heterogeneities could occur in individual patients. Genomic instability processes, as tumor evolves, are the staple of spatial and temporal diversity of cancer and notably play a major role in poor prognosis of lung cancer (de Bruin et al., 2014). Breast cancer is a good example to illustrate spatial heterogeneity. Even with similar histopathological characteristics, lesion samples from multifocal breast cancer could not only harbor multiple mutations with an average of three mutations per patient but also exhibit inter-lesion heterogeneity in mutation (Desmedt et al., 2015). Lesions distant from one another had more different mutations. As tumor progresses due to genomic stability and selective pressure of therapeutic interventions, temporal heterogeneity could occur owing to the accumulation of acquired mutations plus selective pressure leading to a change in the dominant population. Discordance between primary and recurrent breast cancers, for instance, was observed with respect to the expression status of estrogen receptor, progesterone receptor, and human epidermal growth factor receptor 2 (HER2); status changes could include change from negative to positive or vice versa for all three receptors (Aurilio et al., 2014). Such shifting of receptor status was also observed for HER2 between primary gastric cancer and synchronous lymph node metastases (Ieni et al., 2014). Temporal heterogeneity could lead to metastasis and make cancer a moving target; as a result, modifying treatment regimens accordingly throughout the disease course is important. For a patient with the change from human epidermal growth factor receptor 2 (EGFR2) negative in the primary gastric cancer to EGFR2 positive in metastatic lesion, trastuzumab was added to the treatment regimen since this biological product is effective when HER2 is overexpressed (Gumusay et al., 2015).

Temporal and spatial heterogeneity of cancer poses great challenges to curing cancer. Hence, effective treatments would require dynamic implementation of treatment regimens tailored to the molecular features of a patient's cancer at the start of diagnosis, as well as to the observed temporal molecular changes during the course of treatment, in order to prevent metastasis and relapse. Both functional germ line and somatic mutations could affect the clinical outcomes. A recent study revealed that functional germ line mutations may not be associated with CRC risk or survival (Huhn et al., 2014). Accumulation of acquired mutations increases as tumors evolve. Take breast cancer example, the majority of mutations observed in breast cancer genomes occurred following the initial driver event (Stephens et al., 2012). Multiregional sequencing enables detection of the subclonal diversity of primary tumors, thereby allowing the clinicians and patients to make an informed decision of adequate treatment regimens to improve prognosis (Yates et al., 2015). Targeted sequencing and genome-scale sequencing enable identification of new important somatic mutations at a low allelic frequency in addition to common mutations of a high frequency, and of new somatic mutations for aggressive CRCs (Cancer Genome Atlas Network, 2012; Desmedt et al., 2015). With the sequencing data in hand, a crucial step would be to predict the possible mutations that would adversely drive cancer progress.

In the following section, we will discuss computational methods that have been developed to identify driver mutations.

4. COMPUTATIONAL METHODS FOR IDENTIFYING DRIVER MUTATIONS

Advances in sequencing technologies undoubtedly accelerate discovery of somatic mutations in cancer (COSMIC, 2015). The challenge we face is precise identification of driver mutations by sorting through huge cancer genomes so that druggable targets could be selected for drug discovery and development. Notably, among the mutations not every mutation is equally fatal. Driver mutations are nonsynonymous mutations causally linked with cancerogenesis (Stratton et al., 2009). They are usually associated with oncogene or tumor suppressor genes (Garraway & Lander, 2013; Hanahan & Weinberg, 2011). The mutations that are carried along through cell division are passenger mutations and may be involved in cancer progression (McFarland et al., 2013) though they are not critical for cancer evolution.

Characterizing a mutation as driver is not an easy task due to the heterogeneity of the mutational landscape. Many pan-cancer studies have highlighted the complexity of cancer (Kandoth et al., 2013; Lawrence et al., 2013; Watson, Takahashi, Futreal, & Chin, 2013). An exemplifying characteristic is the "long tail" of the mutational landscape of many cancers; i.e., few genes are mutated across many samples while many genes are mutated only in a small number of samples. Those well-known cancer genes, such as *TP53* encoding tumor protein p53 (p53), are on the high-frequency list; most of these genes have already been identified (Lawrence et al., 2014). Sophisticated methods are developed to characterize rare mutations. These methods can be roughly categorized into three main categories: (i) frequency based, (ii) functionally based, and (iii) pathway based. Each method utilizes different distinctive characteristics of driver mutations in their roles in cancer evolution (Table 3).

4.1 Frequency-Based Methods

Driver genes are expected to be mutated more frequently in tumor than normal tissues due to the selective pressure of cancer evolution (Stratton et al., 2009), and this is the premise of frequency-based methods. The main challenge for methods in this category is to define a background mutation rate (BMR) for every interrogated gene, i.e., the expected frequencies of passenger mutations. To identify rare mutations, BMRs and modeling different heterogeneity factors would need to be defined. The number of passenger mutations is usually estimated from the number of supposedly selectively neutral mutations such as synonymous mutations and/or mutations in introns and untranslated regions around the gene.

The canonical example for this category is MutSigCV (Lawrence et al., 2013) (*Mut*ation *Sig*nificance with *cov*ariates) which defines three BMRs: patient, mutational mechanism, and region specific. For the first two, the mutational frequency of a gene is computed for every sample and every mutational type (six basic types), while for the regional BMR, genes are clustered together based on their BMRs and other properties (covariates) and then a common frequency is calculated. The three BMRs are then combined, and a final *p*-value is reported based on a projection method that accounts for the two most suspicious mutations of every patient. MutSigCV has been used heavily in the construction of the Cancer Genome Atlas (TCGA), characterizing mutations for various cancers such as lung (Cancer Genome Atlas Research Network et al., 2014), and lately for cutaneous melanoma (The Cancer Genome Atlas Network et al., 2015).

Table 3 A List of Widely Used Methods for the Characterization of Driver Mutations

Scope	Approach	Description	References
Frequency based	CNA Peel-off	Uses "arbitrated peel-off" algorithm to identify regions with significant levels of CNA	Mermel et al. (2011)
	BMR	Computes and combines seven BMRs per gene and also allows for user-defined regions of interest	Dees et al. (2012)
	BMR	Computes BMR using coverage information per patient, mechanism, and region (in a covariate space)	Lawrence et al. (2013)
	Bayesian based	Uses Bayesian approach to estimate BMR, considering eight biological parameters	Hua et al. (2013)
	Permutation based	Computes permutation-based gene BMR per patient and trinucleotide across exons, introns, and UTRs	Hodis et al. (2012)
Function based	Evolutionary	Uses an entropy score to compute the impact induced in the aligned residue of protein homologs	Reva et al. (2011)
	Evolutionary	Uses an alignment-based score to compute and allows for more types of mutations than SNPs	Choi, Sims, Murphy, Miller, and Chan (2012)
	ML	Trains naïve Bayes classifier using both evolutionary and structural information	Adzhubei et al. (2010)
	ML	Random forest classifier trained on mutations from HGMD vs. ESP using SNVBox data	Carter, Douville, Stenson, Cooper, and Karchinn (2013)
	Structural	Searches for clusters of mutations in both the linear and 3D structure of proteins	Ryslik, Cheng, Cheung, Modis, and Zhaoo (2013)
Pathway based	Known sets	Uses belief-propagation algorithm to infer gene activity from gene expression and CNA	Ng et al. (2012)
	Mutually exclusive	Finds subnetworks based on pairs of genes that have mutually exclusive mutations	Ciriello, Cerami, Sander, and Schultzz (2012)
	Mutually exclusive	Simultaneously finds multiple groups of genes with mutually exclusive aberrations	Leiserson, Blokh, Sharan, and Raphael (2013)
	Network	Uses a network diffusion approach to connect genomic perturbations to gene expression changes	Paull et al. (2013)
	Network	Finds significantly mutated subnetworks, via an insulated diffusion process that spreads a mutations effect across the network	Leiserson et al. (2015)
Aggregators	Averaging	Combines SIFT, LogRE, MAPP, polyphen2, and MutationAssessor using a weighted approach	Gonzalez-Perez and Lopez-Bigas (2011)
	ML	Combines SIFT, Pfam-LogRE, and GOSS using a Random Forest classifier to characterize mutations	Kaminker, Zhang, Watanabe, and Zhang (2007)

Note: A catalog of software tools is available at http://omictools.com/ (accessed July, 2015). CNA, copy number aberration; BMR, background mutation rate; ML, machine learning; SNP, single nucleotide polymorphism; HGMD, Human Gene Mutation Database; ESP, Exome Sequencing Project.

4.2 Functional-Based Methods

Functional-based methods aim to pinpoint the mutations based on their functional impact since driver mutations are expected to affect significantly their encoded proteins. There are two main approaches in assessing the impact of a mutation, and they could be used independently or in combination. The first approach is evolutionary based and correlates positively the conservation of an amino acid sequence among homologs with the functional impact of a change of the sequence. The second approach aims to predict the effect of a mutation on the structural characteristics of the encoded protein such as its secondary or 3D structure. These two approaches could be integrated using machine learning (ML) algorithms.

A notable evolutionary-based method is MutationAssessor (Reva et al., 2011), which has been used for TCGA, and is usually used in conjunction with other methods, in studying ovarian cancer (The Cancer Genome Atlas Research Network, 2011) and squamous cell lung cancer (Kannan et al., 2012). MutationAssessor correlates the impact of a mutation with the change in the informational entropy it induces in its corresponding homolog family and its subfamilies. Subfamilies are computationally derived clusters of similar homologs used to characterize differences within a family such as different interaction partners. Different methods in this category may use alternative metrics to calculate the impact of a mutation; PROVEAN (Choi et al., 2012), for example, uses an alignment-based distance which allows assessment of mutations spanning multiple nucleotide bases.

Purely structural-based methods are practically limited to the study of well-known proteins for which high-level data and empirical knowledge about their folding mechanisms are available. However, when combined with evolutionary information, they can increase the accuracy of a predictor. A successful example would be PolyPhen2 (Adzhubei et al., 2010), a naïve Bayes classifier trained on eight sequencing and three structural features.

4.3 Pathway-Based Methods

Pathway-based methods differ from the previously described methods in that they aim to characterize sets of genes instead of individual genes. This approach is inspired both by the "long tail" of the mutational landscape, which indicates that multiple genes are involved, and from the idea that cancer, despite its complexity, is the result of few deregulated processes/pathways (Hanahan & Weinberg, 2011). A key step of pathway-based methods is the selection of gene sets to be investigated, since enumeration of all the possible gene sets is infeasible. There are three main approaches addressing this issue, which can be categorized based on their dependence on *a priori* information.

On the one hand, there are methods that rely heavily on prior information and examine predefined sets of genes. Standard enrichment analysis algorithms such as gene set enrichment analysis (Subramanian et al., 2005) can be used to determine, if a specific set of mutated genes are significant and meaningful in the context of biological functions and pathways. On the other hand, there are methods that do not rely on prior information at all and identify deregulated pathways solely from the combinatorial patterns of the mutational landscape. In particular, these methods aim to identify sets of mutations that cover, as much as possible, a wide range of samples, but, at the same time, do not coexist in the same sample. This approach is inspired by the "mutual exclusivity" property (Yeang, McCormick, & Levine, 2008) of the mutational landscape, i.e., the observation that although possible, it is unlikely for more than one mutations to coexist in a pathway because mutations are relatively rare and most pathways can be deregulated by a single mutation.

Dendrix (*De novo Dri*ver E*x*clusivity) (Vandin, Upfal, & Raphael, 2012) is the prototypical example of this category. It formulates and maximizes inter-, while minimizes intra-, sample coverage through solving a constrained optimization problem. Its derivative, Multi-Dendrix (Leiserson et al., 2013), reformulates to account for multiple pathways simultaneously and solves it through integer linear programming which results in increased accuracy in less time. Multi-Dendrix is not just the core algorithm but also a pipeline to summarize and evaluate the results of the analysis. Both Dendrix and Multi-Dentrix were tested with TCGA data and identified exclusivity patterns in several pathways such as p53 and RTK phosphoinositide 3-kinase–protein kinase B/Akt (PI3K/AKT) pathway.

Finally, there is an intermediate strategy that relies on interaction networks. Methods in this category utilize the network properties together with the mutational information in the samples in order to capture the significantly mutated subnetworks. One of the most modern examples is HotNet2 (Leiserson et al., 2015). In HotNet2, every gene is initially associated with a "temperature," based on its mutational frequency, which eventually diffuses to neighboring nodes. Once the equilibrium is reached, a directed network is retrieved according to the heat flows along the original edges, and all of its complete subnetworks are then evaluated by a two-stage significance test. HotNet2 has been used in a pan-cancer study where it identifies both well-known and less characterized subnetworks. Its predecessor, HotNet (Vandin, Upfal, & Raphael, 2011), which relies on the same principle, has also been used in two TCGA studies for acute myeloid leukemia (The Cancer Genome Atlas Research Network, 2013) and RCC (The Cancer Genome Atlas Research Network, 2013).

4.4 Comparison and Meta-Analysis

In conclusion, the methods described above focus on different aspects of cancer biology; choice of methods is not so much a matter of efficiency as a matter of research scope. In practice, when characterizing the mutational landscape of a cancer, multiple methods are used to elucidate different aspects of the data set. Frequency-based methods are the most unbiased of the methods since they are primarily data driven and adopt a minimum number of assumptions. They are also the only choice available for characterizing noncoding regions. However, their power depends on the sample size which may end up prohibitively large in order to identify rare mutations. As a result, they have been extensively used for massive cancer characterization projects where thousands of tumor samples are available. Apart from their dependence on a large sample size, which may not be sufficient if heterogeneity is not modeled properly, frequency-based methods have also been criticized for their lack of a mechanistic insight into how mutations are actually related to cancer phenotypes. On the other hand, functional-based methods relying on *a priori* knowledge can characterize mutations even from a single sample. However, in this way their scope is limited to relatively well-studied cases for which prior information is available, while for other less characterized cases extrapolation is needed. Pathway-based methods are generally more diverse than the previous two. Methods relying on predefined sets of genes are primarily hypothesis driven, while Dendrix-like methods are primarily data driven. Nonetheless, their scope is limited to cases where the mutual exclusivity principle holds. Other methods that rely on interaction networks would need efforts to balance data and hypotheses. Protein interaction networks available on databases, such as Reactome (Croft et al., 2014), cover a wide range of proteins and interactions, but their quality has been widely criticized (Bonetta, 2010; Cusick et al., 2009; Kirouac et al., 2012).

Finally, precisely because different methods highlight different aspects of cancer biology, meta-analysis methods that aim to systematically integrate their results have also been developed. These methods may rely on simple averaging techniques, such as Condel (Gonzalez-Perez & Lopez-Bigas, 2011) that integrates the results from SIFT (Kumar, Henikoff, & Ng, 2009), LogRE (Clifford, Edmonson, Nguyen, & Buetow, 2004), MAPP (Stone & Sidow, 2005), PolyPhen2, and MutationAssessor, or rely on more sophisticated ML algorithms, such as CanPredict (Kaminker, Zhang, Watanabe, et al., 2007) that uses a random forest classifier trained on data from SIFT, LogRE, and GOSS (Kaminker, Zhang, Waugh, et al., 2007). Cancer-trained methods and general methods seem to perform similarly with respect to accuracy, while predictors integrating multiple methods do not substantially increase accuracy (Gnad, Baucom, Mukhyala, Manning, & Zhang, 2013).

There are still gaps to be filled with data of quality and computational methods of precision; therefore, effective concerted efforts between biologists and computational scientists are needed. Biomedical literature contains knowledge obtained through scientific discoveries in biomedical disciplines including cancer research and is a great resource for research. In the following section, literature mining will be discussed for its utility.

5. LITERATURE MINING IN ONCOLOGY RESEARCH

5.1 Biomedical Literature Mining

PubMed (http://www.ncbi.nlm.nih.gov/pubmed), one of the key biomedical literature resources, currently indexes over 25 million article citations, including 3.5 million full-text in PubMed Central (http://www.ncbi.nlm.nih.gov/pmc/), and continuously expands in volume. The field of biomedical literature mining, a computer-assisted information, and knowledge extraction from biomedical text data has become increasingly important to cope with the huge volume of electronically available biomedical literature. Literature mining has four main areas (Fleuren & Alkema, 2015; Zhu et al., 2013): (1) information retrieval (IR), (2) named entity recognition (NER), (3) relation extraction (RE), and (4) knowledge discovery (KD). First, IR aims to properly identify and retrieve text data relevant to a desired topic from document repositories. The best-known system is the Entrez query and retrieval system in PubMed (Baxevanis, 2008; Jensen, Saric, & Bork, 2006). Second, NER identifies genes, proteins, chemical compounds, metabolites, diseases, etc., from the literature. The methods and algorithms used in NER include rule-based approaches and ML-based approaches. The former relies on patterns of systematic variations in names (Fukuda, Tamura, Tsunoda, & Takagi, 1998; Hur, Schuyler, States, & Feldman, 2009; Jensen et al., 2006; Krauthammer, Rzhetsky, Morozov, & Friedman, 2000; Tanabe & Wilbur, 2002). The latter utilizes statistical and probabilistic models such as the hidden Markov model (Nobata, Collier, & Tsujii, 1999; Zhou, Zhang, Jian, Shen, & Tan, 2004), the support vector machine (Donaldson et al., 2003; Kazama, Makino, Ohta, & Tsujii, 2002), and the naïve Bayesian learning (Hatzivassiloglou, Duboue, & Rzhetsky, 2001). Third, RE identifies the relationship such as interaction and regulation between biological entities and is applied to identify physical protein–protein interactions (PPIs) (Cooper & Kershenbaum, 2005; Donaldson et al., 2003; Li, Hu, Lin, & Yang, 2010; Ramani, Bunescu, Mooney, & Marcotte, 2005). For example, STRING (http://string-db.org/) and the Michigan Molecular Interactions (MiMI, http://mimi.ncibi.org/) incorporate literature-derived PPI information to augment the PPI database. Lastly, KD aims at discovering hidden information in the literature.

5.2 Literature Mining for Integrating Molecular Events at the Cellular Level

With the advent of next-generation sequencing technology, the NCBI Genome database continuously expands and currently contains the sequences of over 13,000 genomes (National Center for Biotechnology Information, 2010). To handle these data, computational approaches, relying on protein or DNA sequence similarities with known species and literature-mined association between protein families and functional keywords or Gene Ontology (GO), have been extensively used (Abascal & Valencia, 2003; Andrade & Valencia, 1998; Arnaud, Costanzo, Shah, Skrzypek, & Sherlock, 2009; Ashurst & Collins, 2003; Lovering, Dimmer, & Talmud, 2009; Tamames, Ouzounis, Casari, Sander, & Valencia, 1998). Text mining approaches facilitate analysis of gene expression data by providing highly relevant functional terms to genes or clusters of genes (Blaschke, Oliveros, & Valencia, 2001; Shatkay, Edwards, Wilbur, & Boguski, 2000). In addition, tools like PubGene (http://pubgene.org; Jenssen, Laegreid, Komorowski, & Hovig, 2001), Genomatix Pathway System (Genomatix Software GmbH, Munich, Germany; https://www.genomatix.de/; (Lim, Hwang, Aw, & Sun, 2008; Vock, Gleissner, Klapper, & Doring, 2008), MetaCore from GeneGO (Thomson Reuters; https://portal.genego.com/), and Ingenuity Pathway Analysis (QIAGEN, California; http://www.ingenuity.com/products/ipa) allow navigation and visualization of the potential functional connections among genes in various domains using the gene co-citation networks constructed from

PubMed (Hur et al., 2011). Importantly, identifying PPI using literature mining is an active research area as PPIs govern the majority of cellular events (Raman, 2010), and is employed to associate cellular events to clinical phenotypes with respect to adverse drug reactions (Hur, Guo, Loh, Feldman, & Bai, 2014; Hur, Zhao, & Bai, 2015). Experimentally verified PPIs are estimated to represent only a portion of the complete PPI networks (Hart, Ramani, & Marcotte, 2006; Stumpf et al., 2008). Thus, literature mining potential PPIs critically augments the current PPI compendium using approaches such as semantic proposition (Rindflesch, Hunter, & Aronson, 1999), rules and patterns (Ng & Wong, 1999; Ono, Hishigaki, Tanigami, & Takagi, 2001), and ML-based methods (Donaldson et al., 2003; Yang, Lin, & Li, 2010).

5.3 Literature Mining to Construct Informed Knowledge Bases for Cancer Research

Creating knowledge base through text mining and curation has long facilitated biomedical research. An open access knowledge base for precision cancer medicine (PCM; http://rubinlab.bgu.ac.il/pcm/) incorporates literature-based associations among drug, gene, and disease (Geifman, Haviv, Kurzrock, & Rubin, 2014). CancerResource (http://bioinf-data.charite.de/cancerresource/) is a database integrating cancer-relevant relationships of drugs and targets obtained through text mining of the literature and compiling other external resources (Ahmed et al., 2011). Other resources include Pharmacogenomics Knowledge Base (PharmGKB; https://www.pharmgkb.org/), DrugBank (http://drugbank.ca/), KEGG pathways (http://www.genome.jp/kegg/pathway.html), and NCI60 cancer cell lines.

Li et al. constructed a miRNA-cancer association network (miCancerna; http://micancerna.appspot.com/) based on over 1000 miRNA and cancer associations identified from the PubMed abstract by literature mining, which includes 226 miRNA families and 20 common cancers (Li, Hu, et al., 2014). With the cutoff of top 5% significant association, miCancerna allows for prioritization of cancer-related miRNAs. *p53* and *E2F1* genes were identified as therapeutic targets of human NSCLC, from a molecular network of genes and their products inhuman and mouse, which was constructed from literature-derived orthologus interactions (Li, Bi, et al., 2014).

High-throughput DNA/RNA-Seq produces a large number of genes associated with cancer, and utilizing these results to identify and prioritize candidate genes for further studies is important. Lee et al. developed OncoSearch (http://oncosearch.biopathway.org), a Web-based cancer gene search engine (Lee, Dang, Lee, & Park, 2014). OncoSearch searches PubMed abstracts for differentially expressed genes in cancer, for related cancer progression in patients, and for the plausible roles of these genes. Various tools have also been developed to prioritize cancer genes and pathways using keyword-, vector space modeling-, ontology structure-, and statistical modeling-based literature mining approaches. These tools aim at identifying candidate genes and pathways that are strongly associated with the pathogenic mechanisms of various cancers (Luo, Riedlinger, & Szolovits, 2014). In brief, literature mining constitutes an important part of our efforts on developing effective cancer therapy by providing biological context with scientific evidences using text mining algorithms and critically supports the computational methods discussed above.

In addition to identifying the druggable targets, computational methods are also developed to facilitate research in genome editing (Stoimenov, Ali, Pandzic, & Sjoblom, 2015) for the purpose of developing research tools and for treating diseases. In the following section, we will highlight genomic editing technologies and its use in cancer research.

6. PRECISE GENOME EDITING

In the context of precision medicine, characterization of the cancer genomes in individual patients (for example by high-throughput sequencing) would enable individualized treatment. Recent advances in genome editing technologies may, for the first time, allow the editing of the human genome at the patient level for the purpose of maximizing a therapeutic outcome. On this front, two technologies have emerged during the past few years overcoming the limitations of the most commonly used zinc finger nucleases (ZFNs) (Kim, Cha, & Chandrasegaran, 1996). These are (i) the transcription activator-like effector nucleases (TALENs) (Boch et al., 2009; Moscou & Bogdanove, 2009; Reyon et al., 2012) and, more recently, (ii) the clustered regularly interspaced short palindromic repeats (CRISPR)/Cas nuclease systems (Sander & Joung, 2014; Sato et al., 2015; Takahashi et al., 2015).

Precise genome editing may correct disease-causing mutations in a specific DNA region through specific site recognition for targeted DNA double-strand breaks and repairs with the healthy pieces of DNA fragments. For example, in the work by Sebastiano et al. (2011), the authors used a ZFN to correct the mutation that causes sickle cell anemia in induced pluripotent stem cells (iPSCs) derived from a patient with the disease. This strategy could be used to generate engineered, patient-derived cells that could be transplanted back without fear of rejection or the use of immunosuppressive drugs (http://directorsblog.nih.gov/2013/01/22/copy-editing-the-genome-extreme-personalized-medicine/).

6.1 Genome Editing Using TALENs

TALENs are artificial restriction enzymes that recognize specific regions in the genome and create DNA double-strand breaks. The double-stranded breaks (DSBs) created by TALENs are repaired by one of multiple mechanisms used in eukaryotic cells to maintain genomic stability during the cell cycle (Takata et al., 1998), and the two common mechanisms are (i) nonhomologous end joining (NHEJ) and (ii) homology-directed repair (HDR) (Roth, Porter, & Wilson, 1985). NHEJ, handled by a NHEJ complex, is the most common repair mechanism (Reid & Rothenberg, 2015). During NHEJ, direct ligation could bypass minor errors at the DSB ends, while adaptive processing-dependent resolution could correct the errors at the DSB termini so that the correct sequence can be aligned and ligated (Waters et al., 2014). In addition to the canonical pathway DNA–PK-dependent pathway of NHEJ (D-NHEJ), a quasi-backup function (B-NHEJ) pathway could operate as an alternative if D-NHEJ is deficient (Iliakis et al., 2004; Mladenov & Iliakis, 2011). For HDR in mammalian cells, several transferases such as XRCC3, Rad51, Rad54, XRCC2, and BRCA1 (Johnson & Jasin, 2001; Pierce, Johnson, Thompson, & Jasin, 1999) are involved in mammalian cells.

TALEN technology is based on the TAL effector proteins, naturally secreted by the plant pathogen, *Xanthomonas*, that bind onto a host DNA. TAL effectors can be engineered to bind onto any specific DNA sequence, and TALEs are then fused, by design, to nonspecific DNA-cleaving nucleases (Joung & Sander, 2013), to construct TALENs; a specific TALEN designed could cleave a DNA at the desired location, opening prospects for use in research and biotechnology. Advances in TALEs are evidenced from the results of *in vitro* genomic editing of human *NTF3* and *CCR5* genes (Miller et al., 2011), to *in vivo* editing of DNAs in somatic tissues of zebrafish (Bedell et al., 2012), to multiple knock-in edits of human genome in human iPSC and neural stem cell (Cerbini et al., 2015), to reverse the phenotype of a transgenic mouse model of HPV (human papillomavirus)-driven neoplasms (Hu et al., 2015), and to selectively correct mitochondrial mutations in the germ line (Reddy et al., 2015). Reportedly, efficacy of genomic editing could reach 100% for some loci (Bedell et al., 2012).

Most recently, TALENs have been used for high-throughput genome editing by leveraging the fast ligation-based automatable solid-phase high-throughput (FLASH) platform (Reyon et al., 2012). Such a high-throughput construction of TALEN pairs followed by a large-scale assessment will optimize the utility of genomic editing for battling cancers.

6.2 Genome Editing Using CRISPR/Cas9 Systems

CRISPR/Cas9 is a RNA-guided site-specific DNA cleavage technology for genomic engineering and is based on a prokaryotic immune mechanism that protects bacteria and archaea from foreign DNA (Cong et al., 2013; Sander & Joung, 2014). Through CRISPR, bacteria and archaea continue to evolve for survival by acquiring sequence spacer from foreign DNA, and use the acquired spacer to target and cleave invasive DNA from virus or plasmid (Deveau, Garneau, & Moineau, 2010; Horvath & Barrangou, 2010). Its adaptation for genome editing incorporates a piece of RNA called guide RNA (gRNA), about 20 nt long, coupled to a Cas9 nuclease. The Cas9 nuclease induces a DSB at exactly 3 bp upstream from a PAM (protospacer adjacent motif). The 3-bp sequence is in the form of 5′-NGG-3′ where N is any nucleotide and G is guanine nucleotide. By designing the gRNA to be complementary to the target position in the genome, one can guide the nuclease to cut anywhere in the genome (as long as there is a PAM close by). The CRISPR/Cas9 system is characterized by high specificity and efficacy, making it attractive for *in vivo* applications in addition to generating mutated cell lines for use in pharmaceutical research. Moreover, novel methods such as the GONAD (Genome-editing via Oviductal Nucleic Acids Delivery) system overcome previous limitations and enable the delivery of nucleic acids (eGFP mRNA or Cas9 mRNA/sgRNAs) into the preimplantation embryo in the intact mouse oviduct (Takahashi et al., 2015). Recently, Sato et al. (2015) edited mouse spermatogonial cells by using both TALEN and CRISPR/Cas9. In another *in vivo* application, Swiech et al. (2015) used CRISPR/Cas9 to target the *Mecp2* gene as well as multiple genes (*Dnmt1*, *Dnmt3a*, and *Dnmt3b*) in the adult mouse brain *in vivo* and characterized the effects of genome modifications in postmitotic neurons using biochemical, genetic, electrophysiological, and behavioral readouts.

Finally, Liang et al. (2015) achieved for the first time the CRISPR/Cas9-mediated gene editing in human tripronuclear zygotes, opening the door for future success. However, as noted by the Liang et al., editing efficiency was low and off-target cleavage led to undesired mutations. Therefore, there is a need to improve this technology in order to achieve clinical application of precision genome editing technologies. Though the utility of genome editing in battling cancer is so far limited to developing preclinical *in vivo* research tools, such technologies are a stepping stone to our ultimate goal of curing cancer.

7. CONCLUSION

With the claims databases and EMRs available for computation and mining, we could investigate population-level outcomes across various cancers. Though at the bedside, we are learning, one patient at a time, how heterogeneous a type of cancer can be and how tumor heterogeneity affects treatment outcome. With the aids of omics technologies and computational methods, we have a large number of mutations identified and continuously compile into databases the mutations that are newly discovered. Spatial heterogeneity creates opportunities for simultaneous evolution of multiple driver genes, while temporal heterogeneity further diversifies mutations and cancer characteristics. The landscape of cancer genomes is ever expanding. Though advances in computational methods and computer sciences are enabling us to separate driver from passenger mutations and compile databases of cancer mutations. In fact, application of the driver mutations identified computationally in drug discovery, and development is scarce based on literature. Implementation of many theoretical findings in the realm of drug discovery is likely far more challenging than we would hope.

In addition to cancer genomes, tissue environment and patient physiology could affect the outcome in an individual patient. In the microenvironment of a tumor tissue, cancer evolution takes place. Computational methods and computer sciences could presumably be utilized to address the issue how a unique tissue environment in a specific patient physiological condition affects the treatment outcome. Conceivably, one could incorporate the information from all the technologies, create virtual cancer models considering an array of mutations coexisting in a cancer, and then simulate to assess all possible scenarios. Even if it is possible to computationally create all the cancer models, simulating the specific microenvironment in cancer tissues in individual patients, which could be critical to cancer evolution, would require extensive knowledge and expertise in many areas. It would need concerted efforts from a group of experts including biologists, geneticists, computational scientists, clinicians, and statisticians to be more realistically simulate cancer evolution in a specific cancer microenvironment.

Though decipher cancer complexity for the purpose of curing cancer is very challenging, cutting-edge technologies, such as genomic editing, are being used to create cancer models and to develop novel treatments. Research efforts in searching for effective drugs to treat various cancers are simultaneously advancing on all the critical fronts of battling cancer. It is hopeful that one day we could stay ahead of cancer mutation before it takes place with precision, rather than chase after mutation.

Disclaimer: The views expressed in this chapter do not represent the views of the U.S. Food and Drug Administration.

SUPPLEMENTARY MATERIAL

The below data are available on http://dx.doi.org/10.1016/bs.apcsb.2015.09.005:

Spreadsheet 1	Approved small molecular tyrosine kinase inhibitors plus antibodies targeting kinases and their pharmacological targets from their approved labels (Drugs@FDA)
Spreadsheet 2	Uniprot (targets for each small molecular tyrosine kinase inhibitor or antibody obtained from Uniprot)
Spreadsheet 3	Clusters of Targets where targets are clustered together into families with their respective groups of drugs

REFERENCES

Abascal, F., & Valencia, A. (2003). Automatic annotation of protein function based on family identification. *Proteins*, *53*(3), 683–692. http://dx.doi.org/10.1002/prot.10449.

Adzhubei, I. A., Schmidt, S., Peshkin, L., Ramensky, V. E., Gerasimova, A., Bork, P., et al. (2010). A method and server for predicting damaging missense mutations. *Nature Methods*, 7(4), 248–249. http://dx.doi.org/10.1038/nmeth0410-248.

Ahmed, J., Meinel, T., Dunkel, M., Murgueitio, M. S., Adams, R., Blasse, C., et al. (2011). CancerResource: A comprehensive database of cancer-relevant proteins and compound interactions supported by experimental knowledge. *Nucleic Acids Research*, *39*(Database issue), D960–D967. http://dx.doi.org/10.1093/nar/gkq910.

American Cancer Society (2015). *Cancer Facts & Figures 2015*. http://www.cancer.org/acs/groups/content/@editorial/documents/document/acspc-044552.pdf (Accessed June 2015).

Andrade, M. A., & Valencia, A. (1998). Automatic extraction of keywords from scientific text: Application to the knowledge domain of protein families. *Bioinformatics*, *14*(7), 600–607.

Andrejauskas-Buchdunger, E., & Regenass, U. (1992). Differential inhibition of the epidermal growth factor-, platelet-derived growth factor-, and protein kinase C-mediated

signal transduction pathways by the staurosporine derivative CGP 41251. *Cancer Research*, *52*(19), 5353–5358.
Antonescu, C. R., Besmer, P., Guo, T., Arkun, K., Hom, G., Koryotowski, B., et al. (2005). Acquired resistance to imatinib in gastrointestinal stromal tumor occurs through secondary gene mutation. *Clinical Cancer Research*, *11*(11), 4182–4190. http://dx.doi.org/10.1158/1078-0432.CCR-04-2245.
Arnaud, M. B., Costanzo, M. C., Shah, P., Skrzypek, M. S., & Sherlock, G. (2009). Gene ontology and the annotation of pathogen genomes: The case of *Candida albicans*. *Trends in Microbiology*, *17*(7), 295–303. http://dx.doi.org/10.1016/j.tim.2009.04.007.
Ashurst, J. L., & Collins, J. E. (2003). Gene annotation: Prediction and testing. *Annual Review of Genomics and Human Genetics*, *4*, 69–88. http://dx.doi.org/10.1146/annurev.genom.4.070802.110300.
Aurilio, G., Disalvatore, D., Pruneri, G., Bagnardi, V., Viale, G., Curigliano, G., et al. (2014). A meta-analysis of oestrogen receptor, progesterone receptor and human epidermal growth factor receptor 2 discordance between primary breast cancer and metastases. *European Journal of Cancer*, *50*(2), 277–289. http://dx.doi.org/10.1016/j.ejca.2013.10.004.
Baselga, J., Norton, L., Albanell, J., Kim, Y. M., & Mendelsohn, J. (1998). Recombinant humanized anti-HER2 antibody (Herceptin) enhances the antitumor activity of paclitaxel and doxorubicin against HER2/neu overexpressing human breast cancer xenografts. *Cancer Research*, *58*(13), 2825–2831.
Baxevanis, A. D. (2008). Searching NCBI databases using Entrez. *Current Protocols in Bioinformatics. Chapter 1*. http://dx.doi.org/10.1002/0471250953.bi0103s24. Unit 1.3.
Bedell, V. M., Wang, Y., Campbell, J. M., Poshusta, T. L., Starker, C. G., Krug, R. G., 2nd., et al. (2012). In vivo genome editing using a high-efficiency TALEN system. *Nature*, *491*(7422), 114–118. http://dx.doi.org/10.1038/nature11537.
Blaschke, C., Oliveros, J. C., & Valencia, A. (2001). Mining functional information associated with expression arrays. *Functional & Integrative Genomics*, *1*(4), 256–268. http://dx.doi.org/10.1007/s101420000036.
Boch, J., Scholze, H., Schornack, S., Landgraf, A., Hahn, S., Kay, S., et al. (2009). Breaking the code of DNA binding specificity of TAL-type III effectors. *Science*, *326*(5959), 1509–1512. http://dx.doi.org/10.1126/science.1178811.
Bonetta, L. (2010). Protein-protein interactions: Interactome under construction. *Nature*, *468*(7325), 851–854. http://dx.doi.org/10.1038/468851a.
Buchdunger, E., Zimmermann, J., Mett, H., Meyer, T., Muller, M., Druker, B. J., et al. (1996). Inhibition of the Abl protein-tyrosine kinase in vitro and in vivo by a 2-phenylaminopyrimidine derivative. *Cancer Research*, *56*(1), 100–104.
Cancer Genome Atlas Network. (2012). Comprehensive molecular characterization of human colon and rectal cancer. *Nature*, *487*(7407), 330–337. http://dx.doi.org/10.1038/nature11252.
Cancer Genome Atlas Research Network, Collisson, E. A., Campbell, J. D., Brooks, A. N., Berger, A. H., Lee, W., et al. (2014). Comprehensive molecular profiling of lung adenocarcinoma. *Nature*, *511*(7511), 543–550. http://dx.doi.org/10.1038/nature13385.
Carter, H., Douville, C., Stenson, P. D., Cooper, D. N., & Karchin, R. (2013). Identifying Mendelian disease genes with the variant effect scoring tool. *BMC Genomics*, *14*(Suppl. 3), S3. http://dx.doi.org/10.1186/1471-2164-14-S3-S3.
Cejas, P., Lopez-Gomez, M., Aguayo, C., Madero, R., de Castro Carpeno, J., Belda-Iniesta, C., et al. (2009). KRAS mutations in primary colorectal cancer tumors and related metastases: A potential role in prediction of lung metastasis. *PLoS One*, *4*(12). e8199. http://dx.doi.org/10.1371/journal.pone.0008199.
Cerbini, T., Funahashi, R., Luo, Y., Liu, C., Park, K., Rao, M., et al. (2015). Transcription activator-like effector nuclease (TALEN)-mediated CLYBL targeting enables enhanced transgene expression and one-step generation of dual reporter human induced

pluripotent stem cell (iPSC) and neural stem cell (NSC) lines. *PLoS One*, *10*(1). e0116032. http://dx.doi.org/10.1371/journal.pone.0116032.

Choi, Y., Sims, G. E., Murphy, S., Miller, J. R., & Chan, A. P. (2012). Predicting the functional effect of amino acid substitutions and indels. *PLoS One*, 7(10). e46688. http://dx.doi.org/10.1371/journal.pone.0046688.

Ciriello, G., Cerami, E., Sander, C., & Schultz, N. (2012). Mutual exclusivity analysis identifies oncogenic network modules. *Genome Research*, *22*(2), 398–406. http://dx.doi.org/10.1101/gr.125567.111.

Clifford, R. J., Edmonson, M. N., Nguyen, C., & Buetow, K. H. (2004). Large-scale analysis of non-synonymous coding region single nucleotide polymorphisms. *Bioinformatics*, *20*(7), 1006–1014. http://dx.doi.org/10.1093/bioinformatics/bth029.

Collins, F. S., & Varmus, H. (2015). A new initiative on precision medicine. *The New England Journal of Medicine*, *372*(9), 793–795. http://dx.doi.org/10.1056/NEJMp1500523.

Cong, L., Ran, F. A., Cox, D., Lin, S., Barretto, R., Habib, N., et al. (2013). Multiplex genome engineering using CRISPR/Cas systems. *Science*, *339*(6121), 819–823. http://dx.doi.org/10.1126/science.1231143.

Cooper, J. W., & Kershenbaum, A. (2005). Discovery of protein-protein interactions using a combination of linguistic, statistical and graphical information. *BMC Bioinformatics*, *6*, 143. http://dx.doi.org/10.1186/1471-2105-6-143.

COSMIC (Catalogue of Somatic Mutations in Cancer). http://cancer.sanger.ac.uk/cosmic (Accessed on July 8, 2015).

Croft, D., Mundo, A. F., Haw, R., Milacic, M., Weiser, J., Wu, G., et al. (2014). The Reactome pathway knowledgebase. *Nucleic Acids Research*, *42*(Database issue), D472–D477. http://dx.doi.org/10.1093/nar/gkt1102.

Cusick, M. E., Yu, H., Smolyar, A., Venkatesan, K., Carvunis, A. R., Simonis, N., et al. (2009). Literature-curated protein interaction datasets. *Nature Methods*, *6*(1), 39–46. http://dx.doi.org/10.1038/nmeth.1284.

de Bruin, E. C., McGranahan, N., Mitter, R., Salm, M., Wedge, D. C., Yates, L., et al. (2014). Spatial and temporal diversity in genomic instability processes defines lung cancer evolution. *Science*, *346*(6206), 251–256. http://dx.doi.org/10.1126/science.1253462.

Dees, N. D., Zhang, Q., Kandoth, C., Wendl, M. C., Schierding, W., Koboldt, D. C., et al. (2012). MuSiC: Identifying mutational significance in cancer genomes. *Genome Research*, *22*(8), 1589–1598. http://dx.doi.org/10.1101/gr.134635.111.

Desmedt, C., Fumagalli, D., Pietri, E., Zoppoli, G., Brown, D., Nik-Zainal, S., et al. (2015). Uncovering the genomic heterogeneity of multifocal breast cancer. *The Journal of Pathology*, *236*(4), 457–466. http://dx.doi.org/10.1002/path.4540.

Deveau, H., Garneau, J. E., & Moineau, S. (2010). CRISPR/Cas system and its role in phage-bacteria interactions. *Annual Review of Microbiology*, *64*, 475–493. http://dx.doi.org/10.1146/annurev.micro.112408.134123.

Donaldson, I., Martin, J., de Bruijn, B., Wolting, C., Lay, V., Tuekam, B., et al. (2003). PreBIND and Textomy—Mining the biomedical literature for protein-protein interactions using a support vector machine. *BMC Bioinformatics*, *4*, 11.

Drugs@FDA. http://www.accessdata.fda.gov/scripts/cder/drugsatfda/index.cfm (Accessed July 2015).

Edkins, S., O'Meara, S., Parker, A., Stevens, C., Reis, M., Jones, S., et al. (2006). Recurrent KRAS codon 146 mutations in human colorectal cancer. *Cancer Biology & Therapy*, *5*(8), 928–932. 3251 [pii].

Evaluation of Genomic Applications in Practice and Prevention (EGAPP) Working Group, Calonge, N., Fisher, N. L., Berg, A. O., Campos-Outcalt, D., Djulbegovic, B., et al. (2013). Recommendations from the EGAPP Working Group: Can testing of tumor tissue for mutations in EGFR pathway downstream effector genes in patients with metastatic colorectal cancer improve health outcomes by guiding decisions regarding anti-EGFR therapy? *Genetics in Medicine*, *15*(7), 517–527. http://dx.doi.org/10.1038/gim.2012.184.

Ferguson, B. D., Carol Tan, Y. H., Kanteti, R. S., Liu, R., Gayed, M. J., Vokes, E. E., et al. (2015). Novel EPHB4 receptor tyrosine kinase mutations and kinomic pathway analysis in lung cancer. *Scientific Reports*, *5*, 10641. http://dx.doi.org/10.1038/srep10641.

Fleuren, W. W., & Alkema, W. (2015). Application of text mining in the biomedical domain. *Methods*, *74*, 97–106. http://dx.doi.org/10.1016/j.ymeth.2015.01.015.

Fukuda, K., Tamura, A., Tsunoda, T., & Takagi, T. (1998). Toward information extraction: Identifying protein names from biological papers. *Pacific Symposium on Biocomputing*, 707–718.

Garraway, L. A., & Lander, E. S. (2013). Lessons from the cancer genome. *Cell*, *153*(1), 17–37. http://dx.doi.org/10.1016/j.cell.2013.03.002.

Geifman, N., Haviv, I., Kurzrock, R., & Rubin, E. (2014). Promoting precision cancer medicine through a community-driven knowledgebase. *Journal of Personalized Medicine*, *4*(4), 475–488. http://dx.doi.org/10.3390/jpm4040475.

Giuliani, J., Martelli, S., Remo, A., & Bonetti, A. (2015). Primary TKI resistance in advanced non-small cell lung cancer with EGFR mutation: An open question. *Tumori*, *101*(4), e115–e117. http://dx.doi.org/10.5301/tj.5000317.

Gnad, F., Baucom, A., Mukhyala, K., Manning, G., & Zhang, Z. (2013). Assessment of computational methods for predicting the effects of missense mutations in human cancers. *BMC Genomics*, *14*(Suppl. 3), S7. http://dx.doi.org/10.1186/1471-2164-14-S3-S7.

Gonzalez-Perez, A., & Lopez-Bigas, N. (2011). Improving the assessment of the outcome of nonsynonymous SNVs with a consensus deleteriousness score, Condel. *American Journal of Human Genetics*, *88*(4), 440–449. http://dx.doi.org/10.1016/j.ajhg.2011.03.004.

Gotlib, J., Cools, J., Malone, J. M., 3rd., Schrier, S. L., Gilliland, D. G., & Coutre, S. E. (2004). The FIP1L1-PDGFRalpha fusion tyrosine kinase in hypereosinophilic syndrome and chronic eosinophilic leukemia: Implications for diagnosis, classification, and management. *Blood*, *103*(8), 2879–2891. http://dx.doi.org/10.1182/blood-2003-06-1824.

Gumusay, O., Benekli, M., Ekinci, O., Baykara, M., Ozet, A., Coskun, U., et al. (2015). Discordances in HER2 status between primary gastric cancer and corresponding metastatic sites. *Japanese Journal of Clinical Oncology*, *45*(5), 416–421. http://dx.doi.org/10.1093/jjco/hyv020.

Hanahan, D., & Weinberg, R. A. (2011). Hallmarks of cancer: The next generation. *Cell*, *144*(5), 646–674. http://dx.doi.org/10.1016/j.cell.2011.02.013.

Hart, G. T., Ramani, A. K., & Marcotte, E. M. (2006). How complete are current yeast and human protein-interaction networks? *Genome Biology*, 7(11), 120. http://dx.doi.org/10.1186/gb-2006-7-11-120.

Hatzivassiloglou, V., Duboue, P. A., & Rzhetsky, A. (2001). Disambiguating proteins, genes, and RNA in text: A machine learning approach. *Bioinformatics*, *17*(Suppl. 1), S97–S106.

Heinrich, M. C., Maki, R. G., Corless, C. L., Antonescu, C. R., Harlow, A., Griffith, D., et al. (2008). Primary and secondary kinase genotypes correlate with the biological and clinical activity of sunitinib in imatinib-resistant gastrointestinal stromal tumor. *Journal of Clinical Oncology*, *26*(33), 5352–5359. http://dx.doi.org/10.1200/JCO.2007.15.7461.

Hodis, E., Watson, I. R., Kryukov, G. V., Arold, S. T., Imielinski, M., Theurillat, J. P., et al. (2012). A landscape of driver mutations in melanoma. *Cell*, *150*(2), 251–263. http://dx.doi.org/10.1016/j.cell.2012.06.024.

Horvath, P., & Barrangou, R. (2010). CRISPR/Cas, the immune system of bacteria and archaea. *Science*, *327*(5962), 167–170. http://dx.doi.org/10.1126/science.1179555.

Hu, Z., Ding, W., Zhu, D., Yu, L., Jiang, X., Wang, X., et al. (2015). TALEN-mediated targeting of HPV oncogenes ameliorates HPV-related cervical malignancy. *The Journal of Clinical Investigation*, *125*(1), 425–436. http://dx.doi.org/10.1172/JCI78206.

Hua, X., Xu, H., Yang, Y., Zhu, J., Liu, P., & Lu, Y. (2013). DrGaP: A powerful tool for identifying driver genes and pathways in cancer sequencing studies. *American Journal of Human Genetics*, *93*(3), 439–451. http://dx.doi.org/10.1016/j.ajhg.2013.07.003.

Huhn, S., Bevier, M., Pardini, B., Naccarati, A., Vodickova, L., Novotny, J., et al. (2014). Colorectal cancer risk and patients' survival: Influence of polymorphisms in genes somatically mutated in colorectal tumors. *Cancer Causes & Control, 25*(6), 759–769. http://dx.doi.org/10.1007/s10552-014-0379-1.

Hur, J., Guo, A. Y., Loh, W. Y., Feldman, E. L., & Bai, J. P. (2014). Integrated systems pharmacology analysis of clinical drug-induced peripheral neuropathy. *CPT: Pharmacometrics & Systems Pharmacology. 3*. http://dx.doi.org/10.1038/psp.2014.11. e114.

Hur, J., Schuyler, A. D., States, D. J., & Feldman, E. L. (2009). SciMiner: Web-based literature mining tool for target identification and functional enrichment analysis. *Bioinformatics, 25*(6), 838–840. http://dx.doi.org/10.1093/bioinformatics/btp049.

Hur, J., Sullivan, K. A., Pande, M., Hong, Y., Sima, A. A., Jagadish, H. V., et al. (2011). The identification of gene expression profiles associated with progression of human diabetic neuropathy. *Brain, 134*(Pt 11), 3222–3235. http://dx.doi.org/10.1093/brain/awr228.

Hur, J., Zhao, C., & Bai, J. P. (2015). Systems pharmacological analysis of drugs inducing Stevens-Johnson syndrome and toxic epidermal necrolysis. *Chemical Research in Toxicology, 28*(5), 927–934. http://dx.doi.org/10.1021/tx5005248.

Ieni, A., Barresi, V., Caltabiano, R., Caleo, A., Bonetti, L. R., Lanzafame, S., et al. (2014). Discordance rate of HER2 status in primary gastric carcinomas and synchronous lymph node metastases: A multicenter retrospective analysis. *International Journal of Molecular Sciences, 15*(12), 22331–22341. http://dx.doi.org/10.3390/ijms151222331.

Iliakis, G., Wang, H., Perrault, A. R., Boecker, W., Rosidi, B., Windhofer, F., et al. (2004). Mechanisms of DNA double strand break repair and chromosome aberration formation. *Cytogenetic and Genome Research, 104*(1–4), 14–20. http://dx.doi.org/10.1159/000077461.

Jensen, L. J., Saric, J., & Bork, P. (2006). Literature mining for the biologist: From information retrieval to biological discovery. *Nature Reviews. Genetics*, 7(2), 119–129. http://dx.doi.org/10.1038/nrg1768.

Jenssen, T. K., Laegreid, A., Komorowski, J., & Hovig, E. (2001). A literature network of human genes for high-throughput analysis of gene expression. *Nature Genetics, 28*(1), 21–28. http://dx.doi.org/10.1038/88213.

Johnson, R. D., & Jasin, M. (2001). Double-strand-break-induced homologous recombination in mammalian cells. *Biochemical Society Transactions, 29*(Pt 2), 196–201.

Joung, J. K., & Sander, J. D. (2013). TALENs: A widely applicable technology for targeted genome editing. *Nature Reviews. Molecular Cell Biology, 14*(1), 49–55. http://dx.doi.org/10.1038/nrm3486.

Kaminker, J. S., Zhang, Y., Watanabe, C., & Zhang, Z. (2007a). CanPredict: A computational tool for predicting cancer-associated missense mutations. *Nucleic Acids Research, 35*(Web Server issue), W595–W598. http://dx.doi.org/10.1093/nar/gkm405.

Kaminker, J. S., Zhang, Y., Waugh, A., Haverty, P. M., Peters, B., Sebisanovic, D., et al. (2007b). Distinguishing cancer-associated missense mutations from common polymorphisms. *Cancer Research, 67*(2), 465–473. http://dx.doi.org/10.1158/0008-5472.CAN-06-1736.

Kandoth, C., McLellan, M. D., Vandin, F., Ye, K., Niu, B., Lu, C., et al. (2013). Mutational landscape and significance across 12 major cancer types. *Nature, 502*(7471), 333–339. http://dx.doi.org/10.1038/nature12634.

Kannan, K., Inagaki, A., Silber, J., Gorovets, D., Zhang, J., Kastenhuber, E. R., et al. (2012). Whole-exome sequencing identifies ATRX mutation as a key molecular determinant in lower-grade glioma. *Oncotarget, 3*(10), 1194–1203. 689 [pii].

Karaman, M. W., Herrgard, S., Treiber, D. K., Gallant, P., Atteridge, C. E., Campbell, B. T., et al. (2008). A quantitative analysis of kinase inhibitor selectivity. *Nature Biotechnology, 26*(1), 127–132. http://dx.doi.org/10.1038/nbt1358.

Kazama, Jun'ichi, Makino, Takaki, Ohta, Yoshihiro, & Tsujii, Jun'ichi (2002). Tuning support vector machines for biomedical named entity recognition. In *Paper presented at the*

Proceedings of the ACL-02 workshop on Natural language processing in the biomedical domain—Volume 3, Pennsylvania, Philadelphia.

Kim, Y. G., Cha, J., & Chandrasegaran, S. (1996). Hybrid restriction enzymes: Zinc finger fusions to Fok I cleavage domain. *Proceedings of the National Academy of Sciences of the United States of America, 93*(3), 1156–1160.

Kirouac, D. C., Saez-Rodriguez, J., Swantek, J., Burke, J. M., Lauffenburger, D. A., & Sorger, P. K. (2012). Creating and analyzing pathway and protein interaction compendia for modelling signal transduction networks. *BMC Systems Biology, 6*, 29. http://dx.doi.org/10.1186/1752-0509-6-29.

Krauthammer, M., Rzhetsky, A., Morozov, P., & Friedman, C. (2000). Using BLAST for identifying gene and protein names in journal articles. *Gene, 259*(1–2), 245–252.

Kroeger, N., Choueiri, T. K., Lee, J. L., Bjarnason, G. A., Knox, J. J., MacKenzie, M. J., et al. (2014). Survival outcome and treatment response of patients with late relapse from renal cell carcinoma in the era of targeted therapy. *European Urology, 65*(6), 1086–1092. http://dx.doi.org/10.1016/j.eururo.2013.07.031.

Kumar, P., Henikoff, S., & Ng, P. C. (2009). Predicting the effects of coding non-synonymous variants on protein function using the SIFT algorithm. *Nature Protocols, 4*(7), 1073–1081. http://dx.doi.org/10.1038/nprot.2009.86.

Law, V., Knox, C., Djoumbou, Y., Jewison, T., Guo, A. C., Liu, Y., et al. (2014). DrugBank 4.0: Shedding new light on drug metabolism. *Nucleic Acids Research, 42*(Database issue), D1091–D1097. http://dx.doi.org/10.1093/nar/gkt1068.

Lawrence, M. S., Stojanov, P., Mermel, C. H., Robinson, J. T., Garraway, L. A., Golub, T. R., et al. (2014). Discovery and saturation analysis of cancer genes across 21 tumour types. *Nature, 505*(7484), 495–501. http://dx.doi.org/10.1038/nature12912.

Lawrence, M. S., Stojanov, P., Polak, P., Kryukov, G. V., Cibulskis, K., Sivachenko, A., et al. (2013). Mutational heterogeneity in cancer and the search for new cancer-associated genes. *Nature, 499*(7457), 214–218. http://dx.doi.org/10.1038/nature12213.

Lee, H. J., Dang, T. C., Lee, H., & Park, J. C. (2014). OncoSearch: Cancer gene search engine with literature evidence. *Nucleic Acids Research, 42*(Web Server issue), W416–W421. http://dx.doi.org/10.1093/nar/gku368.

Leiserson, M. D., Blokh, D., Sharan, R., & Raphael, B. J. (2013). Simultaneous identification of multiple driver pathways in cancer. *PLoS Computational Biology, 9*(5). e1003054http://dx.doi.org/10.1371/journal.pcbi.1003054.

Leiserson, M. D., Vandin, F., Wu, H. T., Dobson, J. R., Eldridge, J. V., Thomas, J. L., et al. (2015). Pan-cancer network analysis identifies combinations of rare somatic mutations across pathways and protein complexes. *Nature Genetics, 47*(2), 106–114. http://dx.doi.org/10.1038/ng.3168.

Lemmon, M. A., & Schlessinger, J. (2010). Cell signaling by receptor tyrosine kinases. *Cell, 141*(7), 1117–1134. http://dx.doi.org/10.1016/j.cell.2010.06.011.

Li, J., Bi, L., Sun, Y., Lu, Z., Lin, Y., Bai, O., et al. (2014). Text mining and network analysis of molecular interaction in non-small cell lung cancer by using natural language processing. *Molecular Biology Reports, 41*(12), 8071–8079. http://dx.doi.org/10.1007/s11033-014-3705-5.

Li, Y., Hu, X., Lin, H., & Yang, Z. (2010). Learning an enriched representation from unlabeled data for protein-protein interaction extraction. *BMC Bioinformatics, 11*(Suppl. 2), S7. http://dx.doi.org/10.1186/1471-2105-11-S2-S7.

Li, L., Hu, X., Yang, Z., Jia, Z., Fang, M., Zhang, L., et al. (2014). Establishing reliable miRNA-cancer association network based on text-mining method. *Computational and Mathematical Methods in Medicine, 2014*, 746979. http://dx.doi.org/10.1155/2014/746979.

Liang, P., Xu, Y., Zhang, X., Ding, C., Huang, R., Zhang, Z., et al. (2015). CRISPR/Cas9-mediated gene editing in human tripronuclear zygotes. *Protein & Cell, 6*(5), 363–372. http://dx.doi.org/10.1007/s13238-015-0153-5.

Lierman, E., Folens, C., Stover, E. H., Mentens, N., Van Miegroet, H., Scheers, W., et al. (2006). Sorafenib is a potent inhibitor of FIP1L1-PDGFRalpha and the imatinib-resistant FIP1L1-PDGFRalpha T674I mutant. *Blood*, *108*(4), 1374–1376. http://dx.doi.org/10.1182/blood-2006-02-004457.

Lierman, E., Michaux, L., Beullens, E., Pierre, P., Marynen, P., Cools, J., et al. (2009). FIP1L1-PDGFRalpha D842V, a novel panresistant mutant, emerging after treatment of FIP1L1-PDGFRalpha T674I eosinophilic leukemia with single agent sorafenib. *Leukemia*, *23*(5), 845–851. http://dx.doi.org/10.1038/leu.2009.2.

Lim, K. H., Huang, M. J., Chen, L. T., Wang, T. E., Liu, C. L., Chang, C. S., et al. (2008). Molecular analysis of secondary kinase mutations in imatinib-resistant gastrointestinal stromal tumors. *Medical Oncology*, *25*(2), 207–213. http://dx.doi.org/10.1007/s12032-007-9014-2.

Lim, C. K., Hwang, W. Y., Aw, S. E., & Sun, L. (2008). Study of gene expression profile during cord blood-associated megakaryopoiesis. *European Journal of Haematology*, *81*(3), 196–208. http://dx.doi.org/10.1111/j.1600-0609.2008.01104.x.

Lovering, R. C., Dimmer, E. C., & Talmud, P. J. (2009). Improvements to cardiovascular gene ontology. *Atherosclerosis*, *205*(1), 9–14. http://dx.doi.org/10.1016/j.atherosclerosis.2008.10.014.

Luo, Y., Riedlinger, G., & Szolovits, P. (2014). Text mining in cancer gene and pathway prioritization. *Cancer Informatics*, *13*(Suppl. 1), 69–79. http://dx.doi.org/10.4137/CIN.S13874.

McFarland, C. D., Korolev, K. S., Kryukov, G. V., Sunyaev, S. R., & Mirny, L. A. (2013). Impact of deleterious passenger mutations on cancer progression. *Proceedings of the National Academy of Sciences of the United States of America*, *110*(8), 2910–2915. http://dx.doi.org/10.1073/pnas.1213968110.

Mermel, C. H., Schumacher, S. E., Hill, B., Meyerson, M. L., Beroukhim, R., & Getz, G. (2011). GISTIC2.0 facilitates sensitive and confident localization of the targets of focal somatic copy-number alteration in human cancers. *Genome Biology*, *12*(4), R41. http://dx.doi.org/10.1186/gb-2011-12-4-r41.

Miller, J. C., Tan, S., Qiao, G., Barlow, K. A., Wang, J., Xia, D. F., et al. (2011). A TALE nuclease architecture for efficient genome editing. *Nature Biotechnology*, *29*(2), 143–148. http://dx.doi.org/10.1038/nbt.1755.

Mladenov, E., & Iliakis, G. (2011). Induction and repair of DNA double strand breaks: The increasing spectrum of non-homologous end joining pathways. *Mutation Research*, *711*(1–2), 61–72. http://dx.doi.org/10.1016/j.mrfmmm.2011.02.005.

Moscou, M. J., & Bogdanove, A. J. (2009). A simple cipher governs DNA recognition by TAL effectors. *Science*, *326*(5959), 1501. http://dx.doi.org/10.1126/science.1178817.

National Center for Biotechnology Information (2010). *Entrez Genome*. from, http://www.ncbi.nlm.nih.gov/sites/genome.

Ng, S., Collisson, E. A., Sokolov, A., Goldstein, T., Gonzalez-Perez, A., Lopez-Bigas, N., et al. (2012). PARADIGM-SHIFT predicts the function of mutations in multiple cancers using pathway impact analysis. *Bioinformatics*, *28*(18), i640–i646. http://dx.doi.org/10.1093/bioinformatics/bts402.

Ng, S. K., & Wong, M. (1999). Toward routine automatic pathway discovery from on-line scientific text abstracts. *Genome Informatics. Workshop Genome Informatics*, *10*, 104–112.

Nobata, C., Collier, N., & Tsujii, J. (1999). Automatic term identification and classification in biology texts. In *Proceedings of the Natural Language Pacific Rim Symposium*.

Ono, T., Hishigaki, H., Tanigami, A., & Takagi, T. (2001). Automated extraction of information on protein-protein interactions from the biological literature. *Bioinformatics*, *17*(2), 155–161.

Paull, E. O., Carlin, D. E., Niepel, M., Sorger, P. K., Haussler, D., & Stuart, J. M. (2013). Discovering causal pathways linking genomic events to transcriptional states using Tied

Diffusion Through Interacting Events (TieDIE). *Bioinformatics*, *29*(21), 2757–2764. http://dx.doi.org/10.1093/bioinformatics/btt471.

Pereira, A. A., Rego, J. F., Morris, V., Overman, M. J., Eng, C., Garrett, C. R., et al. (2015). Association between KRAS mutation and lung metastasis in advanced colorectal cancer. *British Journal of Cancer*, *112*(3), 424–428. http://dx.doi.org/10.1038/bjc.2014.619.

Pierce, A. J., Johnson, R. D., Thompson, L. H., & Jasin, M. (1999). XRCC3 promotes homology-directed repair of DNA damage in mammalian cells. *Genes & Development*, *13*(20), 2633–2638.

Raman, K. (2010). Construction and analysis of protein-protein interaction networks. *Automated Experimentation*, *2*(1), 2. http://dx.doi.org/10.1186/1759-4499-2-2.

Ramani, A. K., Bunescu, R. C., Mooney, R. J., & Marcotte, E. M. (2005). Consolidating the set of known human protein-protein interactions in preparation for large-scale mapping of the human interactome. *Genome Biology*, *6*(5), R40. http://dx.doi.org/10.1186/gb-2005-6-5-r40.

Reddy, P., Ocampo, A., Suzuki, K., Luo, J., Bacman, S. R., Williams, S. L., et al. (2015). Selective elimination of mitochondrial mutations in the germline by genome editing. *Cell*, *161*(3), 459–469. http://dx.doi.org/10.1016/j.cell.2015.03.051.

Reid, D. A., & Rothenberg, E. (2015). Repair of chromosomal breaks by NHEJ. *Oncotarget*, *6*(18), 15730–15731. 4593 [pii].

Reva, B., Antipin, Y., & Sander, C. (2011). Predicting the functional impact of protein mutations: Application to cancer genomics. *Nucleic Acids Research*. *39*(17). e118. http://dx.doi.org/10.1093/nar/gkr407.

Reyon, D., Tsai, S. Q., Khayter, C., Foden, J. A., Sander, J. D., & Joung, J. K. (2012). FLASH assembly of TALENs for high-throughput genome editing. *Nature Biotechnology*, *30*(5), 460–465. http://dx.doi.org/10.1038/nbt.2170.

Rindflesch, T. C., Hunter, L., & Aronson, A. R. (1999). Mining molecular binding terminology from biomedical text. *Proceedings of the AMIA Symposium*, *127–131*. D005564 [pii].

Roth, D. B., Porter, T. N., & Wilson, J. H. (1985). Mechanisms of nonhomologous recombination in mammalian cells. *Molecular and Cellular Biology*, *5*(10), 2599–2607.

Ryslik, G. A., Cheng, Y., Cheung, K. H., Modis, Y., & Zhao, H. (2013). Utilizing protein structure to identify non-random somatic mutations. *BMC Bioinformatics*, *14*, 190. http://dx.doi.org/10.1186/1471-2105-14-190.

Sadovnik, I., Lierman, E., Peter, B., Herrmann, H., Suppan, V., Stefanzl, G., et al. (2014). Identification of Ponatinib as a potent inhibitor of growth, migration, and activation of neoplastic eosinophils carrying FIP1L1-PDGFRA. *Experimental Hematology*, *42*(4), 282–293. http://dx.doi.org/10.1016/j.exphem.2013.12.007. e284.

Salemi, S., Yousefi, S., Simon, D., Schmid, I., Moretti, L., Scapozza, L., et al. (2009). A novel FIP1L1-PDGFRA mutant destabilizing the inactive conformation of the kinase domain in chronic eosinophilic leukemia/hypereosinophilic syndrome. *Allergy*, *64*(6), 913–918. http://dx.doi.org/10.1111/j.1398-9995.2009.01943.x.

Sander, J. D., & Joung, J. K. (2014). CRISPR-Cas systems for editing, regulating and targeting genomes. *Nature Biotechnology*, *32*(4), 347–355. http://dx.doi.org/10.1038/nbt.2842.

Sato, T., Sakuma, T., Yokonishi, T., Katagiri, K., Kamimura, S., Ogonuki, N., et al. (2015). Genome editing in mouse spermatogonial stem cell lines using TALEN and double-nicking CRISPR/Cas9. *Stem Cell Reports*, *5*(1), 75–82. http://dx.doi.org/10.1016/j.stemcr.2015.05.011.

Sebastiano, V., Maeder, M. L., Angstman, J. F., Haddad, B., Khayter, C., Yeo, D. T., et al. (2011). In situ genetic correction of the sickle cell anemia mutation in human induced pluripotent stem cells using engineered zinc finger nucleases. *Stem Cells*, *29*(11), 1717–1726. http://dx.doi.org/10.1002/stem.718.

Shatkay, H., Edwards, S., Wilbur, W. J., & Boguski, M. (2000). Genes, themes and microarrays: Using information retrieval for large-scale gene analysis. *Procedings of the International Conference on Intelligent Systems for Molecular Biology*, *8*, 317–328.

Stephens, P. J., Tarpey, P. S., Davies, H., Van Loo, P., Greenman, C., Wedge, D. C., et al. (2012). The landscape of cancer genes and mutational processes in breast cancer. *Nature*, *486*(7403), 400–404. http://dx.doi.org/10.1038/nature11017.

Stoimenov, I., Ali, M. A., Pandzic, T., & Sjoblom, T. (2015). Computational and molecular tools for scalable rAAV-mediated genome editing. *Nucleic Acids Research*, *43*(5). e30http://dx.doi.org/10.1093/nar/gku1286.

Stone, E. A., & Sidow, A. (2005). Physicochemical constraint violation by missense substitutions mediates impairment of protein function and disease severity. *Genome Research*, *15*(7), 978–986. http://dx.doi.org/10.1101/gr.3804205.

Stratton, M. R., Campbell, P. J., & Futreal, P. A. (2009). The cancer genome. *Nature*, *458*(7239), 719–724. http://dx.doi.org/10.1038/nature07943.

Stumpf, M. P., Thorne, T., de Silva, E., Stewart, R., An, H. J., Lappe, M., et al. (2008). Estimating the size of the human interactome. *Proceedings of the National Academy of Sciences of the United States of America*, *105*(19), 6959–6964. http://dx.doi.org/10.1073/pnas.0708078105.

Subramanian, A., Tamayo, P., Mootha, V. K., Mukherjee, S., Ebert, B. L., Gillette, M. A., et al. (2005). Gene set enrichment analysis: A knowledge-based approach for interpreting genome-wide expression profiles. *Proceedings of the National Academy of Sciences of the United States of America*, *102*(43), 15545–15550. http://dx.doi.org/10.1073/pnas.0506580102.

Swiech, L., Heidenreich, M., Banerjee, A., Habib, N., Li, Y., Trombetta, J., et al. (2015). In vivo interrogation of gene function in the mammalian brain using CRISPR-Cas9. *Nature Biotechnology*, *33*(1), 102–106. http://dx.doi.org/10.1038/nbt.3055.

Takahashi, G., Gurumurthy, C. B., Wada, K., Miura, H., Sato, M., & Ohtsuka, M. (2015). GONAD: Genome-editing via Oviductal Nucleic Acids Delivery system: A novel microinjection independent genome engineering method in mice. *Scientific Reports*, *5*, 11406. http://dx.doi.org/10.1038/srep11406.

Takata, M., Sasaki, M. S., Sonoda, E., Morrison, C., Hashimoto, M., Utsumi, H., et al. (1998). Homologous recombination and non-homologous end-joining pathways of DNA double-strand break repair have overlapping roles in the maintenance of chromosomal integrity in vertebrate cells. *The EMBO Journal*, *17*(18), 5497–5508. http://dx.doi.org/10.1093/emboj/17.18.5497.

Tamames, J., Ouzounis, C., Casari, G., Sander, C., & Valencia, A. (1998). EUCLID: Automatic classification of proteins in functional classes by their database annotations. *Bioinformatics*, *14*(6), 542–543.

Tanabe, L., & Wilbur, W. J. (2002). Tagging gene and protein names in biomedical text. *Bioinformatics*, *18*(8), 1124–1132.

The Cancer Genome Atlas Network, Akbani, R., Akdemir, K. C., Aksoy, B. A., Albert, M., Ally, A., et al. (2015). Genomic classification of cutaneous melanoma. *Cell*, *16*(7), 1681–1696. http://dx.doi.org/10.1016/j.cell.2015.05.044.

The Cancer Genome Atlas Research Network, Bell, D., Berchuck, A., Birrer, M., Chien, J., Cramer, D., et al. (2011). Integrated genomic analyses of ovarian carcinoma. *Nature*, *474*(7353), 609–615. http://dx.doi.org/10.1038/nature10166.

The Cancer Genome Atlas Research Network, Ley, T. J., Miller, C., Ding, L., Raphael, B. J., Mungall, A. J., et al. (2013). Genomic and epigenomic landscapes of adult de novo acute myeloid leukemia. *The New England Journal of Medicine*, *368*(22), 2059–2074. http://dx.doi.org/10.1056/NEJMoa1301689.

Vandin, F., Upfal, E., & Raphael, B. J. (2011). Algorithms for detecting significantly mutated pathways in cancer. *Journal of Computational Biology*, *18*(3), 507–522. http://dx.doi.org/10.1089/cmb.2010.0265.

Vandin, F., Upfal, E., & Raphael, B. J. (2012). De novo discovery of mutated driver pathways in cancer. *Genome Research*, *22*(2), 375–385. http://dx.doi.org/10.1101/gr.120477.111.

Vock, C., Gleissner, M., Klapper, M., & Doring, F. (2008). Oleate regulates genes controlled by signaling pathways of mitogen-activated protein kinase, insulin, and hypoxia. *Nutrition Research (New York)*, *28*(10), 681–689. http://dx.doi.org/10.1016/j.nutres.2008.06.010.

Waters, C. A., Strande, N. T., Pryor, J. M., Strom, C. N., Mieczkowski, P., Burkhalter, M. D., et al. (2014). The fidelity of the ligation step determines how ends are resolved during nonhomologous end joining. *Nature Communications*, *5*, 4286. http://dx.doi.org/10.1038/ncomms5286.

Watson, I. R., Takahashi, K., Futreal, P. A., & Chin, L. (2013). Emerging patterns of somatic mutations in cancer. *Nature Reviews. Genetics*, *14*(10), 703–718. http://dx.doi.org/10.1038/nrg3539.

Wellcome Trust Sanger Institute, Genome Research Limited. (Accessed June 2015). Catalogue of Somatic Mutations in Cancer http://cancer.sanger.ac.uk/cosmic.

Yang, Z., Lin, H., & Li, Y. (2010). BioPPISVMExtractor: A protein-protein interaction extractor for biomedical literature using SVM and rich feature sets. *Journal of Biomedical Informatics*, *43*(1), 88–96. http://dx.doi.org/10.1016/j.jbi.2009.08.013.

Yates, L. R., Gerstung, M., Knappskog, S., Desmedt, C., Gundem, G., Van Loo, P., et al. (2015). Subclonal diversification of primary breast cancer revealed by multiregion sequencing. *Nature Medicine*, *21*(7), 751–759. http://dx.doi.org/10.1038/nm.3886.

Yeang, C. H., McCormick, F., & Levine, A. (2008). Combinatorial patterns of somatic gene mutations in cancer. *The FASEB Journal*, *22*(8), 2605–2622. http://dx.doi.org/10.1096/fj.08-108985.

Zhou, G., Zhang, J., Su, J., Shen, D., & Tan, C. (2004). Recognizing names in biomedical texts: A machine learning approach. *Bioinformatics*, *20*, 1190.

Zhu, F., Patumcharoenpol, P., Zhang, C., Yang, Y., Chan, J., Meechai, A., et al. (2013). Biomedical text mining and its applications in cancer research. *Journal of Biomedical Informatics*, *46*(2), 200–211. http://dx.doi.org/10.1016/j.jbi.2012.10.007.

CHAPTER SEVEN

Molecular Dynamics: New Frontier in Personalized Medicine

P. Sneha, C. George Priya Doss[1]
Medical Biotechnology Division, School of Biosciences and Technology, VIT University, Vellore, Tamil Nadu, India
[1]Corresponding author: e-mail addresses: georgecp77@yahoo.co.in; georgepriyadoss@vit.ac.in

Contents

Advances in Protein Chemistry and Structural Biology, Volume 102
ISSN 1876-1623
http://dx.doi.org/10.1016/bs.apcsb.2015.09.004

Abstract

The field of drug discovery has witnessed infinite development over the last decade with the demand for discovery of novel efficient lead compounds. Although the development of novel compounds in this field has seen large failure, a breakthrough in this area might be the establishment of personalized medicine. The trend of personalized medicine has shown stupendous growth being a hot topic after the successful completion of Human Genome Project and 1000 genomes pilot project. Genomic variant such as SNPs play a vital role with respect to inter individual's disease susceptibility and drug response. Hence, identification of such genetic variants has to be performed before administration of a drug. This process requires high-end techniques to understand the complexity of the molecules which might bring an insight to understand the compounds at their molecular level. To sustenance this, field of bioinformatics plays a crucial role in revealing the molecular mechanism of the mutation and thereby designing a drug for an individual in fast and affordable manner. High-end computational methods, such as molecular dynamics (MD) simulation has proved to be a constitutive approach to detecting the minor changes associated with an SNP for better understanding of the structural and functional relationship. The parameters used in molecular dynamic simulation elucidate different properties of a macromolecule, such as protein stability and flexibility. MD along with docking analysis can reveal the synergetic effect of an SNP in protein–ligand interaction and provides a foundation for designing a particular drug molecule for an individual. This compelling application of computational power and the advent of other technologies have paved a promising way toward personalized medicine. In this in-depth review, we tried to highlight the different wings of MD toward personalized medicine.

1. INTRODUCTION

Traditionally drug discovery process involves discovery of new candidate drugs with the aid of pharmacological and chemistry-based drug discovery approaches (Katara, 2013). Although significant steps have taken place in the field of drug development to meet the market's demand, still constant failure have been observed. This failure might be due to the continuous variations seen in the target protein. Drug designing involves genes that can provide help in the metabolism of a drug (Meyer, 2000). A change at the genetic level can lead to toxicity of a drug (Okazaki, Javle, Tanaka, Abbruzzese, & Li, 2010), which may directly influence the drug's action toward an unfavorable condition. SNPs (single nucleotide polymorphisms) are also known to influence this process in which, a nucleotide base pair change might lead to a change in the amino acid sequence of a protein. SNPs are often confused with point mutations when taken at DNA level. Both point mutation and SNPs are same that produce a "variation" in the genetic code that may or may not have a phenotypic change. The difference lies in the frequency of the variation observed in a population. Based on the position of these SNPs, the intensity of a disease condition or drug interaction varies and also known to be associated with destabilizing the target itself (Kimura-Kataoka et al., 2012). SNPs present in drug metabolizing enzymes, drug transporters, and receptors affect ADME (absorption, distribution, metabolism, and excretion) of a drug leading to interindividual drug response. Hence, drug designing for these special conditions might be a possible way of treating the disorders that can lead to advanced personalized drug development (Ginsburg & Mccarthy, 2001). In this regard, two branches have emerged to understand the drug response; pharmacogenetics and pharmacogenomics. Pharmacogenetics investigates the correlation between a drug's response and genetic differences. Whereas pharmacogenomics uses a genome-wide approach to dissecting the entire range of genes involved in drug response (Katara, 2013; Shastry, 2006). Identification of the genetic variations (biomarkers) is another scope to develop this field that could predict the patient's ability to respond or not to a particular drug.

The need for the development of unique lead molecule from thousands of compounds that are isolated might double with an increase in population and disease complications (Harvey, 2008). This process is known to consume almost 15 years to deliver a novel compound to the market after passing a serious *in vitro* and *in vivo* experiment to ensure with no adverse effects. The experimental analysis for lead compound becomes tedious, time consuming, and highly impossible as it includes ethical rights. The compounds may be diverse from principal foundation such as plant source, animal source, microbial source, or any synthetically derived compounds. This process becomes tedious, because of its low flexibility, lower availability of assessment procedures, environmental factors, poor drug efficacy, bioavailability, or instability in the target molecule. However, the evaluation of toxicity, binding affinity, sustainability, interactions between the target, and lead molecule becomes comparatively challenging by experimental methods. In recent years, two major branches, bioinformatics and pharmacogenomics have created a positive impact in drug discovery process.

The advent of new sequencing technologies and completion of Human Genome Project and 1000 genome project resulted in the identification and deposition of millions of SNPs in public repositories like NCBI. Major interest in medical genomics is to identify, characterize, and prioritize the functional mutations that bring about a change in protein function by altering the protein stability, dynamics, and flexibility. Knowledge over this paved the way for the development of numerous *in silico* methods to distinguish the deleterious from neutral SNPs in a pool. These predictions are made based on sequence and structure-based features such as biochemical properties of an amino acid, hydrogen bonding, and solvent accessibility. A few tools that provide insight over the impact of SNP are BALL-SNP (Mueller et al., 2015), SIFT (Ng & Henikoff, 2003), PolyPhen2 (Adzhubei, Jordan, & Sunyaev, 2013), HANSA (Acharya & Nagarajaram, 2012), PredictSNP (Bendl et al., 2014), SNPs&Go (Capriotti et al., 2013), and MutPred (Li et al., 2009). The recent progress made in the development of sophisticated algorithms for SNP analysis prompted the researchers to test the prediction power and provide a template for identification of disease-associated SNPs (Alanazi et al., 2011; De Alencar & Lopes, 2010; Doss & Rajith, 2012; George Priya Doss & Rajith, 2013; George Priya Doss et al., 2008; Gillard, Van Der Perren, Moguilevsky, Massingham, & Chatelain, 2002; Hussain et al., 2012; Magesh & George Priya Doss, 2012).

Also, mapping the mutation onto 3D structure allow us to understand the structure–function relationship in a detailed manner. Understanding the mechanism at atomic level measures the protein stability and dynamics and gives a better prediction at the structural level. Recent progress in structural genomic consortiums brought molecular docking and molecular dynamics (MD) simulation into the spotlight to reduce the overload of experimental techniques. This might further lead to the discovery of valid drug compounds in a short span of time (Lee, 2001; Lyne, 2002). Computational microscope MD analysis involves the use of force fields for atoms present in a macromolecule. This accurate prediction helps us understand its motion and elucidates the small difference occurred due to a variation (Levitt, Hirshberg, Sharon, & Daggett, 1995; Skopalik, Anzenbacher, & Otyepka, 2008). Therefore, a molecule's complete description is better understood when defined at an atomic level and becomes an inevitable method that may bypass the difficulties faced by experimental methods. MD also explains the significant change in the binding affinity of a macromolecule upon mutation (SNP) and drug response (Aleksandrov & Simonson, 2010; Doss, Chakraborty, Chen, & Zhu, 2014; Doss & NagaSundaram, 2012; Doss, Nagasundaram, Chakraborty, Chen, & Zhu, 2013; Doss, Chakraborty, et al., 2014; Doss, Rajith, et al., 2014; Kalaiarasan et al., 2015; Zhang & Zhong, 2010).

Exponential growth in the field of computational biology and experimental protein structure determinations with reduced extensive labor work have drastically increased the identification of novel drug (De la Iglesia, Cachau, García-Remesal, & Maojo, 2013; Hillisch, Pineda, & Hilgenfeld, 2004; Marrink & Tieleman, 2002; McInnes, 2007; Zhao & Iyengar, 2012). This process not only aims to save money and time by experimental analysis but also aids in determining the efficacy, safety, and stability of the drug toward the molecule. With the advancement of high-throughput screening, computational approaches have also grown along. Thus, computational approach became a prime important feature in efficiently analyzing and integrating all the knowledge available (Csermely, Korcsmaros, Kiss, London, & Nussinov, 2013; Sharma & Sarkar, 2013; Zhu et al., 2012). Various strategies have to be noted while designing a new drug: hydrogen bond interaction, fluctuations, conformational changes, environmental factors, and distribution of atoms in the molecule. Furthermore, the drug has to abide by the Lipinski rule of five (Lipinski, Lombardo, Dominy, & Feeney, 2012). The field of science that involves a combination of biological system and computational approach to study the drug and their target interaction effects being termed as "system pharmacology" (Berger & Iyengar, 2011).

With this availability of vast knowledge, drug discovery might ultimately lead to the development of personalized medications constructed upon the genotype of every individual. Although personalized medicine requires tedious diagnosis, this field has the potential for better treatment with reduced adverse drug response. In recent years, several researchers have illustrated the entry of personalized medicine in individual medical care (Derks & Diosdado, 2015; Hayes, Markus, Leslie, & Topol, 2014; Katsios & Roukos, 2010; Kream, 2015; Pokorska-Bocci et al., 2014; Servant et al., 2014; Wang, Russell, & Yan, 2014; Xie et al., 2014). A very crucial analysis is required to match each variant with drug response that is not feasible by experimental methods. Although several reports have continuously stated the importance of computational approach in identifying the impact of SNP over the protein's function (George Priya Doss, Chakraborty, Narayan, & Thirumal Kumar, 2014), drug response (Zhang, Miteva, Wang, & Alexov, 2012), and personalized medicine (Alexov, 2014). This review brings the influence of MD to identify the SNP-linked drug response to aid in personalized medicine. Further, this might fetch attention of experimental biologist to predict or understand the molecular mechanism in pursuing toward personalized medicine using MD.

2. SINGLE NUCLEOTIDE POLYMORPHISMS IN DRUG RESPONSE

Evolution on earth has begun based on genetic variations that continuously happen and give the recipient an increased or decreased ability to survive. These genetic variation transfers to their respective off-springs of next generations and also have the increasing chance of survival for the entire population. SNPs are the most common form of genetic variation and occur in about every 300 nucleotides. SNPs, as stated earlier, may be found at any region on the DNA. Some SNPs occur at the coding and few at the non-coding regions. A polymorphism present within an active site, substrate binding site, a DNA binding site, and protein domain affects the function of the encoded protein (Bell et al., 2002; Giacomini et al., 2007). An SNP may not certainly be deleterious, but if the changes are known to alter the conformation and affect the functionality of a protein, designated as non-synonymous SNPs (nsSNPs). SNPs present in the exonic region (Johnson, Wang, & Sadee, 2005; Wang, Johnson, Papp, Kroetz, & Sadee, 2005), promoter region (Kimura-Kataoka et al., 2012), *transcription* factors (Lockwood et al., 2014), and micro-RNA binding sites (Kim, Yoo, Choi, & Park, 2010; Peng et al., 2010; Xu, Feng, & Li, 2010; Zhou et al., 2011) identified as the key factor in inducing various diseases in human. Furthermore, SNPs observed in the genes or proteins involved in the pharmacokinetic or pharmacodynamics of a drug are found to be hindering the drug response (Ma & Lu, 2011). Drug-metabolizing enzymes and transporters play a crucial role in drug response; any genetic polymorphisms identified may directly affect the kinetics ADME of the drug. The target for a particular drug may widely vary based on the disease condition and the organ targeted. The drug may act in one or more targets based on the mode of action and have more affinity toward the targets belonging to the same family. For instance, histamine H1 receptor acts as a key mechanistic target for drugs like hydroxyzine and cetirizine (Tashiro et al., 2002; Van Ruitenbeek, Vermeeren, & Riedel, 2010). All these drugs also show better binding with other G-protein coupled receptors (GPCRs) *in vitro* assays (Gillard et al., 2002). D receptors or Dopamine receptors that belong to the GPCRs family are also taken as targets for disease involving neurons. This phenomenon explains that the drug may act through multiple mechanisms, and henceforth unrelated targets may also involve. Another example isozymes belonging to Cytochrome P450 family, widely known to be participating in the biotransformation of several drugs. Genetic variations in few subclasses of isozymes of this family show poor drug metabolism (Ogu, & Maxa, 2000). SNPs present in Cytochrome P450 family genes have been widely found to be lowering the hepatic drug metabolism (Flück, Mullis, & Pandey, 2010; Lockwood et al., 2014). Variants associated with any of the proteins involved in the mechanism of drug response might reduce the efficacy of the target. A challenge in identification of these approaches has been clearly stated by Fang et al. (2012). Lack of statistical power and widely available hypothesis have constrained the sample size that led to the false prediction of these deleterious SNPs.

3. DRUG DISCOVERY

Pharmaceutical companies with huge capital investment have largely involved in the development of a novel drug molecule for past three decades. Identification of the target and validation of a drug is the foremost and complex process in the formulation of a drug. Drugs are designed to interact with the protein and few genes involved in metabolizing the drug inside the living system. However, there is a consistent failure witnessed in this field (Couch & Mott, 2012; Heinemann, Douillard, Ducreux, & Peeters, 2013). The mechanism of drug response is highly complex and various factors are involved in deciding the fate of a drug. The binding or interaction of a molecule toward its target to alter a process has to be strong enough to overcome all the biological hurdles. Although the metabolism of a drug involves various factors, a large attribution toward the failure may be because of the deformities at the genetic level (Lewis, 2005; Pirmohamed, 2014). Accurately understanding the genes that directly or indirectly associated with a disease condition becomes a fundamental aspect of any novel drug discovery process. The genetic variants also play a crucial role as they may directly influence the acceptance of a particular drug for a disease condition. In case of an SNP or copy number variant condition, a change in a nucleotide system can directly have a significant influence on the protein produced. To overcome this, understanding the whole gene along with their regulatory regions might provide a basis for understanding a disease at the molecular level. Along with the huge time-consuming process, pharmaceutical industries also face expenses that are inevitable. The costs include commercialization of the drug, litigation expenses, drug recall costs, leading the drug into the market (Katara, 2013).

Hence to understand all the factors mentioned above and cost-effectiveness in mind, bioinformatics has shown a significant impact (Katara, 2013). The understanding of the drug interaction and binding patterns requires much developed methodology. "Functional genomics" study has indicated to help in identifying the genes that may be involved with drug metabolism. These genes are so much necessary as they become the first step in the discovery of prognostic biomarkers to help classify patients for the discrete personalized therapeutic approach. Hence, the existence of personalized medicine also known as genetic medicine comes into the field of drug discovery.

4. ADVENT OF PERSONALIZED MEDICINE

The Human Genome Project has paved the way for personalized medicine and become a sizzling topic in drug discovery process. Personalized medicine, as defined by Van der Greef, Hankemeier, and McBurney (2006) is a customized medical care for each patient's unique condition with respect to their genetic makeup (Van der Greef et al., 2006). Personalized medicine is the next-generation treatment strategy in health care research to overcome the failure. Development of new diagnostic tests and expanding the use of biomarkers will empower the identification of a disease at the molecular level. Ultimately, it will be the support for the development of novel targeted treatments. There is also the remarkable development of personalized medicine that has established itself in various crucial disorders such as schizophrenia, where clozapine is a drug molecule to which a particular subtype has shown to react. This finding was considered to be the first systematic evaluation of an antipsychotic response in a given genetic subtype of schizophrenia (Butcher et al., 2015). Personalized medicine has already entered the huge cancer world, where Zeidan et al. (2015) describes the use of clinical proteomics in breast cancer. Hence, this leads to the development of medicine and novel biomarkers for the prognosis of the disease condition (Zeidan, Townsend, Garbis, Copson, & Cutress, 2015). Chronic myeloid leukemia is another experimental research area that supports personalized medication therapy (Lee & Swanton, 2012; Zeidan et al., 2015). To accomplish success in the development of personalized medicine, understanding the response and resistance to a therapy becomes a primary feature (Fisher, Larkin, & Swanton, 2011). Thus, the adoption of personalized medicine might take the drug development strategy to a more successful platform and also minimizes the variability in every phase of drug development process.

With the availability of complete human genome sequences, information availability in the public databases has increased largely (Mizuno, Niwa, Yotsumoto, & Sugiyama, 2003). Many computational studies have also stood along to meet this scenario for identification of new drug molecules. To recognize the disease at a molecular level, analyzing gene, its product, and targeted molecular testing along with computational analysis have paved a better route at this stage (Kumar, 2007; Oscarson, 2003). Although personalized medicine is a breakthrough in the modern era, there are challenges that continuously interrupt the growth of this field. The diagnosis for the establishment of personalized medicine requires sequencing of the genome, comparison of the genome, identification of a particular biomarker, and pharmacogenomics study (Salari, Watkins, & Ashley, 2012).

The challenges commonly faced are processing of the whole genome data, interpretation and comparison of the data obtained with the available data. Further the data incorporates with wide systems and information to understand the complexity associated. These hectic diagnostic measures become a tedious millstone in case of nsSNP, which may have a deleterious effect on the resulting protein. As stated earlier, the impact of SNPs on the drug metabolism has led to the failure of a drug response. This further explains the need for identifying genetic variations before drug administration. High-end molecular methods support the fact that failure of drug response involves genetic variation. MD might be helpful in predicting a personalized therapy for a particular subtype with a visible genetic variation. A diagrammatic representation explains the difference between the traditional and personalized medicine (Fig. 1).

5. EVOLUTION OF MOLECULAR DYNAMICS IN THE FIELD OF MACROMOLECULE AND BINDING INTERACTION ANALYSIS

Macromolecules such as proteins are more flexible and dynamic in nature. They have the ability to change its conformation in response to external factors such as temperature, pH, charge, ion concentration, phosphorylation, or binding of a ligand. These changes also account for the conversion from an agonist to antagonist behavior (Krarup, Christensen, Hovgaard, & Frokjaer, 1998; Nichols, Swift, & Amaro, 2012; Teague, 2003). Therefore, analyzing the conformational change of a macromolecule

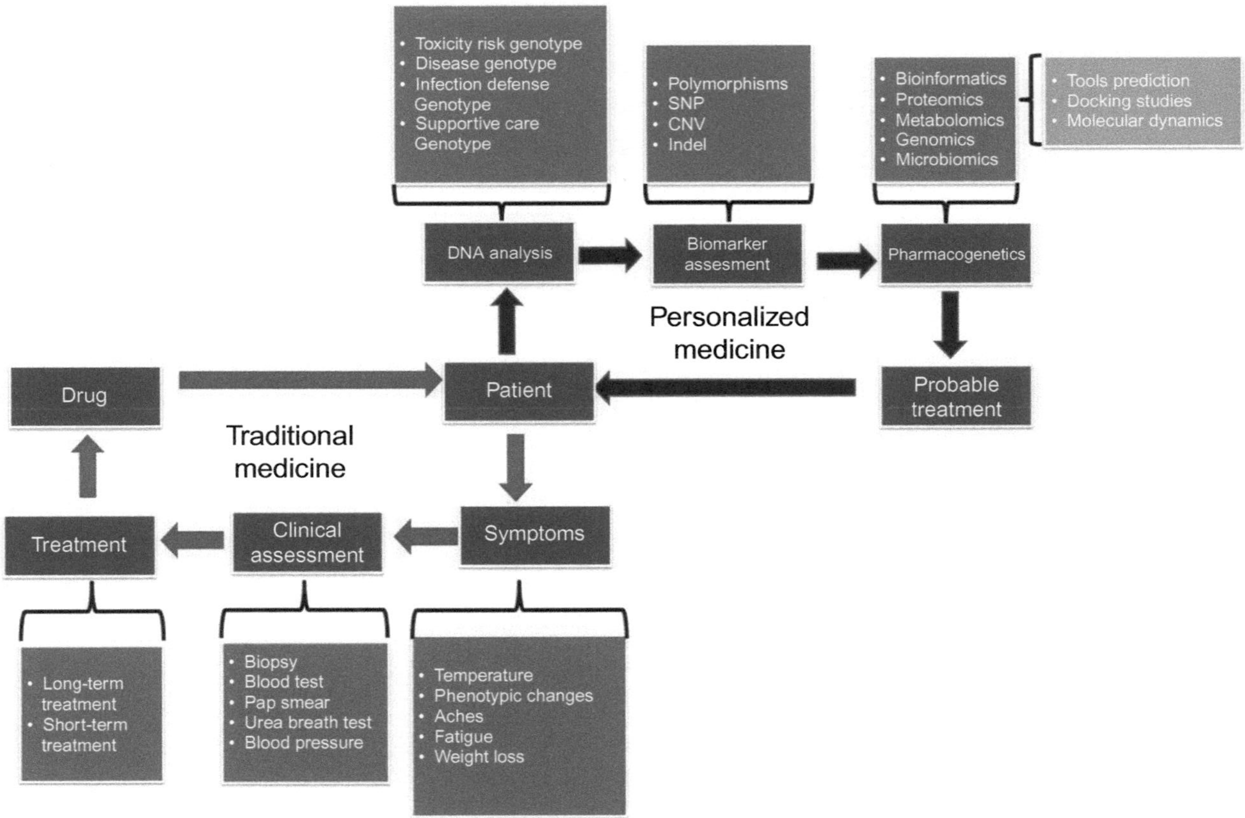

Figure 1 Pathologic states and clinical observations are used to evaluate and adjust treatments as in traditional medicine. The personal genomics connect the genotype to phenotype and provide insight into disease. Pink (gray in the print version) color directions represent the route of traditional medicine and blue (black in the print version) represents the route of personalized medicine.

at an atomistic level becomes a crucial step in studying the molecular aspects of a whole protein. Also understanding the binding interactions within and between the molecules has proved to be a significant part of drug designing (Berjanskii & Wishart, 2005; Goh, Milburn, & Gerstein, 2004; Onufriev, Bashford, & Case, 2004; Sotomayor & Schulten, 2007). Studying the interacting and binding patterns of the drug at the molecular level might help us understand the mechanism of the drug action. Cost effective sequencing methods resulted in the drastic information of genomic sequences. Most of the proteins do not have solved structures in protein data bank (PDB). A large gap exists between the solved structures by NMR, X-ray crystallography, and protein sequences in databases. To minimize the gap, affordable homology modeling techniques showed positive approach (Guex, Peitsch, & Schwede, 2009; Fiser & Sali, 2003; Zhang, 2008). Homology modeling, searches template sequences for the protein where no 3D structures available in PDB. Along with this CADD (Computer-Aided Drug Discovery) have been evolved largely to overcome the current situation, where docking analysis plays a very crucial role in drug designing. The binding affinity of the drug toward the target explains the specificity of a drug. Docking is usually a multistep process in which each step introduces one or more additional degrees of complexity to understanding the binding affinities (Gohlke & Klebe, 2002). Drug discovery process also involves screening of large compounds using methods such as visual screening that has evolved recently to aid the difficulties in screening of large compounds (Nichols et al., 2012). As previously noted, the hydrogen bonding, charges are a crucial crust in interactions between the drug and the molecule (Gohlke & Klebe, 2002). Docking clearly predicts all the parameters required for the successful interaction of the drug with a particular target. To analyze the motions of a drug along with the target, MD approach elucidates the detailed movement of all the atoms present. Furthermore, MD can be used in reefing the 3D structures determined by X-ray and NMR techniques.

6. MOLECULAR DYNAMICS—A BOON IN PERSONALIZED MEDICINE

The initiation and advancement in MD technique have led to the development of better analysis of mutation instability and loss of interaction. MD simulation explained about the protein with a minimum of nano to microsecond variation and proved to be a boon in elucidating the dynamic nature of macromolecules. They also served as the only method to understand the complex experimental phenomena (Beierlein et al., 2013; Dien, Deane, & Knapp, 2014; Dodson, Lane, & Verma, 2008; Fang & Weng, 2000). Measuring the change of confirmation at every nanosecond enables us to understand the impact of mutations and their adverse effects on the drug metabolism. Since deviation in a macromolecule takes place at atomistic level (Beierlein et al., 2013), explaining them becomes more tedious. Although there is a still much room for improvement in computational methods to make far more significant contributions aiding in drug development that has been impossible in the past (Lybrand, 1995). Increasing demands monitored in the case of structure-based drug design utilizes various computational methods. MD plays an essential role in monitoring the stability and specificity of the drug and the protein (Karatrantos, Composto, Winey, & Clarke, 2014). The robustness of this method has produced a significant impact on minimizing the risks, labor intensity, time, and tedious experimental analysis. Thus validating a target or explain the incapability of drug binding in mutated targets (Adcock & McCammon, 2006). MD simulations at equilibrium have also been known to identify correctly the mode of binding of few chemical compounds to their receptors such as binding of dimethyl sulfoxide in the aromatase FKBP (Huang & Caflisch, 2011). Rapid mutations are frequently observed in the viral genome has become a great trouble in designing a sustainable drug molecule. In the case of H5N1 virus, ostelmavir was reported to be prescribed but yet did not show successful results. This failure of the drug toward the target was further understood using MD that, those mutations at H274Y or N294S may be a reason that leads to resistance toward drug action. MD simulation explains the spontaneous binding of benzamidine to trypsin, explaining the crystal-structure-defined pose and disclosing the binding pathway (Borhani & Shaw, 2012). A further failure in drug response is the ability of the drug to bypass through the bi-lipid layer. Li et al. (2014) designed three inhibitors using axitinib as the reference drug against (VEGFR) receptor with the influence of MD simulations (Li et al., 2014). MD approach has largely evolved with different constraints and parameters are developed to study the details of the protein at a molecular level. MD has found to predict the permeability coefficient so as to understand the solvent and behavior pattern of the solvent toward the drug molecule (Huang & Caflisch, 2011). MD simulations is a multistep that includes, force field application, neutralizing the system, energy minimization, equilibration as explained below.

6.1 Force Fields

A recent development in computational approaches, in analyzing the potential energy functions (force fields) within a framework increases the knowledge about a macromolecule (Cornell et al., 1995). External forces are exerted to induce rotation without any significant disturbance in the nature of the protein's structure (Kutzner, Czub, & Grubmüller, 2011). A force field is an expression that mathematically describes the energy dependence of a system on all the coordinates of its particles. Force fields also explain the interaction between the atoms present in a system (Åqvist, Luzhkov, & Brandsdal, 2002; Guvench & MacKerell, 2008). These force fields, also known as potential energy, largely used to assess a protein at an atomistic level. The possible motions that take place within a macromolecule inside a biological system aids in calculating the potential energy to analyze and understand the protein and its function (Karplus & McCammon, 2002; Yin & MacKerell, 1998). Force fields are constructed using either experimental results such as X-ray, electron diffraction, NMR spectroscopy, IR spectroscopy or derived from *ab initio* and semi-empirical quantum mechanical calculation. Further, acts as an efficient helping aid in drug designing, chemoinformatics, nanoinformatics, and other fields that involve the study of macromolecules (Tosco, Stiefl, & Landrum, 2014). Various strategies have been applied to improve and validate the force fields to balance the atoms (Weiner & Kollman, 1981). Of the broadly available, most importantly used force fields are CHARMM, AMBER, GROMOS, OPLS, and COMPASS (Hansson, Oostenbrink, & van Gunsteren, 2002; Karplus & McCammon, 2002). The first three force fields were quite general and exclusively used in the simulation of macromolecules (such as protein) whereas the latter were designed to run condensed matter. Most popularly used force fields in MD simulation of proteins and nucleic acids (NA) are described in Table 1.

6.2 Solvation

Solvation model play a significant role in describing the electrostatics of the aqueous environment stability, sustainability and also the interactions between the macromolecules (Berjanskii & Wishart, 2005; Galiceanu, Reis, & Dolgushev, 2014; Onufriev et al., 2004; Seeliger & de Groot, 2010). In a biological system, there may be a huge variation in the solvent system that handles certain reactions to take place to maintain the sustainability of the whole system. The drug interaction differs with the solvent

Table 1 Various Force Fields Used for Molecular Dynamics Simulation

Force Fields	Interpretation	References
AMBER	Polar hydrogen bonds were explicitly represented. The hydrogen bonds were excluded during the earlier development of force fields	Ponder and Case (2003)
Amber ff94	All the torsion parameters were adjusted	Halgren (1999)
AMBER99sb	Designed for better balanced secondary structure elements of a protein. Proved to be an exception among force fields to show better hydrogen bonding and better accuracy	Weiner et al. (1984), Hornak et al. (2006), and Lange, Van der Spoel, and de Groot (2010)
CHARMM	The CHARMM (Chemistry at HARvard Molecular Mechanics) all atom energy forces was originally developed in early 1980s with no explicit hydrogen Designed to simulate rigid molecules. NAMD (nanoscale molecular dynamics program) simulation program was written using Charmm++ parallel program	Showalter and Bruschweiler (2007), Cournia, Vaiana, Ullmann, and Smith (2004), Phillips et al. (2005), and Kale et al. (1999)
CHARMM19	Able to obtain a balanced interaction between solute–water, water–water energies and useful for protein and peptide simulations	Brooks et al. (1983)
CHARMM22	Developed in the year 1997 which includes unsaturated hydrocarbons	
CHARMM27	Was compatible for proteins, nucleic acids, lipids, and many other small molecules	Cournia et al. (2004)
OPLS	Optimized potentials for liquid simulations. The early	Jorgensen and Tirado-Rives (1988), Kaminski and Friesner

Table 1 Various Force Fields Used for Molecular Dynamics Simulation—cont'd

Force Fields	Interpretation	References
	application of OPLS was for the rigid molecule. Later, AMBER/OPLS was used only for polar-hydrogen representation. The major difference between the AMBER and AMBER/OPLS is that, the force fields are empirical and able to reproduce properties of organic liquids. The functional group charges are transferable between the use of neutral subunits for large molecules and also to predict beta-sheet structure. Movement in the atoms and bond stretching was standardized using the force field	(2001), and Jorgensen, Maxwell, and Tirado-rives (1996)
GROMOS	GROMACS an acronym of GROningen molecular simulation package which was developed in 1978. GROMACS is an open source and free software available at http://www.gromacs.org	Christen et al. (2005)
GROMACS 45A3/4	Three force fields were exceptionally designed for calculating the interaction between the protein molecules with *n*-alkanes, cyclo-, iso-, neoalkanes, and branched aliphatics	Pronk et al. (2013)
GROMACS 53A5/6, and 54A7	Cyclohexane was used to reproduce a thermodynamic property of small molecules that present in pure liquids and solvation free enthalpies	Schuler, Daura, and van Gunsteren (2001)
GROMOS87, GROMOS96, GROMOS05	Developed to meet the requirement for calculating the energy	Jardón-Valadez, Bondar, and Tobias (2014) and Oostenbrink, Villa, Mark, and van Gunsteren (2004)

Continued

Table 1 Various Force Fields Used for Molecular Dynamics Simulation—cont'd

Force Fields	Interpretation	References
MMFF	MMFF (Merck molecular force field) cannot be applied for bulk phase simulation and performs very poorly when used to model organic liquids	Halgren (1996, 1999)
MMFF94 and MMFF94s	This uses fixed atomic center point charge to describe electrostatic energies. Hence, development in polarizable force fields may help in wider use. It is very difficult to generate "gold standards." Developed to show consistency when compared with other force fields	Lobanov, Bogatyreva, and Galzitskaya (2008)

system, as there are larger changes for the molecule to produce strong interactions with the target in the presence of water molecules (Karatrantos et al., 2014). Not only molecules have an effect because of the solvent system, but also the hydrogen bonding has an impact on the surrounding environment (Nogovitsyn, Kolesnikov, & Budkov, 2014). Hence for the same reason, MD solvation system is designed purely based on the requirement of the compound. For simulation of a compound, solute, solvent, and the total solvent–solute system were calculated and further help in deciding the fate of the compound. In some instance, alcohol and water were tested to understand the solvation. In designing these solubility parameters, the free energy binding plays a significant role as described by Duan et al. (2003).

6.3 Energy Minimization and Periodic Boundary

Force fields applied to the atoms in systems that are essential to finding a stable point to begin dynamics. The net force on a particular atom vanishes when a molecule attains minimum potential energy. More than one minimum point will be present in a given polymer, biopolymer, or liquid under the periodic boundary conditions. Also, constraints are imposed during a simulation process based on the experimental details of the structures. Minimized structures greatly help as average thermodynamic calculation and entropy estimations for a large number of structures are difficult to analyze. Thus, minimization provides detailed information that is complementary to MD simulation. To minimize a structure, it requires a particular function that is specified by the force fields. The magnitude of the first derivative is most important as it determines the directions and also helps in characterizing the convergence of the derivatives. Energy has converged or minimized when the derivative has come closer to zero. A minimization iteration process takes place that explains the complete cycle of differentiation and steps involved during minimization. MD simulation requires a large number of iteration to reach convergence. Minimization process involves several protocols of which Steepest descent, Conjugate gradient, and Newton–Raphson are the major protocols in practice today. PBCs (periodic boundary constraints) are considered only while calculating the nonbonded interactions between atoms that are belonging to different molecules. If the potential range is not too long, then the minimum image convention will be adopted. This convention means that each atom in a system interacts only with the nearest atom in the periodic array. Macromolecules such as protein when studied in solution, these constraints are not quite sufficient for simulation. In a PBC, the linear momentum is conserved whereas the angular momentum is not conserved (Kuzkin, 2014). Thus, these constraints play a major role during simulation of macromolecules.

6.4 Temperature and Pressure Differences

Temperature plays a crucial role in the kinetics of protein. Temperature difference leads to structural variation. For instance, increase in temperature has been observed in unfolding of the protein (Day, Bennion, Ham, & Daggett, 2002). Thus, simulation at different temperature may give an overview of the protein's behavior with respect to temperature. Similarly, pressure gradients are another set of parameters in understanding the motion of protein. Difference regions of proteins react differently to pressure gradient. Under pressure, the hydrophobic interactions tend to be disturbed thereby affecting the protein folding (Grigera & McCarthy, 2010). Thereby both temperature and pressure have a direct impact on the protein folding mechanism. MD has shown to be an important method to evaluate these differences also. The molecule was prepared by energy minimizing and equilibrating with required parameters. Equilibrated files used for performing MD results in a set of trajectory file used for further analysis. Various steps involved in MD simulation are illustrated in Fig. 2.

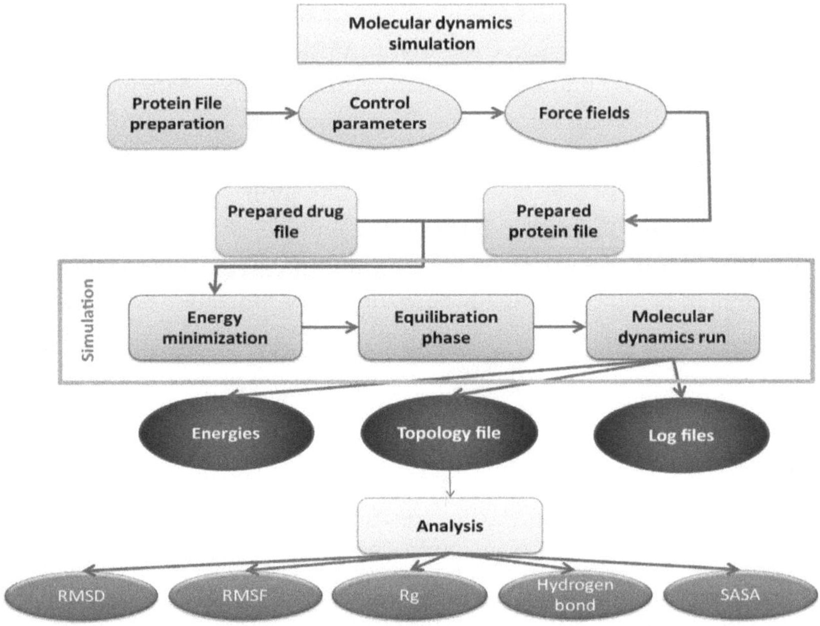

Figure 2 Different steps involved in molecular dynamics simulation.

7. MD TRAJECTORIES ANALYSIS

Structural property calculation of proteins from the trajectory files performed with the built-in functions of GROMACS software. Xmgrace program, a widely used program to analyze the trajectories obtained from MD Simulation (Turner, 2005). The trajectory files were analyzed through the various GROMACS utilities as discussed below.

7.1 Root Mean Square Deviation

Root mean square deviation (RMSD) is the deviation observed between two heavy atoms to predict the stability of a protein. RMSD of the protein explains the protein folding nature, where the RMSD of N set of atoms at time t, is calculated using

$$\mathrm{RMSD}(t) = \sqrt{\frac{\sum_{i=0}^{N} \left| \vec{r}_i(t) - \vec{r}_i^{\,0} \right|^2}{N}}.$$

Graphical representation of the RMSD values explains the possible variations that occur throughout the simulation. Thus, RMSD of all the Cα atoms are calculated which is considered as the central criterion to measure the convergence (Gargallo, Oliva, Querol, & Aviles, 2000; Wittayanarakul et al., 2005). Furthermore, RMSD analysis proved to be an important parameter in providing insight over the protein folding *in vivo* studies (Lucent, Vishal, & Pande, 2007). Wittayanarakul et al. (2005) calculated the RMSD values between each subsite present in the complex structure to the relative initial structures. A clear picture of the flexibility of Saquinavir, which clearly explained that all the subsites were more or less unchanged except that of one site (P2) (Wittayanarakul et al., 2005). Thus, RMSD plays a role in determining the convergence as well as explaining the protein folding. g_rms program fetches the 2D RMSD graph of the simulated protein. NMSim is an online tool to carry out simulation process consists of three steps to retrieve trajectory of the protein molecules (Krüger, Ahmed, & Gohlke, 2012). Apart from these online servers to calculate the RMSD, various standalone tools such as SPDBViewer (Guex & Peitsch, 1997) and PyMol (DeLano, 2002) were widely used to calculate deviations between two proteins.

7.2 Root Mean Square Fluctuation

RMSF (root mean square fluctuation) is the fluctuation observed between the residues or atoms present in a macromolecule. The atomic fluctuation explains the level of flexibility a protein during a simulation. RMSF of a protein can be calculated using

$$\text{RMSF}_i = \sqrt{\left\langle \left(\vec{r}_i - \langle \vec{r}_i \rangle\right)^2 \right\rangle}.$$

RMSF calculations are also determined to understand the flexibility and motility differences in the atoms between the mutant and the native (Roccatano, Colombo, Fioroni, & Mark, 2002). g_rmsf is further used to obtain a graphical representation of RMSF plot using Xmgrace program for the obtained MD trajectories. Other tools such as NAMD, visual molecular dynamics (VMD) plugin are also used widely to calculate RMSF.

7.3 Radial Distribution Function

Radial distribution function (RDF) helps to understand the binding process that was aided by the solvent. It is also been extensively used to understand the interactions of the atoms present for small molecules such as germanium (Ding & Andersen, 1986). The increase in the intensity of RDF relates to the more loosely packed atomic structure (Kovacs & Laaksonen, 1991). This concept helps to compare that computational approach and more intense experimental analysis. An interaction analysis was also reported by Bellissent-Funel, Lal, Bradley, and Chent (1993) with the assistance of RDF. A correlation between the water–protein interaction and RDF were stated and proved that a change in RDF may indicate the changes in the water–protein interactions (Bellissent-Funel et al., 1993). Using Xmgrace, the g_rdf program is used to obtain the plot radial distribution function.

7.4 Hydrogen Bonds

MD simulations reveal any stronger or weaker interaction that occurred due to a change in the environment inside a biological system or a change in the macromolecule itself due to random mutations. Hydrophobic interactions due to hydrogen bonds play a crucial role in stabilizing the macromolecule. The binding affinity and drug efficacy are associated directly with hydrophobic interactions (Antunes et al., 2014; Panigrahi, 2008). Mutations are known to change the bonding pattern thereby disturbing the stability and interacting patterns of the molecule. The unstable protein as a target may hinder the process of drug metabolism and drug response may not have been observed at all (Jardón-Valadez et al., 2014; Sands & Sansom, 2007). Lesser the hydrogen bonds, the more deviations can be observed which correlates with the results of RMSD. The positively charged molecules have strong hydrogen bonds, and poor binding has negatively charged molecules. This bonding can become a positive approach toward drug designing and also understanding the fate of the drug (Feng et al., 2015). Thus, increase in the number of hydrophobic atoms at the point of the active site of the target/macromolecule may further enhance the activity of the lead compound in a biological environment (Panigrahi, 2008). Binding of the drug to the target depends upon the binding site and its stability. Cryptic binding sites are occult sites present on the target (e.g., receptor) that are not obviously seen in NMR or X-ray crystallographic structures, sometimes they poses vital druggable sites. Molecular simulation plays an important role to find these cryptic binding sites (Durrant & McCammon, 2011a, 2011b; Frembgen-Kesner & Elcock, 2006). The stability of a particular binding site can be predicted using MD demonstrated in the case of *NS3* gene to provide for the stability of binding site for Dengue virus protease (De Almeida, Bastos, Ribeiro, Maigret, & Santana, 2013). Xmgrace program is used to obtain inter or intrahydrogen bonds formed in the protein (g_hbond). Hbonanza is a growing tool that explicitly describes the number of hydrogen bonds formed (Durrant & McCammon, 2011b).

7.5 Radius of Gyration

The radius of gyration (Rg) is defined as the distribution of atoms of a protein around its axis. The length that represents the distance between the point when it is rotating and the point where the transfer of energy has the maximum effect gives Rg. This conceptual idea further helps us to identify various polymer types as in the case of a protein. The calculation of Rg and distance calculations are the two most significant indicators that are widely used in predicting the structural activity of a macromolecule (Falsafi-zadeh, Karimi, & Galehdari, 2012). When a ligand/lead compound binds to the protein, there is a conformational change that changes the radius of gyration (Seeliger & de Groot, 2010). The compactness of a protein has a direct relationship with the rate of folding that can be monitored with the advanced computational method for calculating the radius of gyration (Weiner et al., 1984). MD simulation helps to carry out this calculation after the simulation of the protein (Daidone, Amadei, Roccatano, & Nola, 2003; Lobanov et al., 2008). Galiceanu et al. (2014) has extensively used Rg to study the statistic behavior of the protein. Furthermore, this helps to distinguish between the semi-flexible polymer networks with the complex topologies (Galiceanu et al., 2014). Rg also contributes to study the nature of the salt toward the macromolecule and few other conformational analysis of nanoparticles have also been well studied (Seeliger & de Groot, 2010; Karatrantos et al., 2014). It is noticed that a decrease or increase in salt concentration has an indirect proportionality observed in the values of Rg (Nogovitsyn et al., 2014). Understanding the Rg furthermore helps in predicting the compactness and binding patterns of the drug and protein.

7.6 Secondary Structure Analysis

Secondary structure of the proteins can be used to predict the tertiary structure since predicting only with amino acid sequence may not be sufficient. The secondary structure of proteins is determined by the pattern of hydrogen bonding. A large number of server and tools are used to predict the secondary structure analysis. DSSPcont (Carter, Andersen, & Rost, 2003) and STRIDE (Heinig & Frishman, 2004) are online tools used for understanding the secondary structure. In molecular dynamics simulation analysis program DSSP (Dictionary of Protein Secondary Structure) is used to create, visualize secondary structure plot. This enables us to understand the structural change in the protein structure. do_dssp is used to obtain a secondary structure graph in MD simulations. The graph explains the position of all helices; sheets present in the protein for a particular simulation time. Thus, results from MD simulation further elucidate any minor changes in the structure between the native and mutant structures.

7.7 Contact Maps

A contact map represents the distance between all the amino acids found as pairs in a 3D structure of the protein using a matrix, thus serving to understand the structural differences. Methods were already employed to recover the 3Dstructure of a protein from distance contact maps that were based on techniques such as distance geometry and stochastic optimization (Nilges, Clore, & Gronenborn, 1988). Further distance maps were integrated with simulation annealing methods to understand the 3D structure of a protein. A development in this is that a dynamic approach is integrated to generate a structure of the protein that produces a contact map similar to the query structure contact map (Vendruscolo, Kussell, & Domany, 1997). Various online tools are available to derive the contact map of the macromolecules between two chains in a protein molecule. The conserved residues of the protein were taken to identify the contact maps and were specifically designed to identify the point mutations (Chung, Beaver, Scheeff, & Bourne, 2007). COCOMAPS is another open access tool available to study all the interactions between the amino acid pairs present with the two chains of two molecules (Vangone, Spinelli, Scarano, Cavallo & Oliva, 2011). CMView, another PyMol integrated tool along with contact map 3D visualization that promotes much easier determination of the amino acid contacts in the 3D structure of the protein (Shaikh et al., 2013).

7.8 Free Energy

Studying free energy is of prime importance in understanding the function of a macromolecule in a biological system. Free energy gradients drive the biochemical/biophysical processes. In the case of DNA or RNA binding proteins, there are considerable conformational changes that can take place during binding. Thus, computational methods are designed to calculate accurately free energies as explained by Shaikh et al. (2013) $U = U_{\text{bond}} + U_{\text{angle}} + U_{\text{dihedral}} + U_{\text{vdW}} + U_{\text{elec}}$. The free energy calculation may be of three types; solvation free energy, binding free energy, and conformational free energy. MD calculates the binding energies keeping the sites around the ligands more relaxed thereby accurately predicting the binding patterns. Further improvements are carried out to assess accurately these calculations that include technique such as thermodynamic integration, linear interaction, free energy perturbation, and molecular mechanics methods. Though these methods are expensive to be employed, they provide a much easier prediction for a drug that can bind to the molecule that enriches the screening process (Zhao & Caflisch, 2014). Cui and Bastien (2012) described the use of free energy using steered molecular dynamics. A profile was further computed to show an energy barrier near Tyr-31 in the case of yeast aquaporin Aqy1 from *Pichia pastoris* (Cui & Bastien, 2012).

7.9 Covariance Matrix

Eigenvectors explain the different coordinates of movements of a molecule at possible vectors. The angle values predict the dimerization interfaces that suggest the stability of the protein and are found to be indirectly proportional. MD simulations greatly help in calculating these values and thereby helping to predict the stability of the complex. The eigenvalues can be used to describe the movement of the complex molecule. Covariance matrix expresses the negative or positive correlation, also known as cross-correlation matrix or dynamic cross-correlation matrix, where the coefficients measure or calculate the linear correlation between any two residues. Cartesian coordinates of all the Cα atoms present in the conformations with respect to an average structure were taken as the basis for constructing a covariance matrix. First, the covariance matrix is diagonalized for whose eigenvectors represent the principal geometrical axes in a three-dimensional space. Furthermore, the eigenvalues that has the highest values represent the most substantial relationship between dimensions (Andricioaei & Karplus, 2001). Wang, Yang, An, Wang, and Yin (2011) used covariance correlation to define the motions between the loops of the protein. Thereby proving that there is a regulatory relationship observed between the two proteins (Wang et al., 2011).

7.10 Principal Component Analysis

Principal component analysis (PCA) is orthogonal values that are used to convert a set of correlated variables to a set of uncorrelated variables known as a principal component. This parameter is widely utilized in the DNA–protein complex analysis as they incorporate more of domain specific assumptions. These motions also indicate the winding and unwinding of the DNA at specific time intervals based on the movements observed (Sargsyan, Wright, & Lim, 2012; Yang, Eyal, Bahar, & Kitao, 2009). The use of PCs has led to the reduction of the amount of data that are required to describe the conformational changes (Cohen & Moerner, 2007). Porcupine plots are another representation pattern generated using the values generated by eigenvectors which explicitly states the movement of the macromolecule with the use of cones mimicking the possible direction of movement in a biological system. Husby et al. (2012) described the difference in the motion of the complex using the eigenvector values that were obtained using the MD trajectories (Husby et al., 2012). PCA_NEST an online tool is used to understand the PCA that uses as structural input ensembles that are either experimentally characterized structures or computationally generated. The ensemble is primarily subjected to an iterative best-fitting complex and then to PCA, which further renders a series of output data (Yang et al., 2009). GeoPCA is further advancement to the previously utilized tools based on principal component geodesics. The conformational analysis is carried out using circular data of the complex such as bonds, torsion, and also pseudo torsion angles (Sargsyan et al., 2012). Thus, GeoPCA provides a path for visualizing, analyzing and predicting conformations of complex macromolecules. g_covar and g_anaeig are two commands that help in elucidating the motions of the protein from the trajectory file. As stated above, the construction of covariance matrix is carried out using g_covar and further g_anaeig is used to get the PMF of the eigenvectors. The resultant trajectory files are viewed using Xmgrace program. The motion of these coordinates present at vectors can be more easily described using porcupine plots.

7.11 Electrostatic Interactions

Electrostatic interactions are the electric force between any two charged molecules. The flexibility of the nucleic acid (NA) is predominantly based not only on their structure but also on the different solvent molecules. This is because the NA have different groove distribution and also forms extra hydroxyl groups with the surrounded solvent molecules (Sagui & Darden, 1999). Hence, understanding the interaction between protein and NA becomes a crucial strategy. Precise and faster calculation of electrostatic interaction between molecules has been a challenge until recent times. Lazar, Lee, Kim, Chandrasekaran, and Lee (2010) described the use of Particle Mesh Ewald (PME) to calculate the electrostatic interactions and considered best for a periodic or a pseudoperiodic system (Lazar et al., 2010). Kholmirzo showed that calculating the coulombic forces in a periodic system with the help of large reciprocal lattice vectors is comparatively more rapid than the conventional Ewald (PME) method (Kholmurodov, Smith, Yasuoka, Darden, & Ebisuzaki, 2000). This was further improved by using fixed cutoff and B-spline interpolation to permit the use of fast Fourier transforms that enable to calculate the energies efficiently. The implementation proved to have several advantages over the previously used PME (Guo & Gmeiner, 2001). Thus with growing importance of calculation of electrostatic interaction, the more improved methods to calculate these interactions are also evolving. Molsurf is another tool used to understand the interfaces between the macromolecules. This tool comes with a new feature that can compute map Poisson–Boltzmann electrostatic potentials of the interfaces of the molecules. Consrank are used to understand the interactions between the two residues present in the complex structure. Consrank ranks the entire interface based on conserved regions analysis, and the best-conserved interface are depicted in 2D and 3D graphical form (Chermak et al., 2014). POLYVIEW-MM is another highly used tool that combines the visualization with structural and functional annotation by automatically mapping entire functional hot spots and the structural features into thoroughly analyzed models (Porollo & Meller, 2010). Other tools used to calculate the parameters are tabulated in Table 2.

8. APPLICATION OF MD IN SNP ANALYSIS TOWARD DRUG DISCOVERY

MD had played a significant role in analyzing the impact of SNP over the protein's function and also drug response in the presence of a mutation. A mutation's impact on the drug response could be understood by analyzing the number of hydrogen bonds formed between the protein and the drug (Doss, Nagasundaram, et al., 2013). Further SASA explains the impact of

Table 2 Tools Used to Calculate Parameters of MD Trajectory Files

Parameter	Tools
RMSD	NMSim (normal mode analysis) @ http://www.nmsim.de PyMol (Standalone tool) VMD (visual molecular dynamics) @ http://www.ks.uiuc.edu/Research/vmd/ Command: Gromacs- g_rms
RMSF	NMSim @ http://cpclab.uni-duesseldorf.de/nmsim/ NAMD VMD (visual molecular dynamics) @ http://www.ks.uiuc.edu/Research/vmd/ Command: Gromacs- g_rmsf
Hydrogen bond analysis	HBonanza @ http://www.nbcr.net/hbonanza VMD plugin @ (visual molecular dynamics) http://www.ks.uiuc.edu/Research/vmd/ Command: Gromacs- g_hbond
PCA	PCA-ANOVA @ http://lgsun.grc.nia.nih.gov/ANOVA/help.html#PCA PCA-NEST @ http://ignm.ccbb.pitt.edu/oPCA_Online.htm GeoPCA @ http://pca.limlab.ibms.sinica.edu.tw VMD (visual molecular dynamics) @ http://www.ks.uiuc.edu/Research/vmd/ Command: Gromacs-g_covar and g_anaeig
Contact map	CMView—protein contact map visualization and analysis @ http://www.bioinformatics.org/cmview/ CMWeb: an interactive online tool for analyzing residue–residue contacts and contact prediction methods @ http://cmweb.enzim.hu Con-Struct Map: a comparative contact map analysis tool @ http://pdbrs3.sdsc.edu/ConStructMap/viewer_argument_generator/singleArguments COCOMAPS @ https://www.molnac.unisa.it/BioTools/cocomaps/ Command: Gromacs- g_mdmat
Free energy calculation	FEW: a workflow tool for free energy calculations of ligand binding
Secondary structure analysis	DSSPcont @ http://cubic.bioc.columbia.edu/services/DSSPcont STRIDE @ http://webclu.bio.wzw.tum.de/stride/ Command: Gromacs- do_dssp

Table 2 Tools Used to Calculate Parameters of MD Trajectory Files—cont'd

Parameter	Tools
Interface calculation	Molsurfer @ http://projects.villa-bosch.de/dbase/molsurfer/submit.html Consrank @ https://www.molnac.unisa.it/BioTools/consrank/ PolyviewMM
Metadynamics analysis	METAGUI @ http://www.plumed-code.org

mutation present on the surface, over the protein structure (Sudhakar et al., 2015). Free energy calculation and binding energy explain the properties of the bonds and binding patterns of the drug and protein (Kumari, Kumar, & Lynn, 2014; Yu et al., 2015). Secondary structure prediction, cluster analysis, and motion of the protein (PCA) throughout the simulation give an accurate impact of the mutation (Doss, Chakraborty, et al., 2014; Doss, Rajith, et al., 2014; Yang, Qin, Liu, & Yao, 2011). Contact map analysis gives a clearer and broader perspective on mutations (Perryman, Lin, & McCammon, 2004). Understanding the mutational effect of the protein over the drug might provide insight about the drug response. This further becomes an inevitable method to understand genetic variations and drug response by minimizing the cost and time (Doss, Chakraborty, et al., 2014; Doss, Nagasundaram, et al., 2013; Doss, Rajith, et al., 2014; Doss, Chakraborty, et al., 2014; Harris, Gavathiotis, Searle, Orozco, & Laughton, 2001; Krarup et al., 1998; Spacková et al., 2003; Zhao & Caflisch, 2014). Thus, a computational tool like MD may pave the way for better analysis for individualized medicine although disadvantages of this method are continuously encountered. As a concluding remark, MD plays a crucial role in elucidating the differences occurred due to a mutation.

9. PLAUSIBLE WAYS TO OVERCOME THE DISADVANTAGES OF MOLECULAR DYNAMICS

Timescale becomes the major drawback while carrying out MD simulations for large protein molecules. Since biological interactions are known to occur on the microsecond to the millisecond, the timescale is available but further improvement in reduction of time duration required could be minimized to predict the results much faster. Advancement in the computational field can aid in increasing the speed and minimizing the expense for such special computers for MD simulations. Classical MD is known to have simplified potential energy functions, but there is no electric and conformational polarization explained. Force fields are restricted to fix atomic center point charges that are not sufficient to calculate the electrostatic polarization (Xu et al., 2015). Even with continuous progression in this field, there are still certain problems that remain with MD simulations, particularly concerning complex molecule screening applications (Okimoto et al., 2009). Other known limitations in using crystallographic structures in structure-based drug design are that the structures are related to a particular fixed energy minima state. Hence, further refinement of the protocol is required to overcome these issues. When compared with experimental results in the case of explicit solvent simulations require further improvements (Feig & Brooks, 2004). Though this field has been extensively used past few decades to understand the folding, configuration, conformational, structural, and functional relationship, yet improvements are required to standardize features for better implementation in drug discovery process.

Atomistic MD simulations are still computationally expensive for docking of large libraries of compounds. Dynamics and docking analysis for hit discovery and optimization is increasing steadily in the field of drug discovery making it more reliable and efficient (Coe, Levine, & Martinez, 2007). Binding process and also docking reliability of a docked complex or interaction of the molecule toward the ligand is a vital process in designing a drug for a particular target. Hence understanding this concept is aided with the help of MD simulations, where it describes all the thermodynamics and kinetics of this process (Korfi et al., 2015). As MD simulation gives information at an atomistic level, which becomes particularly more useful when the crystal structure of the complex is unavailable. Molecular simulations provide detailed analysis of all the atoms in motion for duration of the period. Understanding this process is important because theoretical estimates of systemic errors inherent in the simulation are still not stated. Since the potentials are approximate and are in control of the user, any altering or removing, these values can be made as per the individual's interest. Further becomes a most important factor toward the use of molecular simulation (Karplus & McCammon, 2002). When a small compound is taken as a drug, the glitches associated with it cannot be clearly understood using *in vivo* studies. MD studies have greatly helped in understanding all the atoms in detail and also examine the favorable positions and orientations available in the active site (Aghaee, Ghasemi, Manouchehri, & Balalaie, 2014). Another gate opening development in this field is pHMD. pH of a solution also has a critical role in protein dynamics, there is prominent development in pH-coupled dynamic phenomena (Mongan, Case, & McCammon, 2004; Williams, de Oliveira, & McCammon, 2010). Force fields are used for running a simulation with Linux interface but to overcome this, a large number of GUI interface are available to perform dynamics. Gromita (Oostenbrink et al., 2004; Sellis, Vlachakis, & Vlassi, 2009), YASARA dynamics (Krieger, Darden, Nabuurs, Finkelstein, & Vriend, 2004), GUI BioPASED (Popov & Vorob, 2010), MDSA (Bakar, Hashim, & Omar, 2013), GUIMACS (Kota, 2006), MDTRA (Popov, Vorobjev, & Zharkov, 2013), and CHARMM-GUI (Jo, Kim, Iyer, & Im, 2008) are few interfaces. Further various features associated with different GUI are explained in Table 3.

Various programs also have extensively developed along with the development of the MD simulations such as VMD. VMD designed by William Hymphrey et al. visualizes the various protein interactions (Humphrey, Dalke, & Schulten, 1996). Proteins that contain heme group such as

Table 3 Graphic User Interface for Molecular Dynamics

GUI Interface	Features
Gromita	Gromita, a user friendly simulation method to break the command line barrier for GROMACS users. Cross-platform, perl/tcl-tk based, interactive front end that guides through each step in the simulation. Interface provides much enhanced functionality and simplified set of tasks for performing dynamics. Gromita is a noncommercial tool and can be downloaded from http://bio.demokritos.gr/gromita/
CHARMM-GUI	A graphical interface to provide a Web-based approach to generate files to use the techniques in CHARMM. Platform helps to identify any problem through visual inspection which could be resolved much easily with the help of GUI CHARMM-GUI http://www.charmm-gui.org
GUIMACS	Standalone application with multiple document interface (MDI) which enables to run and analyze molecular dynamics simulations. Another highlight of these tools is that multiple simulations can be

Continued

Table 3 Graphic User Interface for Molecular Dynamics—cont'd

GUI Interface	Features
	carried out simultaneously. Portals are displayed and simple menu-driven accesses to various programs are provided. The different interface methods to run simulation GUI Sim—graphical interfaces for six major programs—pdb2gmx, editconf, genbox, genion, grompp, and mdrun GUINalyzer—performs a set of parameters—g_rms, g_hbond, g_confrms, g_energy, g_gyrate, g_dist, and g_saltbr Parameter Editor—The basic MDP files of the molecular dynamics simulation can be generated AutoMACS—an automated machine to perform complete molecular dynamics simulation with PDB file, force field information, and time step required for simulation as minimum input
MDSA	GUI includes a toolbar that provides access to the specific tasking form such as changing the current molecular display characteristic. The methodology is much similar to GUIMACS with the use of different interfaces for running simulation; such as GUISim and GUINalyzer
GUIBioPASED	Intended to work with three-dimensional structures of proteins and nucleic acids to analyze and optimize the structure Conformational energy, simulate conformational dynamics, perform statistical analysis of structural dynamics, and calculates free energy of biopolymers
MDTRA—MD trajectory reader and analyzer	The program has been developed to facilitate the process of search and visualization of results. MDTRA can handle trajectories as sets of protein (PDB) files and presents tools and guidelines to convert some other trajectory formats into such sets The various parameters analyzed by MDTRA are interatomic distances, angles, dihedral angles, angles between planes, one-dimensional and two-dimensional root mean square deviation, and solvent-accessible area
YASARA dynamics	This interface developed new force fields unlike other interface; NOVA, YAMBER. Also uses AMBER force fields for performing simulation

cytochrome proteins (CYP) were previously difficult to study because of the complexity that was further solved by using MD simulation. Skopalik et al. (2008) discuss the flexibility and malleability of CYP protein (Hendrychova et al., 2011; Skopalik et al., 2008; Xu et al., 2015). Certain extensive work has been performed in different force fields to understand the binding affinity between the solvents. Understanding this might provide a guideline for analyzing the changes that might have occurred in the hydrogen bond due to SNPs (Sheehan & Sharratt, 1998). Due to advancement in force fields, MD also helps in understanding the SNPs complex molecules such as heme group containing protein, DNA/RNA bound proteins, protein–ligand interactions.

10. CONCLUSION

The increase in the population and disease types has created a large demand for novel compound discovery. Though pharmaceutical companies have invested millions in this field, a constant failure was always witnessed. Genetic variations have found to be a leading cause of the difference in the drug response. Pharmacogenomics and pharmacogenetics shed limelight to understand the drug's response with the genetic differences and the genes associated with drug response. Since every individual responds differently to different drug types, personalized medicine began its journey in drug discovery process. It is not feasible to screen every compound for a variant of experimental analysis. Henceforth, bioinformaticians might play a significant role in cutting the expenditure, time, and labor. Taking all these factors into account, there has been a considerable increase in the evolution of computational methods which includes modeling, docking, and MD simulation. The simulation provides a detailed view of the concentration profiles of the species in a sample and molecular interactions in a bulk at the interface. MD provides flexibility to both the target and drug separately which might provide insight into the impact of mutations on the drug response. The concept of encountering a novel drug compound has protracted its vastness in almost all the fields of biology. Application of MD has shown beneficial effects in other fields and also we also expect this process could be a driving force in personalized medicine. Thus, this review explicits the vastness of MD in drug discovery and personalized medicine to produce a novel drug compound with less expensive, reduced time duration, and also to diminish the labor work.

ACKNOWLEDGMENTS

We sincerely thank the management of VIT University for providing the laboratory facility and support for conducting the study.

REFERENCES

Acharya, V., & Nagarajaram, H. (2012). Hansa: An automated method for discriminating disease and neutral human nsSNPs. *Human Mutation*, *33*(2), 332–337. http://dx.doi.org/10.1002/humu.21642.

Adcock, S. A., & McCammon, J. A. (2006). Molecular dynamics: Survey of methods for simulating the activity of proteins. *Chemical Reviews*, *106*, 1589–1615.

Adzhubei, I., Jordan, D. M., & Sunyaev, S. R. (2013). Predicting functional effect of human missense mutations using PolyPhen-2. *Current Protocols in Human Genetics*, Chapter 7, Unit 7.2. http://dx.doi.org/10.1002/0471142905.hg0720s76.

Aghaee, E., Ghasemi, J. B., Manouchehri, F., & Balalaie, S. (2014). Combined docking, molecular dynamics simulations and spectroscopic studies for the rational design of a dipeptide ligand for affinity chromatography separation of human serum albumin. *Journal of Molecular Modeling*, *20*(10), 2446. http://dx.doi.org/10.1007/s00894-014-2446-7.

Alanazi, M., Abduljaleel, Z., Khan, W., Warsy, A. S., Elrobh, M., Khan, Z., et al. (2011). In silico analysis of single nucleotide polymorphism (snps) in human β-globin gene. *PLoS One*, *6*(10), e25876. http://dx.doi.org/10.1371/journal.pone.0025876.

Aleksandrov, A., & Simonson, T. (2010). Molecular dynamics simulations show that conformational selection governs the binding preferences of Imatinib for several tyrosine kinases. *Journal of Biological Chemistry*, *285*, 13807–13815.

Alexov, E. (2014). Combining genetics and molecular biophysics to pave the way for personalized. *Advances in Human Biology*, *2014*, 1–16.

Andricioaei, I., & Karplus, M. (2001). On the calculation of entropy from covariance matrices of the atomic fluctuations. *The Journal of Chemical Physics*, *115*, 6289. http://dx.doi.org/10.1063/1.1401821.

Antunes, D., Rigo, M. M., Sinigaglia, M., de Medeiros, R. M., Junqueira, D. M., Almeida, S. E. M., et al. (2014). New insights into the in silico prediction of HIV protease resistance to nelfinavir. *PLoS One*, *9*(1), e87520. http://dx.doi.org/10.1371/journal.pone.0087520.

Åqvist, J., Luzhkov, V. B., & Brandsdal, B. O. (2002). Ligand binding affinities from MD simulations. *Accounts of Chemical Research*, *35*(6), 358–365. http://dx.doi.org/10.1021/ar010014p.

Bakar, N. A. A., Hashim, S. Z. M., & Omar, M. S. S. (2013). MDSA—An interactive analysis tool for protein molecular dynamic simulations: Preliminary study.

Beierlein, F. R., Krause, A. M., Jäger, C. M., Fita, P., Vauthey, E., & Clark, T. (2013). Molecular dynamics simulations of liquid phase interfaces: Understanding the structure of the glycerol/water-dodecane system. *Langmuir*, *29*(38), 11898–11907. http://dx.doi.org/10.1021/la4021355.

Bell, P. A., Chaturvedi, S., Gelfand, C. A., Huang, C. Y., Kochersperger, M., Kopla, R., et al. (2002). SNPstream UHT: Ultra-high throughput drug discovery. *Biotechniques*, *74*, 76–77.

Bellissent-Funel, M. C., Lal, J., Bradley, K. F., & Chent, S. H. (1993). Neutron structure factors of in-vivo deuterated amorphous protein C-phycocyanin. *Biophysical Journal*, *64*, 1542–1549.

Bendl, J., Stourac, J., Salanda, O., Pavelka, A., Wieben, E. D., Zendulka, J., et al. (2014). PredictSNP: Robust and accurate consensus classifier for prediction of disease-related mutations. *PLoS Computational Biology*, *10*(1), e1003440. http://dx.doi.org/10.1371/journal.pcbi.1003440.

Berger, S. I., & Iyengar, R. (2011). Role of systems pharmacology in understanding drug adverse events. *Wiley Interdisciplinary Reviews: Systems Biology and Medicine*, *3*, 129–135. http://dx.doi.org/10.1002/wsbm.114.Role. 1–11.

Berjanskii, M. V., & Wishart, D. S. (2005). A simple method to predict protein flexibility using secondary chemical shifts. *Journal of the American Chemical Society*, *127*(43), 14970–14971. http://dx.doi.org/10.1021/ja054842f.

Borhani, D. W., & Shaw, D. E. (2012). The future of molecular dynamics simulations in drug discovery. *Journal of Computer-Aided Molecular Design*, *26*(1), 15–26. http://dx.doi.org/10.1007/s10822-011-9517-y.

Brooks, B. R., Bruccoleri, R. E., Olafson, B. D., States, D. J., Swaminathan, S., & Karplus, M. (1983). CHARMM: A program for macromolecular energy, minmimization, and dynamics calculations. *Journal of Combinatorial Chemistry*, *4*, 187–217.

Butcher, N. J., Fung, W. L. A., Fitzpatrick, L., Guna, A., Andrade, D. M., Lang, A. E., et al. (2015). Response to clozapine in a clinically identifiable subtype of schizophrenia. *The British Journal of Psychiatry: The Journal of Mental Science*, *206*, 484–491. http://dx.doi.org/10.1192/bjp.bp.114.151837.

Capriotti, E., Calabrese, R., Fariselli, P., Martelli, P. L., Altman, R. B., & Casadio, R. (2013). WS-SNPs&GO: A web server for predicting the deleterious effect of human protein variants using functional annotation. *BMC Genomics*, *14*(3), S6. http://dx.doi.org/10.1186/1471-2164-14-S3-S6.

Carter, P., Andersen, C. a. F., & Rost, B. (2003). DSSPcont: Continuous secondary structure assignments for proteins. *Nucleic Acids Research*, *31*(13), 3293–3295. http://dx.doi.org/10.1093/nar/gkg626.

Chermak, E., Petta, a., Serra, L., Vangone, a., Scarano, V., Cavallo, L., et al. (2014). CONSRANK: A server for the analysis, comparison and ranking of docking models based on inter-residue contacts. *Bioinformatics*, *31*, 1481–1483. http://dx.doi.org/10.1093/bioinformatics/btu837.

Christen, M., Hünenberger, P. H., Bakowies, D., Baron, R., Bürgi, R., Geerke, D. P., et al. (2005). The GROMOS software for biomolecular simulation: GROMOS05. *Journal of Computational Chemistry*, *26*(16), 1719–1751. http://dx.doi.org/10.1002/jcc.20303.

Chung, J. L., Beaver, J. E., Scheeff, E. D., & Bourne, P. E. (2007). Con-Struct Map: A comparative contact map analysis too. *Bioinformatics*, *23*(18), 2491–2492. http://dx.doi.org/10.1093/bioinformatics/btm356.

Coe, J. D., Levine, B. G., & Martinez, T. J. (2007). Ab initio molecular dynamics of excited-state intramolecular proton transfer using multireference perturbation theory. *Journal of Physical Chemistry A*, *111*, 11302–11310.

Cohen, A. E., & Moerner, W. E. (2007). Principal-components analysis of shape fluctuations of single DNA molecules. *Proceedings of the National Academy of Sciences of the United States of America*, *104*(31), 12622–12627. http://dx.doi.org/10.1073/pnas.0610396104.

Cornell, W. D., Cieplak, P., Bayly, C. I., Gould, I. R., Merz, K. M., Ferguson, D. M., et al. (1995). A second generation force field for the simulation of proteins, nucleic acids, and organic molecules. *Journal of the American Chemical Society*, *117*(19), 5179–5197. http://dx.doi.org/10.1021/ja00124a002.

Couch, R. D., & Mott, B. T. (2012). Personalized medicine: Changing the paradigm of drug development. *Methods in Molecular Biology (Clifton, N.J.)*, *823*, 367–378. http://dx.doi.org/10.1007/978-1-60327-216-2_24.

Cournia, Z., Vaiana, A. C., Ullmann, G. M., & Smith, J. C. (2004). Derivation of a molecular mechanics force field for cholesterol. *Pure and Applied Chemistry*, *76*(1), 189–196. http://dx.doi.org/10.1351/pac200476010189.

Csermely, P., Korcsmaros, T., Kiss, H. J. M., London, G., & Nussinov, R. (2013). Structure and dynamics of molecular networks: A novel paradiagram of drug discovery: A comprehensive review. *Pharmacology and Therapeutics*, *138*, 333–408.

Cui, Y., & Bastien, D. (2012). Molecular dynamics simulation and bioinformatics study on yeast aquaporin Aqy1 from Pichia pastoris. *International Journal of Biological Sciences*, *8*(7), 1026–1035. http://dx.doi.org/10.7150/ijbs.4703.

Daidone, I., Amadei, A., Roccatano, D., & Nola, A. D. (2003). Molecular dynamics simulation of protein folding by essential dynamics sampling: Folding landscape of horse heart cytochrome c. *Biophysical Journal*, *85*(5), 2865–2871.

Day, R., Bennion, B. J., Ham, S., & Daggett, V. (2002). Increasing temperature accelerates protein unfolding without changing the pathway of unfolding. *Journal of Molecular Biology*, *322*(1), 189–203. http://dx.doi.org/10.1016/S0022-2836(02)00672-1.

De Alencar, S., & Lopes, J. C. D. (2010). A Comprehensive in silico analysis of the functional and structural impact of SNPs in the IGF1R gene. *Journal of Biomedicine and Biotechnology*, *2010*, 715139. http://dx.doi.org/10.1155/2010/715139.

De Almeida, H., Bastos, I. M. D., Ribeiro, B. M., Maigret, B., & Santana, J. M. (2013). New binding site conformations of the dengue virus NS3 protease accessed by molecular dynamics simulation. *PLoS One*, *8*(8), e72402. http://dx.doi.org/10.1371/journal.pone.0072402.

De la Iglesia, D., Cachau, R. E., García-Remesal, M., & Maojo, V. (2013). Nanoinformatics knowledge infrastructures: Bringing efficient information management to nanomedical research. *Computational Science & Discovery*, *6*(1), 014011. http://dx.doi.org/10.1088/1749-4699/6/1/014011.

DeLano, W. L. (2002). *The PyMOL molecular graphics system, version 0.99*. San Carlos, CA: DeLano Scientific.

Derks, S., & Diosdado, B. (2015). Personalized cancer medicine: Next steps in the genomic era. *Cellular Oncology*, *38*, 1–2. http://dx.doi.org/10.1007/s13402-015-0221-0.

Dien, H., Deane, C. M., & Knapp, B. (2014). Gro2mat: A package to efficiently read gromacs output in MATLAB. *Journal of Computational Chemistry*, *35*(20), 1528–1531. http://dx.doi.org/10.1002/jcc.23650.

Ding, K., & Andersen, H. (1986). Molecular-dynamics simulation of amorphous germanium. *Physical Review B*, *34*(10), 6987–6991. http://dx.doi.org/10.1103/PhysRevB.34.6987.

Dodson, G. G., Lane, D. P., & Verma, C. S. (2008). Molecular simulations of protein dynamics: New windows on mechanisms in biology. *EMBO Reports*, *9*(2), 144–150. http://dx.doi.org/10.1038/sj.embor.7401160.

Doss, C. G. P., Chakraborty, C., Chen, L., & Zhu, H. (2014). Integrating in silico prediction methods, molecular docking, and molecular dynamics simulation to predict the impact of ALK missense mutations in structural perspective. *BioMed Research International*, 895831. http://dx.doi.org/10.1155/2014/895831.

Doss, C. G. P., & NagaSundaram, N. (2012). Investigating the structural impacts of I64T and P311S mutations in APE1-DNA complex: A molecular dynamics approach. *PLoS One*, 7(2), 1–11. http://dx.doi.org/10.1371/journal.pone.0031677.

Doss, C. G. P., Nagasundaram, N., Chakraborty, C., Chen, L., & Zhu, H. (2013). Extrapolating the effect of deleterious nsSNPs in the binding adaptability of flavopiridol with CDK7 protein: A molecular dynamics approach. *Human Genomics*, 7(1), 10. http://dx.doi.org/10.1186/1479-7364-7-10.

Doss, C. G. P., & Rajith, B. (2012). Computational refinement of functional single nucleotide polymorphisms associated with ATM gene. *PLoS One*, 7(4), 1–11. http://dx.doi.org/10.1371/journal.pone.0034573.

Doss, G. P. C., Rajith, B., Chakraborty, C., NagaSundaram, N., Ali, S. K., & Zhu, H. (2014). Structural signature of the G719S-T790M double mutation in the EGFR kinase domain and its response to inhibitors. *Scientific Reports*, *4*, 5868. http://dx.doi.org/10.1038/srep05868.

Duan, Y., Wu, C., Chowdhury, S., Lee, M. C., Xiong, G., Zhang, W., et al. (2003). A point-charge force field for molecular mechanics simulations of proteins based on condensed-phase quantum mechanical calculations. *Journal of Computational Chemistry*, *24*(16), 1999–2012. http://dx.doi.org/10.1002/jcc.10349.

Durrant, J. D., & McCammon, J. A. (2011a). Molecular dynamics simulations and drug discovery. *BMC Biology*, *9*, 71. http://dx.doi.org/10.1186/1741-7007-9-71.

Durrant, J. D., & McCammon, J. A. (2011b). HBonanza: A computer algorithm for molecular-dynamics- trajectory hydrogen-bond analysis. *Journal of Molecular Graphics and Modelling*, *31*, 5–9. http://dx.doi.org/10.1016/j.jmgm.2011.07.008.

Falsafi-zadeh, S., Karimi, Z., & Galehdari, H. (2012). VMD DisRg: New user-friendly implement for calculation distance and radius of gyration in VMD program. *Bioinformation*, *8*, 7–9.

Fang, G., Haznadar, M., Wang, W., Yu, H., Steinbach, M., Church, T. R., et al. (2012). High-order SNP combinations associated with complex diseases: Efficient discovery, statistical power and functional interactions. *PLoS One*, 7(4), e33531. http://dx.doi.org/10.1371/journal.pone.0033531.

Fang, T.-H., & Weng, C. I. (2000). Three-dimensional molecular dynamics analysis of processing using a pin tool on the atomic scale. *Nanotechnology*, *11*(3), 148–153. http://dx.doi.org/10.1088/0957-4484/11/3/302.

Feig, M., & Brooks, C. L. (2004). Recent advances in the development and application of implicit solvent models in biomolecule simulations. *Current Opinion in Structural Biology*, *14*(2), 217–224. http://dx.doi.org/10.1016/j.sbi.2004.03.009.

Feng, Z., Alqarni, M. H., Yang, P., Tong, Q., Chowdhury, A., Wang, L., et al. (2015). Modeling, molecular dynamics simulation, and mutation validation for structure of cannabinoid receptor 2 based on known crystal structures of GPCRs. *Journal of Chemical Information and Modeling*, *54*(9), 2483–2499.

Fiser, A., & Sali, A. (2003). Modeller: Generation and refinement of homology-based protein structure models. *Methods in Enzymology*, *374*, 461–491. http://dx.doi.org/10.1016/S0076-6879(03)74020-8. PMID:14696385.

Fisher, R., Larkin, J., & Swanton, C. (2011). Delivering preventive, predictive and personalised cancer medicine for renal cell carcinoma: The challenge of tumour heterogeneity. *The EPMA Journal*, *3*(1), 1. http://dx.doi.org/10.1007/s13167-011-0137-3.

Flück, C. E., Mullis, P. E., & Pandey, A. V. (2010). Reduction in hepatic drug metabolizing CYP3A4 activities caused by P450 oxidoreductase mutations identified in patients with disordered steroid metabolism. *Biochemical and Biophysical Research Communications*, *401*(1), 149–153. http://dx.doi.org/10.1016/j.bbrc.2010.09.035.

Frembgen-Kesner, T., & Elcock, A. H. (2006). Computational sampling of a cryptic drug binding site in a protein receptor: Explicit solvent molecular dynamics and inhibitor docking to p38 MAP kinase. *Journal of Molecular Biology*, *359*(1), 202–214. http://dx.doi.org/10.1016/j.jmb.2006.03.021.

Galiceanu, M., Reis, A. S., & Dolgushev, M. (2014). Dynamics of semiflexible scale-free polymer networks. *The Journal of Chemical Physics*, *141*(14), 144902. http://dx.doi.org/10.1063/1.4897563.

Gargallo, R., Oliva, B., Querol, E., & Aviles, F. X. (2000). Effect of the reaction field electrostatic term on the molecular dynamics simulation of the activation domain of procarboxypeptidase *B. Protein Engineering*, *13*, 21–26.

George Priya Doss, C., Chakraborty, C., Narayan, V., & Thirumal Kumar, D. (2014). Computational approaches and resources in single amino acid substitutions analysis toward clinical research. *Advances in Protein Chemistry and Structural Biology*, *94*, 365–423. http://dx.doi.org/10.1016/B978-0-12-800168-4.00010-X.

George Priya Doss, C., Rajasekaran, R., Sudandiradoss, C., Ramanathan, K., Purohit, R., & Sethumadhavan, R. (2008). A novel computational and structural analysis of nsSNPs in CFTR gene. *Genomic Medicine*, *2*(1–2), 23–32. http://dx.doi.org/10.1007/s11568-008-9019-8.

George Priya Doss, C., & Rajith, B. (2013). A new insight into structural and functional impact of single-nucleotide polymorphisms in PTEN gene. *Cell Biochemistry and Biophysics*, *66*(2), 249–263. http://dx.doi.org/10.1007/s12013-012-9472-9.

Giacomini, K. M., Brett, C. M., Altman, R. B., Benowitz, N. L., Dolan, M. E., Da, F., et al. (2007). The pharmacogenetics research network: From SNP discovery to clinical drug response. *Clinical Pharmacology and Therapeutics*, *81*(3), 328–345. http://dx.doi.org/10.1038/sj.clpt.6100087.

Gillard, M., Van Der Perren, C., Moguilevsky, N., Massingham, R., & Chatelain, P. (2002). Binding characteristics of cetirizine and levocetirizine to human H(1) histamine receptors: Contribution of Lys(191) and Thr(194). *Molecular Pharmacology*, *61*(2), 391–399. http://dx.doi.org/10.1124/mol.61.2.391.

Ginsburg, G. S., & Mccarthy, J. J. (2001). Personalized medicine: Revolutionizing drug discovery and patient care. *Trends in Biotechnology*, *19*(12), 491–496.

Goh, C.-S., Milburn, D., & Gerstein, M. (2004). Conformational changes associated with protein-protein interactions. *Current Opinion in Structural Biology*, *14*(1), 104–109. http://dx.doi.org/10.1016/j.sbi.2004.01.005.

Gohlke, H., & Klebe, G. (2002). Approaches to the description and prediction of the binding affinity of small-molecule ligands to macromolecular receptors. *Angewandte Chemie (International Ed in English)*, *41*(15), 2644–2676. http://dx.doi.org/10.1002/1521-3773(20020802)41:15<2644::AID-ANIE2644>3.0.CO;2-O.

Grigera, J. R., & McCarthy, A. N. (2010). The behavior of the hydrophobic effect under pressure and protein denaturation. *Biophysical Journal*, *98*(8), 1626–1631. http://dx.doi.org/10.1016/j.bpj.2009.12.4298.

Guex, N., & Peitsch, M. C. (1997). SWISS-MODEL and the Swiss-PdbViewer: An environment for comparative protein modelling. *Electrophoresis*, *18*, 2714–2723.

Guex, N., Peitsch, M. C., & Schwede, T. (2009). Automated comparative protein structure modeling with SWISS-MODEL and Swiss-PdbViewer: A historical perspective. *Electrophoresis*, *30*(Suppl. 1), S162–S173. http://dx.doi.org/10.1002/elps.200900140.

Guo, J. X., & Gmeiner, W. H. (2001). Molecular dynamics simulation of the human U2B" protein complex with U2 snRNA hairpin IV in aqueous solution. *Biophysical Journal*, *81*(2), 630–642.

Guvench, O., & MacKerell, A. D. (2008). Comparison of protein force fields for molecular dynamics simulations. *Methods in Molecular Biology (Clifton, NJ)*, *443*, 63–88. http://dx.doi.org/10.1007/978-1-59745-177-2_4.

Halgren, T. A. (1996). Merck molecular force field. II. MMFF94 van der Waals and electrostatic parameters for intermolecular interactions. *Journal of Computational Chemistry*, *17*(5–6), 520–552. http://dx.doi.org/10.1002/(SICI)1096-987X(199604)17:5/6<520::AID-JCC2>3.0.CO;2-W.

Halgren, T. A. (1999). MMFF VII. Characterization of MMFF94, MMFF94s, and other widely available force fields for conformational energies and for intermolecular-interaction energies and geometries. *Journal of Computational Chemistry*, *20*(7), 730–748. http://dx.doi.org/10.1002/(SICI)1096-987X(199905)20:7<730::AID-JCC8>3.0.CO;2-T.

Hansson, T., Oostenbrink, C., & van Gunsteren, W. (2002). Molecular dynamics simulations. *Current Opinion in Structural Biology*, *12*(2), 190–196. http://dx.doi.org/10.1016/S0959-440X(02)00308-1.

Harris, S. A., Gavathiotis, E., Searle, M. S., Orozco, M., & Laughton, C. A. (2001). Cooperativity in drug—DNA recognition: A molecular dynamics study. *Journal of the American Chemical Society*, *123*(50), 12658–12663. http://dx.doi.org/10.1021/ja016233n.

Harvey, A. L. (2008). Natural products in drug discovery. *Drug Discovery Today*, *13*(19–20), 894–901. http://dx.doi.org/10.1016/j.drudis.2008.07.004.
Hayes, D. F., Markus, H. S., Leslie, R. D., & Topol, E. J. (2014). Personalized medicine: Risk prediction, targeted therapies and mobile health technology. *BMC Medicine*, *12*, 37. http://dx.doi.org/10.1186/1741-7015-12-37.
Heinemann, V., Douillard, J. Y., Ducreux, M., & Peeters, M. (2013). Targeted therapy in metastatic colorectal cancer—An example of personalised medicine in action. *Cancer Treatment Reviews*, *39*(6), 592–601. http://dx.doi.org/10.1016/j.ctrv.2012.12.011.
Heinig, M., & Frishman, D. (2004). STRIDE: A web server for secondary structure assignment from known atomic coordinates of proteins. *Nucleic Acids Research*, *32*, 500–502. http://dx.doi.org/10.1093/nar/gkh429.
Hendrychova, T., Anzenbacherova, E., Hudecek, J., Skopalik, J., Lange, R., Hilderbrandt, P., et al. (2011). Flexibility of human cytochrome P450 enzymes: Molecular dynamics and spectroscopy reveal important function-related variations. *Biochemica et Biophysica Acta*, *1814*, 58–86.
Hillisch, A., Pineda, L. F., & Hilgenfeld, R. (2004). Utility of homology models in the drug discovery process. *Drug Discovery Today*, *9*(15), 659–669. http://dx.doi.org/10.1016/S1359-6446(04)03196-4.
Hornak, V., Abel, R., Okur, A., Strockbine, B., Roitberg, A., & Simmerling, C. (2006). Comparison of multiple amber force fields and development of improved protein backbone parameters. *Proteins*, *65*(3), 712–725. http://dx.doi.org/10.1002/prot. 21123.
Huang, D., & Caflisch, A. (2011). The free energy landscape of small molecule unbinding. *PLoS Computational Biology*. 7(2). e1002002http://dx.doi.org/10.1371/journal.pcbi.1002002.
Humphrey, W., Dalke, A., & Schulten, K. (1996). VMD: Visual molecular dynamics. *Journal of Molecular Graphics*, *7855*, 33–38.
Husby, J., Todd, A. K., Haider, S. M., Zinzalla, G., Thurston, D. E., & Neidle, S. (2012). Molecular dynamics studies of the STAT3 homodimer: DNA complex: Relationships between STAT3 mutations and protein-DNA recognition. *Journal of Chemical Information and Modeling*, *52*(5), 1179–1192. http://dx.doi.org/10.1021/ci200625q.
Hussain, M. R. M., Shaik, N. A., Al-Aama, J. Y., Asfour, H. Z., Subhani Khan, F., Masoodi, T. A., et al. (2012). In silico analysis of single nucleotide polymorphisms (SNPs) in human BRAF gene. *Gene*, *508*(2), 188–196. http://dx.doi.org/10.1016/j.gene.2012.07.014.
Jardón-Valadez, E., Bondar, A. N., & Tobias, D. J. (2014). Electrostatic interactions and hydrogen bond dynamics in chloride pumping by halorhodopsin. *Biochimica et Biophysica Acta*, *1837*(12), 1964–1972. http://dx.doi.org/10.1016/j.bbabio.2014.09.006.
Jo, S., Kim, T., Iyer, V. G., & Im, W. (2008). CHARMM-GUI: A web-based graphical user interface for CHARMM. *Journal of Computational Chemistry*, *29*(11), 1859–1865. http://dx.doi.org/10.1002/jcc.20945.
Johnson, A. D., Wang, D., & Sadee, W. (2005). Polymorphisms affecting gene regulation and mRNA processing: Broad implications for pharmacogenetics. *Pharmacology and Therapeutics*, *106*(1), 19–38.
Jorgensen, W. L., Maxwell, D. S., & Tirado-rives, J. (1996). Development and testing of the OPLS all-atom force field on conformational energetics and properties of organic liquid. *Journal of the American Chemical Society*, *7863*(15), 11225–11236.
Jorgensen, W. L., & Tirado-Rives, J. (1988). The OPLS [optimized potentials for liquid simulations] potential functions for proteins, energy minimizations for crystals of cyclic peptides and crambin. *Journal of the American Chemical Society*, *110*(6), 1657–1666. http://dx.doi.org/10.1021/ja00214a001.
Kalaiarasan, P., Kumar, B., Chopra, R., Gupta, V., Subbarao, N., & Bamezai, R. N. K. (2015). In silico screening, genotyping, molecular dynamics simulation and activity

studies of SNPs in pyruvate kinase M2. *Plos One*, *10*(3), e0120469. http://dx.doi.org/10.1371/journal.pone.0120469.

Kale, L., Skeel, R., Bhandarkar, M., Brunner, R., Gursoy, A., Krawetz, N., et al. (1999). NAMD2: Greater scalability for parallele molecular dynamics. *Journal of Computational Physics*, *151*, 283–312.

Kaminski, G. A., & Friesner, R. A. (2001). Evaluation and reparametrization of the OPLS-AA force field for proteins via comparison with accurate quantum chemical calculations on peptides. *Journal of Physical Chemistry B*, *105*, 6474–6487.

Karatrantos, A., Composto, R. J., Winey, K. I., & Clarke, N. (2014). Polymer conformations in polymer nanocomposites containing spherical nanoparticles. *Soft Matter*, *11*, 382–388. http://dx.doi.org/10.1039/C4SM01980F.

Karplus, M., & McCammon, J. A. (2002). Molecular dynamics simulations of biomolecules. *Nature Structural Biology*, *9*(9), 646–652. http://dx.doi.org/10.1038/nsb0902-646.

Katara, P. (2013). Role of bioinformatics and pharmacogenomics in drug discovery and development process. *Network Modeling Analysis in Health Informatics and Bioinformatics*, *2*(4), 225–230. http://dx.doi.org/10.1007/s13721-013-0039-5.

Katsios, C., & Roukos, D. H. (2010). Individual genomes and personalized medicine: Life diversity and complexity. *Personalized Medicine*, 7(4), 347–350. http://dx.doi.org/10.2217/pme.10.30.

Kholmurodov, K., Smith, W., Yasuoka, K., Darden, T., & Ebisuzaki, T. (2000). A smooth-particle mesh Ewald method for DL_POLY molecular dynamics simulation package on the Fujitsu VPP700. *Journal of Computational Chemistry*, *21*(13), 1187–1191. http://dx.doi.org/10.1002/1096-987X(200010)21:13<1187::AID-JCC7>3.0.CO;2-7.

Kim, M. J., Yoo, S. S., Choi, Y. Y., & Park, J. Y. (2010). A functional polymorphism in the pre-microRNA-196a2 and the risk of lung cancer in a Korean population. *Lung Cancer*, *69*, 127–129.

Kimura-Kataoka, K., Yasuda, T., Fujihara, J., Toga, T., Ono, R., Otsuka, Y., et al. (2012). Genetic and expression analysis of SNPs in the human deoxyribonuclease II: SNPs in the promoter region reduce its in vivo activity through decreased promoter activity. *Electrophoresis*, *33*(18), 2852–2858.

Korfi, K., Mandal, A., Furney, S., Wiseman, D., Somervaille, T. C. P., & Marais, R. (2015). A personalised medicine approach for ponatinib-resistant chronic myeloid leukaemia. *Annals of Oncology*, *26*(6), 1180–1187. http://dx.doi.org/10.1093/annonc/mdv110.

Kota, P. (2006). GUIMACS—A Java based front end for GROMACS. *In Silico Biology*, 7, 0008.

Kovacs, H., & Laaksonen, A. (1991). Molecular dynamic simulation and NMR study of water-acetonitrile mixtures. *Journal of American Chemical Society*, *113*(15), 5596–5605.

Krarup, L. H., Christensen, I. T., Hovgaard, L., & Frokjaer, S. (1998). Predicting drug absorption from molecular surface properties based on molecular dynamics simulations. *Pharmaceutical Research*, *15*(7), 972–978. http://dx.doi.org/10.1023/A:1011905522110.

Kream, R. M. (2015). Personalized- and one- medicine: Bioinformatics foundation in health and its economic feasibility. *Medical Science Monitor*, *21*, 201–204. http://dx.doi.org/10.12659/MSM.893207.

Krieger, E., Darden, T., Nabuurs, S. B., Finkelstein, A., & Vriend, G. (2004). Making optimal use of empirical energy functions: Force-field parameterization in crystal space. *Proteins*, *57*(4), 678–683. http://dx.doi.org/10.1002/prot.20251. PMID: 15390263.

Krüger, D. M., Ahmed, A., & Gohlke, H. (2012). NMSim web server: Integrated approach for normal mode-based geometric simulations of biologically relevant conformational transitions in proteins. *Nucleic Acids Research*, *40*(W1), 310–316. http://dx.doi.org/10.1093/nar/gks478.

Kumar, D. (2007). Genomic medicine: A new frontier of medicine in the twenty first century. *Genomic Medicine*, *1*, 3–7. http://dx.doi.org/10.1007/s11568-007-9003-8.

Kumari, R., Kumar, R., & Lynn, A. (2014). G-mmpbsa—A GROMACS tool for high-throughput MM-PBSA calculations. *Journal of Chemical Information and Modeling*, *54*(7), 1951–1962. http://dx.doi.org/10.1021/ci500020m.

Kutzner, C., Czub, J., & Grubmüller, H. (2011). Keep it flexible: Driving macromolecular rotary motions in atomistic simulations with GROMACS. *Journal of Chemical Theory and Computation*, 7(5), 1381–1393. http://dx.doi.org/10.1021/ct100666v.

Kuzkin, V. A. (2014). On angular momentum balance for particle systems with periodic boundary conditions. *ZAMM—Journal of Applied Mathematics and Mechanics/Zeitschrift für Angewandte Mathematik und Mechanik*. http://dx.doi.org/10.1002/zamm.201400045.

Lange, O. F., Van der Spoel, D., & de Groot, B. L. (2010). Scrutinizing moleuclar mechanics force fields on the submicrosecond timescale with NMR data. *Biophysical Journal*, *99*(647–655).

Lazar, P., Lee, Y., Kim, S., Chandrasekaran, M., & Lee, K. W. (2010). Molecular dynamics simulation study for ionic strength dependence of RNA-host factor interaction in Staphylococcus aureus Hfq. *Bulletin of the Korean Chemical Society*, *31*(6), 1519–1526. http://dx.doi.org/10.5012/bkcs.2010.31.6.1519.

Lee, J. (2001). Drug to SNP: A pharmacogenomics database for linking drug response to SNPs. *Genome Informatics*, *12*, 482–483.

Lee, A. J. X., & Swanton, C. (2012). Tumour heterogeneity and drug resistance: Personalising cancer medicine through functional genomics. *Biochemical Pharmacology*, *83*(8), 1013–1020. http://dx.doi.org/10.1016/j.bcp.2011.12.008.

Levitt, M., Hirshberg, M., Sharon, R., & Daggett, V. (1995). Potential energy function and parameters for simulations of the molecular dynamics of proteins and nucleic acids in solution. *Computer Physics Communications*, *91*(1-3), 215–231. http://dx.doi.org/10.1016/0010-4655(95)00049-L.

Lewis, L. D. (2005). Personalized drug therapy; the genome, the chip and the physician. *British Journal of Clinical Pharmacology*, *60*(1), 1–4. http://dx.doi.org/10.1111/j.1365-2125.2005.02457.x.

Li, B., Krishnan, V. G., Mort, M. E., Xin, F., Kamati, K. K., Cooper, D. N., et al. (2009). Automated inference of molecular mechanisms of disease from amino acid substitutions. *Bioinformatics (Oxford, England)*, *25*(21), 2744–2750. http://dx.doi.org/10.1093/bioinformatics/btp528.

Li, J., Zhou, N., Luo, K., Zhang, W., Li, X., Wu, C., et al. (2014). In silico discovery of potential VEGFR-2 inhibitors from natural derivatives for anti-angiogenesis therapy. *International Journal of Molecular Sciences*, *15*(9), 15994–16011.

Lipinski, C., Lombardo, F., Dominy, B. W., & Feeney, P. J. (2012). Experimental and computational approaches to estimate solubility and permeability in drug discovery and development settings. *Advanced Drug Delivery Reviews*, *64*, 4–17. http://dx.doi.org/10.1016/j.addr.2012.09.019.

Lobanov, M. Y., Bogatyreva, N. S., & Galzitskaya, O. V. (2008). Radius of gyration as an indicator of protein structure compactness. *Molecular Biology*, *42*(4), 623–628. http://dx.doi.org/10.1134/S0026893308040195.

Lockwood, S. H., Guan, A., Yu, A. S., Zhang, C., Zykovich, A., Korf, I., et al. (2014). The functional significance of common polymorphisms in zinc finger transcription factors. *G3 (Bethesda)*, *4*(9), 1647–1655.

Lucent, D., Vishal, V., & Pande, V. S. (2007). Protein folding under confinement: A role for solvent. *Proceedings of the National Academy of Sciences of the United States of America*, *104*(25), 10430–10434. http://dx.doi.org/10.1073/pnas.0608256104.

Lybrand, T. P. (1995). Ligand-protein docking and rational drug design. *Current Opinion in Structural Biology*, *5*(2), 224–228. http://dx.doi.org/10.1016/0959-440X(95)80080-8.

Lyne, P. D. (2002). Structure-based virtual screening: An overview. *Drug Discovery Today*, 7(20), 1047–1055. http://dx.doi.org/10.1016/S1359-6446(02)02483-2.
Ma, Q., & Lu, A. Y. H. (2011). Pharmacogenetcs, pharmacogenomics and individualised medicine. *Pharmacological Reviews*, *63*, 437–459.
Magesh, R., & George Priya Doss, C. (2012). Computational methods to work as first-pass filter in deleterious SNP analysis of Alkaptonuria. *The Scientific World Journal*, *2012*, 1–9. http://dx.doi.org/10.1100/2012/738423.
Marrink, S. J., & Tieleman, D. P. (2002). Molecular dynamics simulation of spontaneous membrane fusion during a cubic—Hexagonal phase transition. *Biophysical Journal*, *83*, 2386–2392.
McInnes, C. (2007). Virtual screening strategies in drug discovery. *Current Opinion in Chemical Biology*, *11*(5), 494–502. http://dx.doi.org/10.1016/j.cbpa.2007.08.033.
Meyer, U. A. (2000). Pharmacogenetics and adverse drug reactions. *Lancet*, *356*(9242), 1667–1671. http://dx.doi.org/10.1016/S0140-6736(00)03167-6.
Mizuno, N., Niwa, T., Yotsumoto, Y., & Sugiyama, Y. (2003). Impact of drug transporter studies on drug discovery. *Pharmacological Reviews*, *55*(3), 425–461. http://dx.doi.org/10.1124/pr.55.3.1.proteins.
Mongan, J., Case, D. A., & McCammon, J. A. (2004). Constant pH molecular dynamics in generalized born implicit solvent. *Journal of Computational Chemistry*, *25*(16), 2038–2048. http://dx.doi.org/10.1002/jcc.20139.
Mueller, S. C., Backes, C., Kalinina, O. V., Meder, B., Stöckel, D., Lenhof, H.-P., et al. (2015). BALL-SNP: Combining genetic and structural information to identify candidate non-synonymous single nucleotide polymorphisms. *Genome Medicine*, 7(1), 65. http://dx.doi.org/10.1186/s13073-015-0190-y.
Ng, P. C., & Henikoff, S. (2003). SIFT: Predicting amino acid changes that affect protein function. *Nucleic Acids Research*, *31*(13), 3812–3814.
Nichols, E. S., Swift, R. V., & Amaro, R. E. (2012). Rational prediction with molecular dynamics for hit identification. *Current Topics in Medicinal Chemistry*, *12*(18), 2002–2012. http://dx.doi.org/10.2174/156802612804910313.
Nilges, M., Clore, G. M., & Gronenborn, A. M. (1988). Determination of the three-dimensional structures of proteins from interproton distance data by dynamical simulated annealing from a random array of atoms. *FEBS Letters*, *229*, 129–136.
Nogovitsyn, E. A., Kolesnikov, A. L., & Budkov, Y. A. (2014). A coarse-grained model of glycosaminoglycans in aqueous salt solutions. A field-theoretical approach. *The Journal of Physical Chemistry B*, *118*, 13037–13039. http://dx.doi.org/10.1021/jp503749a.
Ogu, C. C., & Maxa, J. L. (2000). Drug interactions due to cytochrome P450. *BUMC Proceedings*, *13*, 421–423.
Okazaki, T., Javle, M., Tanaka, M., Abbruzzese, J. L., & Li, D. (2010). Single nucleotide polymorphisms of gemcitabine metabolic genes and pancreatic cancer survival and drug toxicity. *Clinical Cancer Research*, *16*(1), 320–329. http://dx.doi.org/10.1158/1078-0432.CCR-09-1555.
Okimoto, N., Futatsugi, N., Fuji, H., Suenaga, A., Morimoto, G., Yanai, R., et al. (2009). High-performance drug discovery: Computational screening by combining docking and molecular dynamics simulations. *PLoS Computational Biology*, *5*(10), e1000528. http://dx.doi.org/10.1371/journal.pcbi.1000528.
Onufriev, A., Bashford, D., & Case, D. A. (2004). Exploring protein native states and large-scale conformational changes with a modified generalized born model. *Proteins*, *55*(2), 383–394. http://dx.doi.org/10.1002/prot.20033.
Oostenbrink, C., Villa, A., Mark, A. E., & van Gunsteren, W. F. (2004). A biomolecular force field based on the free enthalpy of hydration and solvation: The GROMOS force-field parameter sets 53A5 and 53A6. *Journal of Computational Chemistry*, *25*(13), 1656–1676. http://dx.doi.org/10.1002/jcc.20090.

Oscarson, M. (2003). Pharmacogenetics of drug metabolising enzymes: Importance for personalised medicine. *Clinical Chemistry and Laboratory Medicine: CCLM/FESCC, 41*(4), 573–580. http://dx.doi.org/10.1515/CCLM.2003.087.

Panigrahi, S. K. (2008). Strong and weak hydrogen bonds in protein-ligand complexes of kinases: A comparative study. *Amino Acids, 34*(4), 617–633. http://dx.doi.org/10.1007/s00726-007-0015-4.

Peng, S., Kuang, Z., Sheng, C., Zhang, Y., Xu, H., & Cheng, Q. (2010). Association of microRNA-196a-2 gene polymorphism with gastric cancer risk in a Chinese population. *Digestive Diseases and Sciences, 55*, 2288–2293.

Perryman, A. L., Lin, J. H., & McCammon, J. A. (2004). HIV-1 protease molecular dynamics of a wild-type and of the V82F/I84V mutant: Possible contributions to drug resistance and a potential new target site for drugs. *Protein Science, 13*(4), 1108–1123. http://dx.doi.org/10.1110/ps.03468904.

Phillips, J. C., Braun, R., Wang, W., Gumbart, J., Tajkhorshid, E., Villa, E., et al. (2005). Scalable molecular dynamics with NAMD. *Journal of Computational Chemistry, 26*(16), 1781–1802. http://dx.doi.org/10.1002/jcc.20289.

Pirmohamed, M. (2014). Personalized pharmacogenomics: Predicting efficacy and adverse drug reactions. *Annual Review of Genomics and Human Genetics, 15*, 349–370. http://dx.doi.org/10.1146/annurev-genom-090413-025419.

Pokorska-Bocci, A., Stewart, A., Sagoo, G. S., Hall, A., Kroese, M., & Burton, H. (2014). "Personalized medicine": What's in a name? *Personalized Medicine, 11*(2), 197–210. http://dx.doi.org/10.2217/pme.13.107.

Ponder, J. W., & Case, David. A. (2003). Force fields for protein simulations. Introduction Molecular dynamics and Monte Carlo simulations of proteins, which directions the field is taking in developing new models. simple (sometimes called "'Class I'") potential energy function,, *Advance in Protein Chemistry, 66*, 27–85.

Popov, A. V., & Vorob, Y. N. (2010). GUI BioPASED: A program for molecular dynamics simulations of biopolymers with a graphical user interface. *Bioinformatics, 44*(4), 648–654. http://dx.doi.org/10.1134/S0026893310040217.

Popov, A. V., Vorobjev, Y. N., & Zharkov, D. O. (2013). MDTRA: A molecular dynamics trajectory analyzer with a graphical user interface. *Journal of Computational Chemistry, 34*(4), 319–325. http://dx.doi.org/10.1002/jcc.23135.

Porollo, A., & Meller, J. (2010). POLYVIEW-MM: Web-based platform for animation and analysis of molecular simulations. *Nucleic Acids Research, 38*(Suppl. 2), 662–666. http://dx.doi.org/10.1093/nar/gkq445.

Pronk, S., Páll, S., Schulz, R., Larsson, P., Bjelkmar, P., Apostolov, R., et al. (2013). GROMACS 4.5: A high-throughput and highly parallel open source molecular simulation toolkit. *Bioinformatics (Oxford, England), 29*(7), 845–854. http://dx.doi.org/10.1093/bioinformatics/btt055.

Roccatano, D., Colombo, G., Fioroni, M., & Mark, A. E. (2002). Mechanism by which 2,2,2-trifluoroethanol/water mixtures stabilize secondary-structure formation in peptides: A molecular dynamics study. *Proceedings of the National Academy of Sciences of the United States of America, 99*(19), 12179–12184. http://dx.doi.org/10.1073/pnas.182199699.

Sagui, C., & Darden, T. A. (1999). Molecular dynamics simulations of biomolecules: Long-range electrostatic effects. *Annual Review of Biophysics and Biomolecular Structure, 28*, 155–179. http://dx.doi.org/10.1146/annurev.biophys.28.1.155.

Salari, K., Watkins, H., & Ashley, E. a. (2012). Personalized medicine: Hope or hype? *European Heart Journal, 33*(13), 1564–1570. http://dx.doi.org/10.1093/eurheartj/ehs112.

Sands, Z., & Sansom, M. S. P. (2007). How does a voltage sensor interact with a lipid bilayer? Simulations of a potassium channel domain. *Structure (London, England: 1993), 15*(2), 235–244. http://dx.doi.org/10.1016/j.str.2007.01.004.

Sargsyan, K., Wright, J., & Lim (2012). GeoPCA: A new tool for multivariate analysis of dihedral angles based on principal component geodesics. *Nucleic Acids Research*, *40*(3), e25. http://dx.doi.org/10.1093/nar/gkr1069.

Schuler, L. D., Daura, X., & van Gunsteren, W. F. (2001). An improved GROMOS96 force field for aliphatic hydrocarbons in the condensed phase. *Journal of Computational Chemistry*, *22*(11), 1205–1218. http://dx.doi.org/10.1002/jcc.1078.

Seeliger, D., & de Groot, B. L. (2010). Conformational transitions upon ligand binding: Holo-structure prediction from apo conformations. *PLoS Computational Biology*. *6*(1). e1000634http://dx.doi.org/10.1371/journal.pcbi.1000634.

Sellis, D., Vlachakis, D., & Vlassi, M. (2009). Gromita: A fully integrated graphical user interface to gromacs 4. *Bioinformatics and Biology Insights*, *3*, 99–102.

Servant, N., Roméjon, J., Gestraud, P., La Rosa, P., Lucotte, G., Lair, S., et al. (2014). Bioinformatics for precision medicine in oncology: Principles and application to the SHIVA clinical trial. *Frontiers in Genetics*, *5*, 152. http://dx.doi.org/10.3389/fgene.2014.00152.

Shaikh, S., Li, J., Enkavi, G., Wen, P. C., Huang, Z., & Tajkhorshid, E. (2013). Visualizing functional motions of membrane transporters with molecular dynamics simulations. *Biochemistry*, *52*(4), 569–587. http://dx.doi.org/10.1021/bi301086x.

Sharma, V., & Sarkar, I. N. (2013). Bioinformatics opportunities for identification and study of medicinal plants. *Briefings in Bioinformatics*, *14*(2), 238–250. http://dx.doi.org/10.1093/bib/bbs021.

Shastry, B. S. (2006). Pharmacogenetics and the concept of individualized medicine. *The Pharmacogenomics Journal*, *6*(1), 16–21. http://dx.doi.org/10.1038/sj.tpj.6500338.

Sheehan, M. E., & Sharratt, P. N. (1998). Molecular dynamics methadology for the study of the solvent effects on a concentrated Diels-Alder reaction and the separation of the post-reaction mixture. *Computers and Chemical Engineering*, *22*, S27–S33.

Showalter, S. A., & Bruschweiler, R. (2007). Quantitative molecualr ensemble interpretation of NMR dipolar couplings without restraints. *Journal of the American Chemical Society*, *129*, 4158–4159.

Skopalik, J., Anzenbacher, P., & Otyepka, M. (2008). Flexibility of human cytochromes P450: Molecular dynamics reveals differences between CYPs 3A4, 2C9 and 2A6 which correlate with their substrate preferences. *Journal of Physical Chemistry B*, *112*, 8165–8173.

Sotomayor, M., & Schulten, K. (2007). Single-molecule experiments in vitro and in silico. *Science (New York, NY)*, *316*(5828), 1144–1148. http://dx.doi.org/10.1126/science.1137591.

Spacková, N., Cheatham, T. E., Ryjácek, F., Lankas, F., Van Meervelt, L., Hobza, P., et al. (2003). Molecular dynamics simulations and thermodynamics analysis of DNA-drug complexes. Minor groove binding between 4',6-diamidino-2-phenylindole and DNA duplexes in solution. *Journal of the American Chemical Society*, *125*(7), 1759–1769. http://dx.doi.org/10.1021/ja025660d.

Sudhakar, N., George Priya Doss, C., Thirumal Kumar, D., Chakraborty, C., Anand, K., & Suresh, M. (2015). Deciphering the impact of somatic mutations in exon 20 and exon 9 of PIK3CA gene in breast tumors among Indian women through molecular dynamics approach. *Journal of Biomolecular Structure & Dynamics*, *13*, 1–13.

Tashiro, M., Mochizuki, H., Iwabuchi, K., Sakurada, Y., Itoh, M., Watanabe, T., et al. (2002). Roles of histamine in regulation of arousal and cognition: Functional neuroimaging of histamine H1 receptors in human brain. *Life Sciences*, *72*(4–5), 409–414. http://dx.doi.org/10.1016/S0024-3205(02)02276-2.

Teague, S. J. (2003). Implications of protein flexibility for drug discovery. *Nature Reviews Drug Discovery*, *2*(7), 527–541. http://dx.doi.org/10.1038/nrd1129.

Tosco, P., Stiefl, N., & Landrum, G. (2014). Bringing the MMFF force field to the RDKit: Implementation and validation. *Journal of Cheminformatics*, *6*(1), 37.

Turner, P. J. (2005). *XMGRACE (version 5.1.19)*. Beaverton: Center for coastal and land-margin research, Oregon Graduate Institute of Science and Technology.

Van der Greef, J., Hankemeier, T., & McBurney, R. N. (2006). Metabolomics-based systems biology and personalized medicine: Moving towards n = 1 clinical trials? *Pharmacogenomics*, 7(7), 1087–1094. http://dx.doi.org/10.2217/14622416.7.7.1087.

Van Ruitenbeek, P., Vermeeren, a., & Riedel, W. J. (2010). Histamine H1 receptor antagonist cetirizine impairs working memory processing speed, but not episodic memory. *British Journal of Pharmacology*, *161*(2), 456–466. http://dx.doi.org/10.1111/j.1476-5381.2010.00907.

Vangone, A., Spinelli, R., Scarano, V., Cavallo, L., & Oliva, R. (2011). COCOMAPS: A web application to analyze and visualize contacts at the interface of biomolecular complexes. *Bioinformatics (Oxford, England)*, *27*(20), 2915–2916. http://dx.doi.org/10.1093/bioinformatics/btr484.

Vendruscolo, M., Kussell, E., & Domany, E. (1997). Recovery of protein structure from contact maps. *Folding and Design*, *2*, 295–306.

Wang, D., Johnson, A. D., Papp, A. C., Kroetz, D. L., & Sadee, W. (2005). Multidrug resistance polypeptide 1 (MDR1, ABCB1) variant 3435C > T affects mRNA stability. *Pharmacogenetics and Genomics*, *15*, 693–704.

Wang, W., Russell, A., & Yan, Y. (2014). Traditional Chinese medicine and new concepts of predictive, preventive and personalized medicine in diagnosis and treatment of suboptimal health. *The EPMA Journal*, *5*(1), 4. http://dx.doi.org/10.1186/1878-5085-5-4.

Wang, S., Yang, S., An, B., Wang, S., & Yin, Y. (2011). Molecular dynamics analysis reveals structural insights into mechanism of nicotine n-demethylation catalyzed by tobacco cytochrome P450 mono-oxygenase. *PLoS One*, *6*(8), e23342. http://dx.doi.org/10.1371/journal.pone.0023342.

Weiner, P. K., & Kollman, P. A. (1981). AMBER: Assisted model building with energy refinement. A general program for modeling molecules and their interactions. *Journal of Combinatorial Chemistry*, *2*, 287–303.

Weiner, S. J., Kollman, P. A., Case, D. A., Singh, U. C., Ghio, C., Alagona, G., et al. (1984). A new force field for molecular mechanical simulation of nucleic acids and proteins. *Journal of the American Chemical Society*, *106*, 765–784.

Williams, S. L., de Oliveira, C. A. F., & McCammon, J. A. (2010). Coupling constant pH molecular dynamics with accelerated molecular dynamics. *Journal of Chemical Theory and Computation*, *6*(2), 560–568. http://dx.doi.org/10.1021/ct9005294.

Wittayanarakul, K., Aruksakunwong, O., Sompornpisut, P., Sanghiran-Lee, V., Parasuk, V., Pinitglang, S., et al. (2005). Structure, dynamics and solvation of HIV-1 protease/saquinavir complex in aqueous solution and their contributions to drug resistance: Molecular dynamic simulations. *Journal of Chemical Information and Modeling*, *45*(2), 300–308. http://dx.doi.org/10.1021/ci049784g.

Xie, L., Ge, X., Tan, H., Xie, L., Zhang, Y., Hart, T., et al. (2014). Towards structural systems pharmacology to study complex diseases and personalized medicine. *PLoS Computational Biology*, *10*(5), e1003554. http://dx.doi.org/10.1371/journal.pcbi.1003554.

Xu, B., Feng, N. H., & Li, P. C. (2010). A functional polymorphism in pre-miR-146a gene is associated with prostate cancer risk and mature miR-146a expression in vivo. *Prostate*, *70*, 467–472.

Xu, P., Wang, J., Xu, Y., Chu, H., Liu, J., Zhao, M., et al. (2015). Advancement of polarizable force field and its use for molecular modeling and design. *Advances in Experimental Medicine and Biology*, *827*, 19–32. http://dx.doi.org/10.1007/978-94-017-9245-5_3.

Yang, L.-W., Eyal, E., Bahar, I., & Kitao, A. (2009). Principal component analysis of native ensembles of biomolecular structures (PCA_NEST): Insights into functional dynamics. *Bioinformatics*, *25*(5), 606–614. http://dx.doi.org/10.1093/bioinformatics/btp023.

Yang, Y., Qin, J., Liu, H., & Yao, X. (2011). IV-1 protease molecular dynamics of a wild-type and of the V82F/I84V mutant: Possible contributions to drug resistance and a potential new target site for drugs. *Journal of Chemical Information and Modeling, 51*(3), 680–692. http://dx.doi.org/10.1021/ci100427j.

Yin, D., & MacKerell, A. D. (1998). Combined ab initio/empirical approach for optimization of Lennard-Jones parameters. *Journal of Computational Chemistry, 19*(3), 334–348.

Yu, M., Chen, Z., Guo, W., Wang, J., Feng, Y., Kong, X., et al. (2015). Specifically targeted delivery of protein to phagocytic macrophages. *International Journal of Nanomedicine, 10*, 1743–1757. http://dx.doi.org/10.2147/IJN.S75950.

Zeidan, B. A., Townsend, P. A., Garbis, S. D., Copson, E., & Cutress, R. I. (2015). Clinical proteomics and breast cancer. *The Surgeon, 13*(5), 271–278. http://dx.doi.org/10.1016/j.surge.2014.12.003.

Zhang, Y. (2008). I-TASSER server for protein 3D structure prediction. *BMC Bioinformatics, 9*(40).

Zhang, Z., Miteva, M. a., Wang, L., & Alexov, E. (2012). Analyzing effects of naturally occurring missense mutations. *Computational and Mathematical Methods in Medicine, 2012*, 805827. http://dx.doi.org/10.1155/2012/805827.

Zhang, N., & Zhong, R. (2010). Structural basis for decreased affinity of Emodin binding to Val66-mutated human CK2α as determined by molecular dynamics. *Journal of Molecular Modeling, 16*, 771–780.

Zhao, H., & Caflisch, A. (2014). Molecular dynamics in drug design. *European Journal of Medicinal Chemistry, 91*, 4–14. http://dx.doi.org/10.1016/j.ejmech.2014.08.004.

Zhao, S., & Iyengar, R. (2012). System pharmocology: Network analysis to identify multiscale mechanisms of drug action. *Annual Review of Pharmacology and Toxicology, 52*, 505–521. http://dx.doi.org/10.1146/annurev-pharmtox-010611-134520.

Zhou, B., Wang, K., Wang, Y., Xi, M., Zhang, Z., Song, Y., et al. (2011). Common genetic polymorphisms in pre-microRNAs and risk of cervical squamous cell carcinoma. *Molecular Carcinogenesis, 50*, 499–505.

Zhu, F., Shi, Z., Qin, C., Tao, L., Liu, X., Xu, F., et al. (2012). Therapeutic target database update 2012: A resource for facilitating target-, oriented drug discovery. *Nucleic Acids Research, 40*, D1128–D1136. http://dx.doi.org/10.1093/nar/gkr797.

CHAPTER EIGHT

Personalized Pharmacoperones for Lysosomal Storage Disorder: Approach for Next-Generation Treatment

S.A. Syed Haneef, C. George Priya Doss[1]
Medical Biotechnology Division, School of Biosciences and Technology, VIT University, Vellore, Tamil Nadu, India
[1]Corresponding author: e-mail addresses: georgecp77@yahoo.co.in; georgepriyadoss@vit.ac.in

Contents

Advances in Protein Chemistry and Structural Biology, Volume 102
ISSN 1876-1623
http://dx.doi.org/10.1016/bs.apcsb.2015.10.001

Abstract

Lysosomal storage disorders (LSDs) are a collection of inborn errors of metabolic disorders affected by mutations in lysosome functional genes, commonly acid hydrolases. From the past decades, many approaches like enzyme replacement therapy, substrate reduction therapy are followed to treat these conditions. However, all these approaches have their own limitations. Proof-of-concept studies on pharmacological chaperone therapy (PCT) is now transformed into clinical practice to treat LSDs. Furthermore, it is narrowed with individuals to chaperone sensitive, specific mutations. Hence, personalizing the PCT will be a new direction to combat LSDs. In this review, we have discussed the available clinical strategies and pointed the light on how pharmacological chaperones can be personalized and hopeful to be a next-generation approach to address LSDs.

1. INTRODUCTION

Lysosomes are membrane-bound organelles first reported by Christian de Duve in 1955 (De Duve, 2005). Lysosomes are dynamic cell-recycling units showing protagonist action on different enzymes. They welcome and degrade macromolecules from various membrane-trafficking pathways like autophagic, endocytic, phagocytic, and secretory mostly for cell substrate breaks down. Lysosomal enzymes carry out precise biochemical reactions in breaking the substrates into smaller components which need to be either recycled or if not required will be eliminated from the cell. If these enzymes are absent in the lysosome, intermediate substrate storage/accumulation occurs eliciting an expansion in the size of the lysosomal compartment with a circumstance known as lysosomal storage disorders (LSDs). LSDs are caused by mutations in lysosomal genes, with subsequent cellular dysfunction. LSDs comprise a group of at least 41 distinct disorders in which accumulation of unwanted substrates in the lysosome of all type of cells in the body (Meikle, Hopwood, Clague, & Carey, 1999). These lysosomal enzymes are soluble acidic hydrolase produced in endoplasmic reticulum (ER), quality checked, and transported to lysosome via Golgi apparatus. Deeper articulation of LSD is due to protein misfolding which disrupts the three-dimensional conformation and creates an unstable enzyme resulting in intracellular retention. Hence, enzymes are not trafficked and found to be missing in the lysosome. Other reasons for LSDs are also credited by any defect in activator, transporter, integral membrane, or nonlysosomal enzyme required for universal lysosomal action. This sanctions the storage of substrates in different organs perceived in the ocular, hematologic, skeletal, visceral, and neurological system of the body (Parenti, Andria, & Valenzano, 2015). Lysosomal enzymes degrade a panoptic group of macromolecules including glycogen, glycolipids, glycosphingolipids, mucopolysaccharides, oligosaccharides, cholesterol, peptides, proteins, and glycoproteins to achieve continual synthesis and growth, as well as to maintain cellular homeostasis (Hopwood & Brooks, 1997). Functional disturbance in lysosome is also reasoned for various neurodegenerative disorders (Todde, Veenhuis, & van der Klei, 2009). LSDs are rare inborn metabolic errors with an autosomal recessive form and some show X-linked inheritance pattern with an overall estimated incidence rate of about 1 in 1500–7000 births (Staretz-Chacham, Lang, LaMarca, Krasnewich, & Sidransky, 2009). The seriousness of LSD is due to the activity of residual enzymes which matures progressively with age and involves unwantedly the action of the central nervous system, ensuring the lack of normal development of intellectual capacities in individuals. LSDs are seen in infantile, juvenile, and adult stages with no, lowered, and medium range of residual enzyme activities, respectively (Fan, 2003). Over the past few years, research is persuaded in aspect with various methods confining diagnostic approaches and increase to manage cell storage with supportive care and treating complications to meliorate LSDs. However, unambiguous treatment is not yet available to the affected individuals with LSDs. The principal cause of the associated disorders is directly or indirectly attributed to the same general mechanism, which is identified as a protein misfolding. As the proteins are decisively facilitating in every cell for comprehensive metabolism, it is very imperative to have our proteins in order, nor it is not a surprise, that protein misfolding will be the cause for many serious pathological conditions. Misfolding not only creates a normal functional loss but also gains toxic functionalities in cases like LSDs. Protein misfolding is known to be influenced by several reasons (Lodish et al., 2000). The underlying reasons could be a genetically inherited condition or nonsynonymous SNP; an error in the amino acid chain of an overly crowded cellular environment. Protein misfolding brings out faulty enzymes. There are molecules associated with ER's quality control system that can sense and tag the misfolded or incompletely folded protein. Such faulty enzymes are retrotranslocated to the cytosol and then rapidly degraded by the endoplasmic reticulum-associated degradation (ERAD) mechanism (Ellgaard & Helenius, 2003; Hegde & Ploegh, 2010; Parenti, 2009). The result of protein misfolding heads for enzyme absence/missing, and proteins are not trafficked to perform site-specific interactions. Misfolding creates an enormous problem, the body cannot perform simple, yet crucial functions without the usual concentration of an enzyme. This scenario is severe in LSDs with Fabry disease as one such example. It is the second most commonly occurring LSD after Gaucher disease. In Fabry disease, affected people lack the principal enzyme-α-galactosidase A. This enzyme helps in catalyzing lipids with terminal α-galactosyl residues. The absence of this enzyme causes glycosphingolipids, specifically, globotriaosylceramide and related α-galactosyl terminating glycol sphingolipids to accumulate in the lysosomes of vascular endothelial cells. In many cases, the misfolding in the final conformation of the protein is relatively negligible and will not have a huge deleterious effect on the protein. They are functionally competent but less stable which leads to inefficient trafficking from the ER. Such proteins underwent ERAD mechanism and destroyed prematurely. There are many diseases that still do not have a cure as the occurrence is extremely rare to carry out an extensive, clinical study. To combat LSDs major therapeutic strategy includes: enzyme replacement therapy (ERT), substrate reduction therapy (SRT), and pharmacological chaperone therapy (PCT). The main aim of this review is to highlight and classify LSDs to provide a clear understanding of the pathophysiology and their large phenotypic spectrum of disease manifestations. Furthermore, we tried to provide an overview of the few promising therapeutic approaches with the main emphasis on personalized PCT (Fig. 1).

2. LSDs CLASSIFICATION

LSDs are classified as, namely, lysosomal glycan degradation disorders, lysosomal lipid degradation disorders, lysosomal protein degradation disorders, lysosomal transporter protein disorders, and lysosomal-trafficking protein disorders (Greiner-Tollersrud & Berg, 2005). Table 1 provides the detailed information on various types of LSDs.

2.1 Lysosomal Glycan Degradation Disorder

2.1.1 Sialidosis (Mucolipidosis Type I)

Sialidosis (OMIM#256550) is a rare autosomal recessive lysosome storage disorder caused by defect in *α-neuraminidase* (Gene Symbol: *NEU1*; UniProt ID: Q99519; Chromosomal Location: 6p21.33) leading to accumulation of sialyl conjugates (validated) of sialic acid-containing oligosaccharides and glycopeptides. This results in visual defects with macular cherry-red spots, ataxia, hyperreflexia, seizures, mental retardation, and hepatosplenomegaly with an approximate frequency of <1/10,00,000 live births.

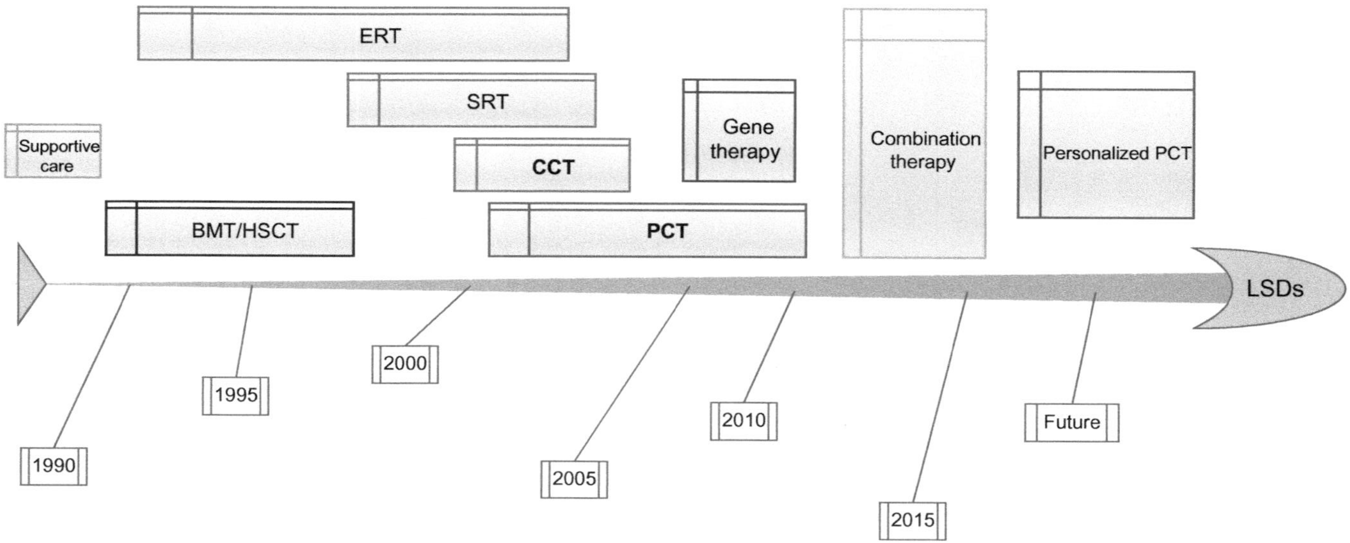

Figure 1 Overview of the few promising therapeutic approaches in LSDs.

Table 1 Classification of Lysosomal Storage Disorders

Disorder	Defective Enzyme	Gene Symbol
Lysosomal glycan degradation disorders		
Sialidosis (mucolipidosis type I)	Sialidase-1	NEU1
Galactosialidosis (GSL)	Lysosomal protective protein	CTSA
α-Mannosidosis	Lysosomal α-mannosidase	MAN2B1
β-Mannosidosis	β-Mannosidase	MANBA
Fucosidosis	α-Fucosidase	FUCA1
Schindler disease	α-*N*-acetylglucosaminidase	NAGA
Aspartylglucosaminuria (AGU)	Glycosylasparaginase	AGA
G_{M1} gangliosidosis	β-Galactosidase	GLB1
Tay–Sachs disease/G_{M2} gangliosidosis 1	β-Hexosaminidase α-subunit	HEXA
Sandhoff disease/G_{M2} gangliosidosis 2	β-Hexosaminidase β-subunit	HEXB
G_{M2} gangliosidosis AB variant	Ganglioside G_{M2}-activator protein	GM2A
Gaucher disease	Glucosylceramidase	GBA
Metachromatic leukodystrophy (MLD)	Arylsulfatase A	ARSA
Multiple sulfatase deficiency	Sulfatase-modifying factor 1	SUMF1
Globoid cell leukodystrophy (GLD)/Krabbe disease	Galactocerebrosidase	GALC
Fabry disease	α-Galactosidase A	GLA
MPS I (Hurler–Scheie syndrome)	α-L-Iduronidase	IDUA
MPS II (Hunter syndrome)	Iduronate 2-sulfatase	IDS
MPS III-Sanfilippo A	*N*-sulfoglucosamine sulfohydrolase	SGSH
MPS III-Sanfilippo B	Alpha-*N*-acetylglucosaminidase	NAGLU

Table 1 Classification of Lysosomal Storage Disorders—cont'd

Disorder	Defective Enzyme	Gene Symbol
MPS III-Sanfilippo C	Heparan-alpha-glucosaminide *N*-acetyltransferase	HGSNAT
MPS III-Sanfilippo D	*N*-acetylglucosamine-6-sulfatase	GNS
MPS IVA-Morquio A	*N*-acetylgalactosamine 6-sulfatase	GALNS
MPS IVB-Morquio B	β-Galactosidase	GLB1
MPS VI (Maroteaux–Lamy syndrome)	Arylsulfatase B	ARSB
MPS VII (Sly syndrome)	β-Glucuronidase	GUSB
MPS IX (Hyaluronidase deficiency)	Hyaluronidase-1	HYAL1
Pompe disease	Lysosomal α-glucosidase	GAA
Defects in lipid degradation		
Niemann–Pick type A (NPDA)	Sphingomyelin phosphodiesterase	SMPD1
Niemann–Pick type B (NPDB)		
Farber lipogranulomatosis (FL)	Acid ceramidase	ASAH1
Wolman disease (WOD)/ cholesteryl ester storage disease (CESD)	Lysosomal acid lipase/ cholesteryl ester hydrolase	LIPA
Defects in protein degradation		
Pycnodysostosis (PKND)	Cathepsin K	CTSK
Ceroid-lipofuscinosis neuronal 1 (CLN1)	Palmitoyl-protein thioesterase 1	PPT1
Ceroid-lipofuscinosis neuronal 2 (CLN2)	Tripeptidyl peptidase	TPP1
Defects in lysosomal transporters		
Cystinosis	Cystinosin	CTNS
Salla disease (SD)/Sialic acid storage disease (SASD)	Sialin	SLC17A5

Continued

Table 1 Classification of Lysosomal Storage Disorders—cont'd

Disorder	**Defective Enzyme**	**Gene Symbol**
Defects in lysosomal-trafficking proteins		
Mucolipidosis type II (I-cell disease)	*N*-acetylglucosamine-1-phosphotransferase subunits alpha/beta	GNPTAB
Mucolipidosis type III (pseudo-Hurler polydystrophy)		
Mucolipidosis type IV (MLIV)	Mucolipin-1	MCOLN1
Danon disease (DAND)	Lysosome-associated membrane glycoprotein 2	LAMP2
Niemann–Pick type C1 (NPC1)	Niemann–Pick C1 protein	NPC1
Ceroid-lipofuscinosis neuronal 3 (CLN3)/Batten disease	Battenin	CLN3
Ceroid-lipofuscinosis neuronal 6 (CLN6)	Ceroid-lipofuscinosis neuronal protein 6	CLN6
Ceroid-lipofuscinosis neuronal 8 (CLN8)	Protein CLN8	CLN8
Chediak–Higashi syndrome (CHS)	Lysosomal-trafficking regulator	LYST
Hermansky–Pudlak syndrome 1 (HPS1)	Hermansky–Pudlak syndrome 1 protein	HPS1
Hermansky–Pudlak syndrome 2 (HPS2)	AP-3 complex subunit beta-1	HPS2

2.1.2 Galactosialidosis

Galactosialidosis (OMIM#256540) is a neurodegenerative LSD caused by defect in *lysosomal protective protein–cathepsin A* (Gene Symbol: *CTSA*; UniProt ID: P10619; Chromosomal Location: 20q13.12) leading to combined deficiency of *β-galactosidase* (*GLB1*) and *neuraminidase* (*NEU1*) genes, responsible for G_{M1} gangliosidosis/MPS IVB (Morquio B) and sialidosis, respectively. It is an autosomal recessive pattern characterized by cherry-red spots with corneal clouding, hydropsfetalis, hepatosplenomegaly, coarse facies, skeletal dysplasia, abnormally large abdominal organs (visceromegaly), myoclonus, ataxia, clusters of enlarged blood vessels forming small, dark red spots on the skin (angiokeratomas), neurological deterioration, growth, psychomotor and mental retardation, and early death. Prevalence of galactosialidosis is unknown with only more than 100 cases reported.

2.1.3 α-Mannosidosis

α-Mannosidosis (OMIM#248500) is a rare autosomal recessive lysosome storage disorder resulting from a deficiency in *lysosomal alpha-mannosidase* (Gene Symbol: *MAN2B1*; UniProt ID: O00754; Chromosomal Location: 19p13.2). This shows a broad range of clinical phenotypes caused by intracellular accumulation of mannose-containing oligosaccharides, which ultimately may lead to immune deficiency, mental retardation, hearing loss, muscular weakness, skeletal abnormalities, and slight strabismus with an overall estimated frequency of 1 in 5,00,000 individuals.

2.1.4 β-Mannosidosis

β-Mannosidosis (OMIM#248510) is an autosomal lysosome storage disorder caused due to *β-mannosidase* deficiency (Gene Symbol: *MANBA*; UniProt ID: O00462; Chromosomal Location: 4q24). Oligosaccharides like sialylated trisaccharide NeuAc(α2–6)Man(β1–4)GlcNAc found to be accumulated in the tissues. Inherited individuals show an increased risk of respiratory and ear infections, hearing loss, speech impairment, swallowing difficulties, poor muscle tone (hypotonia), and reduced sensation or other nervous system abnormalities. So far, only 20 affected individuals have been reported worldwide.

2.1.5 Fucosidosis

Fucosidosis (OMIM#230000) is an autosomal lysosome storage disorder caused due to *α-fucosidase* deficiency (Gene Symbol: *FUCA1*; UniProt ID: P04066; Chromosomal Location: 1p34). Accumulation of lipids (fucose-rich oligosaccharides, sphingolipids, and glycopeptides chains) and the absence of glycosaminoglycans (GAGs) resulting in impaired growth, dysostosis multiplex, seizures, spasticity, angiokeratoma, and visceromegaly. It is a rare condition with only 100 cases reported.

2.1.6 Schindler Disease

Schindler disease (OMIM#609241) is a rare autosomal lysosome storage disorder characterized by the deficient activity of *α-N-acetylglucosaminidase* (Gene Symbol: *NAGA*; UniProt ID: P17050; Chromosomal Location: 22q13.2) with only few reported cases. Accumulation of glycosphingolipid, glycoprotein, and oligosaccharide complexes have terminal *N*-acetylgalactosaminyl residues seen with symptoms like psychomotor retardation, developmental retrogression, hypotonia, spasticity, strabismus, nystagmus, optic atrophy, myoclonic movements, and grand-mal seizures as common.

2.1.7 Aspartylglucosaminuria

Aspartylglucosaminuria (OMIM#208400) is an autosomal lysosome storage disorder with a defect in human *glycosylasparaginase* (Gene Symbol: *AGA*; UniProt ID: P20933; Chromosomal Location: 4q34.3). It is more frequent among Finnish population and globally rare. Excessive glycoasparagine is accumulated in the tissues and excreted in urine. People with this condition have weak bones which are prone to fracture (osteoporosis), hypermobility, loose skin, ocular hypertelorism, small ears, and full lips.

2.1.8 G_{M1} Gangliosidosis

G_{M1} gangliosidosis (OMIM#230500) is rare lysosome storage disorder showing autosomal recessive inheritance pattern caused by a *β-galactosidase* deficiency (Gene Symbol: *GLB1*; UniProt ID: P16278; Chromosomal Location: 3p22.3) with an estimated frequency of 1 in 10,000–2,00,000 live births. Ganglioside substrates get accumulated in lysosome that progressively destroys neurons in the brain and spinal cord leading to developmental regress, hepatosplenomegaly, gingival hypertrophy, and cardiomyopathy.

2.1.9 Tay–Sachs Disease/G_{M2} Gangliosidosis 1

Tay–Sachs disease/G_{M2} gangliosidosis 1 (OMIM#272800) is a rare LSD showing autosomal inheritance pattern by the accumulation of G_{M2} gangliosides due to *β-hexosaminidase α-subunit* deficiency (Gene Symbol: *HEXA*; UniProt ID: P06865; Chromosomal Location: 15q23). Presenting an incidence of 1 in 3,20,000 newborns. Spinocerebellar ataxia, mental capacities, and behavior are affected by early childhood death are seen.

2.1.10 Sandhoff Disease/G_{M2} Gangliosidosis 2

Sandhoff disease/G_{M2} gangliosidosis 2 (OMIM#268800) is a rare lysosome storage disorder caused due to *β-hexosaminidase β-subunit* deficiency (Gene Symbol: *HEXB*; UniProt ID: P07686; Chromosomal Location: 5q13.3) showing an inheritance pattern of autosomal recessive type. High levels of oligosaccharides are excreted in urine with clinical symptoms similar as

Tay–Sachs disease including organomegaly and bone abnormalities with a variable occurrence among different populations.

2.1.11 G_{M2} Gangliosidosis AB Variant

G_{M2} gangliosidosis AB variant (OMIM#272750) is an extremely rare lysosome storage disorder with autosomal inheritance caused due to *ganglioside G_{M2}-activator protein* deficiency (Gene Symbol: *GM2A*; UniProt ID: P17900; Chromosomal Location: 5q33.1). Accumulation of G_{M2} gangliosides is seen with symptoms like seizures, vision and hearing loss, intellectual disability, and paralysis.

2.1.12 Gaucher Disease

Gaucher disease (OMIM#230800) is an autosomal lysosome storage disorder affected due to loss of human *glucosylceramidase* activity (Gene Symbol: *GBA*; UniProt ID: P04062; Chromosomal Location: 1q22) with an estimated prevalence of 1 in 10,000 births globally. In this disorder, accumulation of β-glucocerebrosidase in the cells of the reticuloendothelial system of the liver, spleen, bone marrow presenting organomegaly, bone anomalies, and cytopenia as clinical symptoms.

2.1.13 Metachromatic Leukodystrophy

Metachromatic leukodystrophy (MLD) (OMIM#250100) is a neurodegenerative lysosome storage disorder with autosomal recessive pattern reported 1 in 40,000–1,60,000 individuals worldwide. Caused due to *arylsulfatase A* deficiency (Gene Symbol: *ARSA*; UniProt ID: P15289; Chromosomal Location: 22q13.33) resulting in accumulation of sulfatides (sulfogalactosyl ceramides or sulfogalacto cerebrosides) in the nervous system and kidneys leading to progressive deterioration of intellectual function and motor skills, peripheral neuropathy, incontinence, seizures, paralysis, an inability to speak, blindness, and hearing loss.

2.1.14 Multiple Sulfatase Deficiencies

Multiple sulfatase deficiency (OMIM#272200) is an autosomal lysosome storage disorder resulting in the accumulation of sulfatides, sulfated GAGs, sphingolipids, and steroid sulfates in tissues due to defect in *sulfatase-modifying factor 1* (Gene Symbol: *SUMF1*; UniProt ID: Q8NBK3; Chromosomal Location: 3p26.1). It is estimated to occur 1 in 10,00,000 individuals worldwide showing symptoms like hypotonia, coarse facial features, mild deafness, skeletal anomalies, ichthyosis, hepatomegaly, developmental delay, progressive neurologic deterioration, and hydrocephalus.

2.1.15 Globoid Cell Leukodystrophy/Krabbe Disease

Globoid cell leukodystrophy (GLD)/Krabbe disease (OMIM#245200) is an autosomal recessive lysosome storage disorder caused by *galactocerebrosidase* deficiency (Gene Symbol: *GALC*; UniProt ID: P54803; Chromosomal Location: 14q.31.3). It affects the white matter (myelin) and accumulates multinucleated globoid cells in the central and peripheral nervous system, with an approximate incidence of 1 in 1,00,000–2,50,000 individuals worldwide. Symptoms include irritability, stiffness, poor head control, feeding difficulties, intermittent thumb clasp, episodes of increased temperature, developmental delay, hypertonic episodes occur with opisthotonus, myoclonic seizures, developmental regression, fisting, vision deficits, hypotonia, blindness, and deafness.

2.1.16 Fabry Disease

Fabry disease (OMIM#301500) is an X-linked recessive lysosome storage disorder of glycosphingolipid metabolism leading to the accumulation of globotriaosylceramide in lysosomes, influenced by *α-galactosidase A* deficiency (Gene Symbol: *GLA*; UniProt ID: P06280; Chromosomal Location: Xq22.1). One in 40,000–60,000 males is affected globally and seen less frequently among females. Characteristic features of Fabry disease include acroparesthesias, angiokeratomas, hypohidrosis, corneal opacity, problems with the gastrointestinal system, ringing in the ears (tinnitus), and hearing loss.

2.1.17 MPS I (Hurler–Scheie Syndrome)

MPS I (Hurler–Scheie Syndrome) (OMIM#607015) is an autosomal lysosome storage disorder with problem in breaking down of GAG and getting accumulated in lysosomes due to *α-L-iduronidase* deficiency (Gene Symbol: *IDUA*; UniProt ID: P35475; Chromosomal Location: 4p16.3). One in 1,00,000 individuals is affected by this syndrome with phenotypic characters like facial coarseness, physical problems, developmental delays, mental retardation, hernia, macrocephaly, hydrocephalus, heart-valve abnormalities, hepatosplenomegaly, and macroglossia.

2.1.18 MPS II (Hunter Syndrome)

MPS II (Hunter Syndrome) (OMIM#309900) is rare X-linked recessive lysosome storage disorder leading to progressive accumulation of GAG in all the body cells due to defect in *iduronate 2-sulfatase* (Gene Symbol: *IDS*; UniProt ID: P22304; Chromosomal Location: Xq28). This occurs approximately 1 in 1,00,000 to 1 in 1,70,000 males. Patients with MPS II excrete excessive amount of chondroitin sulfate B (dermatan sulfate) and heparitin sulfate (heparan sulfate) in the urine. Symptoms include macrocephaly, hydrocephalus, hepatosplenomegaly, umbilical hernia or an inguinal hernia, spinal stenosis, and ventricular hypertrophy eventually leading to heart failure.

2.1.19 MPS III

MPS III is the most common type of MPSs, which is also known as Sanfilippo syndrome, presenting autosomal recessive inheritance pattern with four subtypes:

(i) MPS III-Sanfilippo A (OMIM#252900) due to *N-sulfoglucosamine sulfohydrolase* deficiency (Gene Symbol: *SGSH*; UniProt ID: P51688; Chromosomal Location: 17q25.3).

(ii) MPS III-Sanfilippo B (OMIM#252920) due to *alpha-N-acetylglucosaminidase* deficiency (Gene Symbol: *NAGLU*; UniProt ID: P54802; Chromosomal Location: 17q21.2).

(iii) MPS III-Sanfilippo C (OMIM#252930) due to *heparan-alpha-glucosaminide N-acetyltransferase* deficiency (Gene Symbol: *HGSNAT*; UniProt ID: Q68CP4; Chromosomal Location: 8p11.21).

(iv) MPS III-Sanfilippo D (OMIM#252940) due to *N-acetylglucosamine-6-sulfatase* deficiency (Gene Symbol: *GNS*; UniProt ID: P15586; Chromosomal Location: 12q14.3).

A defect in any one of these enzymes disrupts the breakdown of heparin sulfates (HSs) and accumulates in the lysosomes. The estimated incidence of all four types combined is 1 in 70,000 newborns. Clinical symptoms are same as in other MPSs.

2.1.20 MPS IVA-Morquio A

MPS IVA-Morquio A (OMIM#253000) is an autosomal recessive LSD due to *N-acetylgalactosamine 6-sulfatase* deficiency (Gene Symbol: *GALNS*; UniProt ID: P34059; Chromosomal Location: 16q24.3) leading to intracellular accumulation of keratan sulfate and chondroitin-6-sulfate. Key clinical features include short stature, skeletal dysplasia, dental anomalies, and corneal clouding with an estimated incidence of about 1 in 2,00,000–3,00,000 live births.

2.1.21 MPS IVB-Morquio B

MPS IVB-Morquio B (OMIM#253010) is an allelic form of galactosialidosis showing autosomal recessive pattern due to *β-galactosidase* deficiency (Gene Symbol: *GLB1*; UniProt ID: P16278; Chromosomal Location: 3p22.3). Clinical symptoms and prevalence are same as MPS IVA.

2.1.22 MPS VI (Maroteaux–Lamy Syndrome)

MPS VI (Maroteaux–Lamy syndrome) (OMIM#253200) is an autosomal recessive lysosome storage disorder due to *arylsulfatase B* deficiency (Gene Symbol: *ARSB*; UniProt ID: P15848; Chromosomal Location: 5q14.1). Dermatan sulfate is found to be accumulated in the lysosomes with clinical features like short stature, hepatosplenomegaly, dysostosis multiplex, stiff joints, corneal clouding, cardiac abnormalities, and facial dysmorphism. The estimated incidence is found to be 1 in 2,50,000–6,00,000 newborns.

2.1.23 MPS VII (Sly Syndrome)

MPS VII (Sly syndrome) (OMIM#253220) is an autosomal recessive lysosome storage disorder described by the inability to degrade glucuronic acid-containing GAGs due to *β-glucuronidase* deficiency (Gene Symbol: *GUSB*; UniProt ID: P08236; Chromosomal Location: 7q11.21). With an estimated incidence of about 1 in 2,50,000 newborns, showing hydropsfetalis, hepatomegaly, skeletal anomalies, coarse facies, and mental impairment.

2.1.24 MPS IX (Hyaluronidase Deficiency)

MPS IX (hyaluronidase deficiency) (OMIM#601492) is an autosomal lysosome storage disorder with accumulation of hyaluronan (hyaluronic acid) in the tissues due to defect in *hyaluronidase-1* (Gene Symbol: *HYAL1*; UniProt ID: Q12794; Chromosomal Location: 3p21.31). It is a rare condition with less than 1 in 10,00,000 individuals with short stature, lumps with painful swellings around joints, impaired growth, flattened nose bridge, bifid uvula, and cleft palate, respectively.

2.1.25 Pompe Disease

Pompe disease (OMIM#232300) is an autosomal lysosome storage disorder caused by the deficiency of lysosomal *α-glucosidase* (Gene Symbol: *GAA*; UniProt ID: P10253; Chromosomal Location: 17q25.3) also referred as acid maltase deficiency or glycogen storage disease type II as the complex sugars of glycogen stored in the cells. Predominant symptom includes myopathy, hypotonia, hepatomegaly, cardiomegaly with heart defects, affected infants may also fail to gain weight and grow at the expected rate, and some have breathing problems. It is reported to be common in 1 in 40,000 individuals.

2.2 Lysosomal Lipid Degradation Disorder

2.2.1 Niemann–Pick Type A and B

Niemann–Pick type A and B (OMIM#257200; OMIM#607616) are autosomal lysosome storage disorders caused due to defect in *sphingomyelin phosphodiesterase* (Gene Symbol: *SMPD1*; UniProt ID: P17405; Chromosomal Location: 11p15.4) with an estimated combined prevalence of 1 in 2,50,000 individuals holding defect in sphingomyelin degradation and stored in reticuloendothelial cells. This accumulation in the ganglion cells of the central nervous system leads to cell death, early jaundice, enlarging abdomen, poor developmental progress, hepatosplenomegaly, hypotonia, rigidity, and mental retardation.

2.2.2 Farber Lipogranulomatosis

Farber lipogranulomatosis (OMIM#228000) is a rare autosomal lysosome storage disorder involving improper breakdown of sphingolipids and lipids accumulating abundantly around the joints, governed by the defect in *acid ceramidase* (Gene Symbol: *ASAH1*; UniProt ID: Q13510; Chromosomal Location: 8q22). Only 80 cases reported worldwide with distinct clinical symptoms like horse voice, small lumps of fat under the skin (lipogranulomas), swollen and painful joints, breathing difficulty, hepatosplenomegaly, and developmental delays.

2.2.3 Wolman/Cholesteryl Ester Storage Disease

Wolman/cholesteryl ester storage disease (OMIM#278000) is an autosomal lysosome storage disorder caused due to *lysosomal acid lipase/cholesteryl ester hydrolase* deficiency (Gene Symbol: *LIPA*; UniProt ID: P38571; Chromosomal Location: 10q23.31). Excess amount of lipids get accumulated in organs like spleen, liver, bone marrow, small intestine, and kidneys, and calcium deposits in adrenal glands. Estimated prevalence of about 1 in 3,50,000 newborns. Symptoms include jaundice, vomiting, diarrhea, developmental delay, poor weight gain, low muscle tone, anemia, hepatosplenomegaly, and poor absorption of nutrients from food.

2.3 Lysosomal Protein Degradation Disorder

2.3.1 Pycnodysostosis

Pycnodysostosis (OMIM#265800) is a rare autosomal lysosome storage disorder due to *cathepsin K* deficiency (Gene Symbol: *CTSK*; UniProt ID: P43235; Chromosomal Location: 1q21.3), resulting in the accumulation of nondigested phagocytosed collagen in fibroblast with an estimated prevalence of 1–9 in 10,00,000 individuals. Deficiency results in short stature, deformity of the skull, maxilla and phalanges, osteosclerosis, and fragility.

2.3.2 Neuronal Ceroid Lipofuscinosis 1 and 2

Neuronal ceroid lipofuscinosis 1 and 2 (OMIM#256730; OMIM#204500) are autosomal lysosome storage disorders caused due to *palmitoyl-protein thioesterase 1* deficiency (Gene Symbol: *PPT1*; UniProt ID: P50897; Chromosomal Location: 1p34.2) and *tripeptidyl peptidase* deficiency (Gene Symbol: *TPP1*; UniProt ID: O14773; Chromosomal Location: 11q15.4), respectively. In which intracellular accumulation of autofluorescent lipopigment, storage materials are referred as granular osmiophilic deposits (GROD) with an incidence of 1 in 10,00,000 live births having clinical symptoms like motor dysfunction, seizures, spasticity, and loss of vision.

2.4 Lysosomal Protein Degradation Disorder

2.4.1 Cystinosis

Cystinosis (OMIM#219800) is an autosomal recessive lysosome storage disorder characterized by the accumulation of cysteine inside the lysosomes due to defect in *cystinosin* (Gene Symbol: *CTNS*; UniProt ID: O60931; Chromosomal Location: 17p13.2) progressively leading to hypothyroidism, insulin-dependent diabetes, hepatosplenomegaly with portal hypertension. Estimated frequency of 1 in 1,00,000–2,00,000 live births.

2.4.2 Salla/Sialic Acid Storage Disease

Salla/sialic acid storage disease (OMIM#604369) is a very rare autosomal recessive lysosome storage disorder with only few cases reported worldwide caused due to defect in *sialin* (Gene Symbol: *SLC17A5*; UniProt ID: Q9NRA2; Chromosomal Location: 6q13) leading to accumulation of sialic acid in the cells with clinical symptoms like hydropsfetalis, facial coarseness, seizures, bone malformations, hepatosplenomegaly, cardiomegaly, and ascites.

2.5 Lysosomal-Trafficking Protein Degradation Disorder

2.5.1 Mucolipidosis Type II and III

Mucolipidosis type II (I-Cell disease) (OMIM#252500) and Mucolipidosis type III (pseudo-Hurler polydystrophy) (OMIM#252600) are autosomal lysosome storage disorders caused due to *N-acetylglucosamine-1-phosphotransferase subunits alpha/beta* deficiency (Gene Symbol: *GNPTAB*; UniProt ID: Q3T906; Chromosomal Location: 12q23.2) resulting in the accumulation of large molecules within lysosomes. It is a rare condition seen at a frequency of 1 in 1,00,000–4,00,000 individuals. Congenital dislocation of the hip, thoracic deformities, hernia, and hyperplastic gums are evident for this disorder.

2.5.2 Mucolipidosis Type IV

Mucolipidosis type IV (OMIM#252650) is an autosomal recessive neurodegenerative lysosome storage disorder caused due to *mucolipin-1* deficiency (Gene Symbol: *MCOLN1*; UniProt ID: Q9GZU1; Chromosomal Location: 19p13.2) in which phospholipids, gangliosides, and mucopolysaccharides accumulate in lysosomes. It is estimated to occur in 1 in 40,000 individuals with the main symptoms of achlorhydria, anemia, psychomotor delay, hypotonia, and spasticity.

2.5.3 Danon Disease

Danon disease (OMIM#300257) is a rare X-linked dominant lysosome storage disorder especially affecting cardiac muscles due to *lysosome-associated membrane glycoprotein 2* deficiency (Gene Symbol: *LAMP2*; UniProt ID: P13473; Chromosomal Location: Xq24) leading to accumulation of autophagic vacuoles; a condition characterized by cardiomyopathy, myopathy, palpitations, arrhythmia or chest pain, and conduction abnormalities.

2.5.4 Niemann–Pick Type C1

Niemann–Pick type C1 (OMIM#257220) is an autosomal lysosome storage disorder caused due to *Niemann–Pick C1 protein* deficiency (Gene Symbol: *NPC1*; UniProt ID: O15118; Chromosomal Location: 18q11.2), which leads to intracellular accumulation of cholesterol with the main symptoms of cerebellar ataxia, dysarthria, cataplexy, dystonia, vertical supranuclear ophthalmoplegia, seizures, and dementia. Estimated occurrence of about 1–9 in 1,00,000 individuals.

2.5.5 Batten Disease/Neuronal Ceroid Lipofuscinosis 3, 6, and 8

Batten disease/neuronal ceroid lipofuscinosis 3 (OMIM#204200) and 6 (OMIM#601780) and 8 (OMIM#600143) are autosomal lysosome storage disorders caused due to *battenin* deficiency (Gene Symbol: *CLN3*; UniProt ID: Q13286; Chromosomal Location: 16p11.2), *ceroid-lipofuscinosis neuronal protein 6* deficiency (Gene Symbol: *CLN6*; UniProt ID: Q9NWW5; Chromosomal Location: 5q23), and *protein CLN8* deficiency (Gene Symbol: *CLN8*; UniProt ID: Q9UBY8; Chromosomal Location: 8p23.3), respectively. In which intracellular accumulation of autofluorescent lipopigment, storage materials that are referred as GROD with a combined incidence of 1 in 10,00,000 live births having clinical symptoms like motor dysfunction, seizures, spasticity, and loss of vision.

2.5.6 Chediak–Higashi Syndrome

Chediak–Higashi syndrome (OMIM#214500) is a rare autosomal lysosome storage disorder caused due to defect in *lysosomal-trafficking regulator protein* (Gene Symbol: *LYST*; UniProt ID: Q99698; Chromosomal Location: 1q42.3) with only 200 cases reported so far. This condition results in decreased phagocytosis creating symptoms like oculocutaneous albinism, severe immunodeficiency, mild bleeding, neurological dysfunction, and lympho-proliferative disorder.

2.5.7 Hermansky–Pudlak Syndrome 1 and 2

Hermansky–Pudlak syndrome 1 (OMIM#203300) and 2 (OMIM#608233) are autosomal recessive lysosome storage disorders caused due to *Hermansky–Pudlak syndrome 1 protein* deficiency (Gene Symbol: *HPS1*; UniProt ID: Q92902; Chromosomal Location: 10q24.2) and *AP-3 complex subunit beta-1* deficiency (Gene Symbol: *HPS2*; UniProt ID: O00203; Chromosomal Location: 5q14.1), respectively. They show a combined incidence of about 1 in 5,00,000–10,00,000 individuals worldwide characterizing oculocutaneous albinism, bleeding diathesis and, pulmonary fibrosis, or granulomatous colitis resulting from lysosomal ceroid storage.

3. TREATMENT OF LSDs

Supportive care is the only available treatment for most of the LSDs. Recently, two therapeutic ideas were developed, as a clinical strategy for these disorders: substituting the defective enzyme and increasing its activity by ERT (Rohrbach & Clarke, 2007), and making use of substrates to achieve normal catabolism by SRT (Platt & Jeyakumar, 2008). Modernized cell therapy like bone marrow transplantation (BMT) (Krivit, Peters, & Shapiro, 1999) and hematopoietic cell transplantation (Rovelli, 2008) are based on the capabilities of the transferred cells to produce the lacking enzyme and correct the defect in the recipient is an alternative way to combat LSDs. Direct transfer of the normal gene into the defective cells to produce missing enzyme by the process so-called gene therapy (Cheng & Smith, 2003; Hodges & Cheng, 2006; Sands & Davidson, 2006) is also an encouraging approach for LSDs. Another strategy is to assist proper protein folding inside the cellular environment in precluding misfolding with the help of small molecules-chaperones. Chemical chaperone therapy (CCT) (Matsuda et al., 2003) is based on the chemical molecule's tendency to bind and stabilize the misfolded enzyme. A distinctive group of chaperones explicitly called pharmacological chaperones are designed these days especially to bind and stabilize only to the target proteins. PCT (Parenti, 2009; Parenti, Moracci, Fecarotta, & Andria, 2014) is the most recent anticipating therapy for LSDs, which is on the go with rapid development. These pharmacological chaperones are active site-specific chaperones (ASSCs) (Fan, 2003; Fan & Ishii, 2007).

3.1 Enzyme Replacement Therapy

ERT offers the missing enzyme expressed exogenously by repetitive intravenous infusion, and in most cases, this intravenous administration has to be as frequent as weekly or biweekly, due to the short half-life of the enzyme. In this approach, absent or deficient enzyme in an affected individual is replaced with an entirely functional enzyme isolated from an appropriate biological organism or made in the laboratory. The dosage administered and the frequency of the dose differs from disease to disease. ERT is also known to cause unwanted immunologic reactions. ERT does not treat the underlying genetic conditions; it merely increases the concentration of the required functional enzyme ERT with various recombinant enzymes like alglucerase, imiglucerase, velaglucerase, miglustat, and eliglustat for Gaucher disease (Grabowski et al., 2009; Hollak & Weinreb, 2015); agalsidase alpha, agalsidase beta, Amigal, and α-gal A for Fabry disease (Clarke, West, Bultas, & Schiffmann, 2007; Eng et al., 2001; Fervenza, Torra, & Warnock, 2008; Hollak & Weinreb, 2015; Pastores, 2007; Pisani et al., 2012; Schiffmann, Kopp, Austin, et al., 2001); α-glucosidase for Pompe disease (Van den Hout et al., 2000, 2001); laronidase for

MPS I (Dornelles et al., 2014; Harada, Niiyama, Katoh, & Ikeda, 2014; Kakkis et al., 2001); idursulfase for MPS II (Burrow & Leslie, 2008; Dalmau Serra, Vitoria Miñana, CalderónFernández, Cortell Aznar, & Plaza Romero, 2015; Muenzer et al., 2011, 2006); galsulfase for MPS VI (Braunlin et al., 2013; Harmatz et al., 2004, 2010); and recombinant alpha-mannosidase (rhLAMAN) for α-mannosidosis (Borgwardt et al., 2013) have shown to reduce the lysosomal storage. ERT is under the clinical stage for MPS IVA, Niemann–Pick type B, MLD, and Krabbe disease. It proves to be a very efficient method, but the major drawback is the crossing the blood–brain barrier (BBB) which prevents it from administering the disorders associated with CNS (Wang, Bodamer, Watson, Wilcox, & ACMG Work Group on Diagnostic Confirmation of Lysosomal Storage Diseases, 2011). Frequent administration is required due to short half-life of the enzymes, some create unwanted immunological reactions, while some exhaust all the glucosylceramide-based glycolipids lead to other adverse effects, poor distribution in target cells, and cost-effectiveness (Rombach, Hollak, Linthorst, & Dijkgraaf, 2013) are few considerations which makes ERT an unfit approach.

3.2 Substrate Reduction Therapy

In LSD, due to the lack of the particular functional enzyme, the substrate that is usually broken down by the enzyme tends to accumulate owing to the absence of the enzyme. SRT focuses on the oral medication that act and reduces the amount of substrate accumulated in the lysosome. This mitigates the harmful effect caused by the accumulation of substrate to toxic levels. It aims to minimize the production of the particular substrate by the cells, so that no more substrates are left to degrade in the lysosome. Oral therapies are available for few LSDs. Cysteamine is used efficaciously to gain renal function in cystinosis (Gahl et al., 1988; Schneider et al., 1995). *N*-butyl-deoxynojirimycin (NB-DNJ) (*OGT-918, miglustat, Zavesca; Actelion, Basel, Switzerland*) and iminosugar analog used as a glucosylceramide synthase inhibitor which degrades the glucocerebroside storage in Gaucher and Fabry disease (Cox et al., 2000; Dwek, Butters, Platt, & Zitzmann, 2002; Elstein et al., 2004; Motabar, Sidransky, Goldin, & Zheng, 2010; Moyses, 2003; Zimran & Elstein, 2003). *In vitro* inhibition of HS synthesizing genes like *EXTL2* and EXTL3 using siRNAs has shown to decrease in the GAG storage in the fibroblasts of Sanfilippo C patients (Canals, Benetó, Cozar, Vilageliu, & Grinberg, 2015). However, the inhibitory effects of these agents are not very specific, and side effects like gastrointestinal problems are observed. A new compound (*Genz-11638; Genzyme Corporation, Cambridge, MA*) is under development that is likely to produce less side effects. In SRT to reduce lysosomal storage, there must be a residual enzyme activity. Henceforth, combinational therapy using ERT and SRT is suggested as a treatment strategy, where SRT does not reduce substrate turnover rate resulting in cellular depletion, a reason that may finally limit this approach for LSDs.

3.3 Bone Marrow Transplantation

BMT provides the source of the defective enzyme from the engrafted donor cells. BMT is advantageous in crossing the blood–brain barrier (BBB) and helpful in treating disorders related to CNS (Walkley et al., 1994). BMT carried out in patients with Gaucher disease, MPS I, MPS II, MPS VI, MPS VII, GLD, MLD, α-mannosidosis, and fucosidosis showed improvement in the treatment (Hoogerbrugge et al., 1995; Krivit et al., 1999; Mynarek et al., 2012; van Karnebeek et al., 2014). BMT carried out in galactosialidosis murine (Hahn, delPilar Martin, Zhou, Mann, & d'Azzo, 1998; Leimig et al., 2002) and aspartylglucosaminuria mice (Laine, Ahtiainen, Rapola, Richter, & Jalanko, 2004) had reduced the symptoms and also recommended in humans. However, due to long-term uncertain effects related to mortality and graft rejection, it was not advisable later on (Arvio, Saunaaho, & Peippo, 2001). The transplant will not be effective if severe neurological symptoms are seen at the time of transplantation.

3.4 Hematopoietic Stem Cell Transplantation

Hematopoietic stem cell transplantation (HSCT) has been used successfully in the treatment of some LSDs. The concept behind HSCT is that transplanted hematopoietic stem cells from a healthy matched donor can produce the missing enzyme. Thus, produced enzyme will be taken up by mannose-6-phosphate receptors on the cells, endocytosed, and delivered to the lysosome showing the normalized function. LSDs like MPS I, MPS II, MPS III, MPS VI, MPS VII, GLD/Krabbe disease, MLD, Wolman's disease, Mucolipidosis type II (I-cell disease), α-Mannosidosis, Fucosidosis, Niemann–Pick type A and B, Gaucher's disease, ceroid-lipofuscinosis neuronal 3 (CLN3)/Batten disease, Fabry disease, Sandhoff disease, and aspartylglucosaminuria (Grewal et al., 2004; Krivit, 2004;

Ringdén et al., 2006) were treated with HSCT and showed some progressive reports. The major drawback in HSCT is its high morbidity, mortality, and graft failure. Low level of correction and the long-time requirement for integration of the cells into other tissues are some factors that preclude HSCT from being the best approach.

3.5 Gene Therapy

Over the years, cell and gene therapy approaches were tested in many animal models of lysosomal storage diseases. Gene therapy holds the promise of a cure for LSDs (Gritti, 2011; Ioannou, 2000; Rastall & Amalfitano, 2015; Sands & Davidson, 2006). Two major methods are followed: *in vivo* gene transfer by direct infusion of viral vectors (adeno-associated virus and lentivirus vectors) encoding normal enzymes and *ex vivo* gene therapy in which the cells are modified in the culture with viral vectors (retrovirus and lentivirus vectors) encoding normal enzymes and later transplanted. Adeno-associated virus vectors with liver-specific promoter-LSP (AAV-LSPhGAApA) or a hybrid CB promoter (AAV-CBhGAApA) (Franco et al., 2005) and AAV-LSPhGAA (Zhang et al., 2012) to drive GAA expression in Pompe disease are administered in GAA-knockout mice and resulted in normal phenotypic behavior. Administering tissue-specific non-viral promoters like neuron-specific enolase, astrocyte-specific (GFAP) endogenous AGA promoter intrathecally, and adenovirus-mediated glycosylasparaginase (*AGA*) intraventricularly to the aspartylglucosaminuria mouse have expressed normal glycosylasparaginase (*AGA*) and showed partial clearance of lysosomal storage (Peltola et al., 1998; Virta, Rapola, Jalanko, & Laine, 2006). Administering AAV vector of human CLN2 cDNA (AAV2CUhCLN2 or AAV5CUhCLN2) to CLN2-knockout mice reduced the autofluorescent storage in brain regions in ceroid-lipofuscinosis neuronal 2 (CLN2) disorder (Passini et al., 2006). By the means of CNS-directed gene therapy(Sands & Haskins, 2008), AAV5 vector expressing α-L-iduronidase (IDUA) injected in MPS I dogs (Ciron et al., 2006) and AAV1 vector expressing α-Mannosidase in α-mannosidosis cats (Vite et al., 2005) reduced the neurological signs of these disorders intensely. Intravenous injection of RV containing canine β-glucuronidase (cGUSB) cDNA in MPS VII dogs also showed some stable changes (Xu et al., 2002). In sialidosis, NEU1 binds to its chaperone protective protein–cathepsin A (PPCA) for lysosomal activation, in this case, chaperone-mediated gene therapy (Bonten et al., 2013) with a liver-tropic recombinant

AAV2/8 vector expressing PPCA for sialidosis resulted in improvement of the disease phenotype. Recently, many genetherapy-based treatments for human LSDs are in trial phase. RV-α-galactosidase A (NCT00001234) for Fabry disease, RV-glucocerebrosidase (NCT00001234 and NCT00004294) for Gaucher disease, rAAV-CMV-GAA (AAV/NCT00976352) for Pompe disease, LV-ARSA (NCT01560182) for MLD, RV-iduronate-2-sulfatase (NCT00004454) for MPS I and II, rAAV-SGSH and rAAVSUMF1 (NCT01474343) for MPS III A and IVA and VI, rAAV-CUhCLN2 (NCT00151216, NCY01411985, NCT01161576) for Batten disease, and Niemann–Pick disease type C1 disorders were developed to gain normal phenotype (Byrne, Falk, Clément, & Mah, 2012). Proof-of-principle experiments has been evaluated in some animal models of LSDs. However, many hurdles must be overcome before gene therapy can be applied to humans successfully. Few constraints in gene therapy including delivery to the correct cells, random integration, sustained expression, and immune reactions with no apparent short- and long-term toxicity, and most importantly appear to be capable of leading lifelong high-level expression of the recombinant enzymes.

3.6 Chemical Chaperone Therapy

Chemical chaperones are small molecules that bind and stabilize the misfolded proteins in the ER and stimulates regular trafficking that contributes to increase the residual activity. Many low-molecular-weight compounds like glycerol, dimethyl sulfoxide, trimethylamine *N*-oxide, 4-phenylbutyric acid, and heavy water are known to accomplish this beginning (Brown, Hong-Brown, & Welch, 1997). Such compounds that bind selectively at the active site of proteins are called as chemical or pharmacological chaperones (Ulloa-Aguirre, Janovick, Brothers, & Conn, 2004), and can be a potential drug of choice for many specific disorders envisioned as CCT (Sawkar, D'Haeze, & Kelly, 2006). Chemical chaperone of valienamine derivative *N*-octyl-4-epi-β-valienamine (NOEV) stabilizes the *β-galactosidase* activity in galactosialidosis (Hossain et al., 2015). Study on chaperone effect of *N*-(*n*-nonyl) deoxynojirimycin (NN-DNJ) in Gaucher fibroblasts cells showed an increase in the activity of the *β-glucocerebrosidase* (Sawkar et al., 2002). 1-Deoxygalactonojirimycin (DGJ) for Fabry disease, NOEV for G_{M1} gangliosidosis and MPS IVB (Morquio B disease) (Matsuda et al., 2003), and *N*-octyl-β-valienamine, *N*-nonyl-DNJ (deoxynojirimycin), and miglustat-*N*-butyl-DNJ

(NB-DNJ) acts as a chemical chaperones for treating Gaucher disease (Alfonso et al., 2005; Sawkar et al., 2005; Suzuki, 2013; Suzuki, Ogawa, & Sakakibara, 2009). These are few chemical chaperones available for the treatment of LSDs. Clinical experience is limited with unknown long-term effects.

3.7 Pharmacological Chaperone Therapy

Proteins that are chaperone or help in the correct folding of other proteins are known as molecular chaperones (Cooper, 2000). They are not present in the final conformation of the protein but simply help in the stabilization and correct folding of the protein. Molecular chaperones are found even in lower life forms, for example, *Escherichia coli*. DnaK is the equivalent in the *E. coli* for the Hsp70 (Garrido, Paul, Seigneuric, & Kampinga, 2012) in the humans. In the *E. coli*, the DnaK is essential for different processes, namely, protein folding, degradation, translocation, and gene expression (Cellitti et al., 2010). Pharmacological chaperones, like molecular chaperones, are small molecules that selectively bind to the misfolded protein and stabilize them. Such action leads to improved folding and trafficking, thereby increasing the activity (Ulloa-Aguirre et al., 2004). PCT is an investigational approach toward treating disorders like the LSD with an underlying biochemical defect, mostly comprising of misfolded/unstable proteins. Pharmacological chaperones (pharmacoperones) are prototype molecules that help in proper folding of mutated proteins. PCT is an emerging way of treatment in LSDs. Pharmacological chaperones are selectively designed based on the sensitive mutations seen in the target protein. In this context, rescuing the mutated cum misfolded protein with pharmacological chaperones is profitable when compared to the other therapies. Oral small pharmacological chaperones assist in proper folding and trafficking of lysosomal proteins with specific missense mutations. Conceptually, it helps the ribosome to read over nonsense codons and allows a normal translation end at the exact stop codon. Pharmacological chaperones developed are found to increase the cellular activity of the defective enzymes, and many are currently in testing stage for various disorders, may perhaps LSDs patients with nonsense mutations will be benefited (Boyd et al., 2013; Morello, Petäjä-Repo, Bichet, & Bouvier, 2000; Muntau, Leandro, Staudigl, Mayer, & Gersting, 2014; Parenti et al., 2015; Tropak, Blanchard, Withers, Brown, & Mahuran, 2007; Valenzano et al., 2011). The compound DGJ acts as a pharmacological chaperone in Fabry mice (Germain & Fan,

2009; Ishii, 2012; Khanna et al., 2010; Yam, Bosshard, Zuber, Steinmann, & Roth, 2006). Coformulation of the pharmacological chaperone AT1001 (migalastat HCl) DGJ *HCl* with recombinant human α-galactosidase A (ATB100) intravenously to Fabry mice was beneficial (Xu et al., 2015). An iminosugar AT1001 (GR181413A-migalastat hydrochloride) DGJ acts as a pharmacological chaperone in Fabry disease. It binds and stabilizes α-galactosidase A and increased the residual activity (Wu et al., 2011). Patient study of migalastat hydrochloride, DGJ by *Amicus Therapeutics* (Johnson et al., 2013; Giugliani et al., 2013), hiked the levels of *α-galactosidase A* in blood and kidney as well as reduced the levels of globotriaosylceramide-3 in plasma, urine, and tissues with good acceptability in Fabry disease. Isofagomine-based compounds like noncarbohydrate-based inhibitory molecules a 2,4-diamino-5-substituted-quinazoline and a 5-substituted pyridinyl-2-furamide (Tropak et al., 2008) and *N-adamantanyl-4-((3R,4R,5R)-3,4-dihydroxy-5-(hydroxymethyl)piperidin-1-yl)-butanamide* and *N-adamantanyl-4-((3R,4R,5R)-3,4-dihydroxy-5-(hydroxymethyl)piperidin-1-yl)pentanamide* act as pharmacological chaperones (Yu, Sawkar, Whalen, Wong, & Kelly, 2007). Computational analysis of *N*-methylmescaline and shihunine demonstrated to have pharmacological chaperone tendency (Huang, Lee, & Chen, 2014). The glucose analogue, *N*-butyl-deoxynojirimycin (NB-DNJ-miglustat) (Abian et al., 2011) and another compound L-ido nojirimycin derivative, *N-[N′-(4-adamantan-1-ylcarboxamidobutyl) thiocarbamoyl]-1,6-anhydro-L-idonojirimycin* (NAdBTAIJ) along with coenzyme Q10 (*CoQ*) are promising pharmacological chaperones for Gaucher disease (De la Mata et al., 2015). These are few pharmacological chaperone know to increase *glucosylceramidase* activity in Gaucher patient cells. Chaperones like *N-butyl-deoxynojirimycin* (NB-DNJ) and *N-acetylcysteine* (NAC) (Porto et al., 2009, 2012) work as a novel allosteric chaperone for the GAA enzyme involved in Pompe disease. It increases the residual activity of mutated GAA in patients fibroblasts and as well as in COS7 cells expressed with mutated GAA, without disturbing the catalytic activity of the enzyme in Pompe disease. *N-nonyl-deoxygalactonojirimycin* (NN-DGJ) enhanced the *β-galactosidase* activity in patient cell lines containing mutations and suggested as a pharmacological chaperone for treating G_{M1} gangliosidosis (Rigat et al., 2012). OT1001 is a new pharmacological chaperone available for treating G_{M1} gangliosidosis (Knight et al., 2015). Hexosaminidase inhibitor (Tropak, Reid, Guiral, Withers, & Mahuran, 2004), pyrimethamine (*PYR*), a noncarbohydrate pharmacological chaperone (Maegawa et al., 2007) and PYR like iminosugar *NAG-thiazoline* (Bateman, Cherney, Mahuran, Tropak, & James, 2011; Mark et al., 2003) have shown to bind the active site of β-hexosaminidase and increase the residual activity in Tay–Sachs disease and Sandhoff disease. *α-lobeline* inhibits GALC and act as a pharmacological chaperone by boosting its activity in GLD/Krabbe disease (Lee et al., 2010). Pharmacological chaperones like celastrol and MG132 (O'Leary & Igdoura, 2012) in mouse model have shown significant improvements in sialidosis.

4. PHARMACOLOGICAL CHAPERONES VERSUS CHEMICAL CHAPERONES

Pharmacological chaperones have usually been confused with chemical chaperones. They are similar in many aspects but differ in the way they bind and act with proteins. Chemical chaperones protect proteins nonspecifically from various factors that can lead to protein denaturation, by the solvophobic thermodynamic force coming from the interactions between the osmolyte and the peptide backbone. Chemical chaperones are low-molecular-weight substances, usually osmolytes. Large concentrations of chemical chaperones are required to facilitate folding of proteins. Chemical chaperones are more universal. They can bind and stabilize any protein, whereas the pharmacological chaperones are very specific and have effects on particular proteins. Most of the pharmacological chaperones are substrates or active site-directed competitive inhibitors and very efficient at lower concentrations (Bolen & Baskakov, 2001; Fan, 2003; Perlmutter, 2002). However, a pharmacological chaperone cannot bind to all the variants present in a disorder. Oral small pharmacological chaperones assist in proper folding and trafficking of lysosomal proteins with specific missense mutations. Conceptually, it helps the ribosome to read over nonsense codons and allows a normal translation end at the exact stop codon. The pharmacological chaperone might work extremely well with a person having a specific mutation, but it may not with another individual showing the same disorder phenotypes. This is due to the variant difference in the mutant protein. Target proteins, with the help of chaperones, fold correctly and hence pass through the quality control system of the ER and trafficked safely from there to designated location.

5. COMBINATION THERAPY FOR LSDs

Morbific nature of LSDs is very complicated with many interventions that limit the effectiveness of the monotherapy. Concentration on combination therapy and complementary alternative medicine are desirable to tackle LSDs (Balwani, Fuerstman, Desnick, Buckley, & McGovern, 2009; Hawkins-Salsbury, Reddy, & Sands, 2011; Parenti, 2009; Parenti et al., 2015). If these therapies are combined, the upshots will be cohesive and better than monotherapies (Platt & Lachmann, 2009). Combining proteostasis regulators and pharmacological chaperones are approved to treat phenylketonuria with *saproterin dihydrochloride* and amyloidosis with *tafamidis* (Muntau et al., 2014). *Sanofi* has patented a novel chemical inhibitor glucosylceramide synthase (*US 20140255381 A1*) in treating LSDs and cancer, either alone or in combination with ERT (Cox, 2015). *Amicus Therapeutics* has come up with a next-generation approach to, namely, *Chaperone-Advanced Replacement Therapy* (CHART™) which combines pharmacological chaperones with ERT for LSDs. Combining ASSCs and ERT is a contradictory approach for LSDs (Fan, 2003). Recombinant human glucocerebrosidase in combination with *N-butyl-deoxynojirimycin* (NB-DNJ) was applied for Gaucher disease (Butters, Dwek, & Platt, 2000). ERT with migalastat for Fabry disease and ERT with duvoglustat for Pompe disease (Kirkegaard, 2013) are effective strategies. *N-butyl-deoxynojirimycin* (NB-DNJ) along with α-glucosidase (Myozyme™) for Pompe disease shows iminosugar stabilization, resulting in a combination approach (Parenti, Fecarotta, et al., 2014; Parenti, Moracci, et al., 2014). Combined administration of recombinant human α-galactosidase A with pharmacological chaperone AT1001 increases the substrate reduction rate in Fabry mice (Benjamin et al., 2012), and recombinant human α-glucosidase (rhGAA) with an allosteric chaperone *NAC* corrected the enzyme activity in mouse model of Pompe disease (Porto et al., 2012). Mouse model of Sandhoff disease is treated in combination of SRT with BMT and SRT with NSAIDs (*nonsteroidal anti-inflammatory drugs*) (Jeyakumar et al., 2001, 2004), and neuronal stem cell therapy (Lee et al., 2007) are few multimodal strategy to treat LSDs.

6. HOW CAN PCT PERSONALIZED?

Chaperone said to be best, if it shows a high affinity toward an allosteric site in the target protein, creating stability and avoiding inhibition in the protein's activity. Pharmacological chaperones are designed in such a way that are amenable to responsive nonsense mutations present in the targeted protein and clear the way for enhanced enzyme activity. PCT is currently being researched for several diseases plaguing man and utilized to treat or enhance proteostasis of several types of proteins. Research is actively pursued in treating defective lysosomal enzymes; G-protein-coupled receptors, ion channels, transporters, and proteins prone to aggregation leading to toxicity, to name a few. In α-galactosidase A (Fabry disease), mutants like W24C, G35R, A37T, T41I, M42V, H46P, R49C, M51K, M51I, E59K, S65I, E66K, L68F, A73V, D83N, A97V, S102L, R112H, R112C, R118G, R118C, R118L, L120V, S126R, S126G, S126C, S126N, S126I, A135V, D136E, N139S, A143T, A156V, W162G, D165H, C172G, D175N, G183V, S201F, W204C, P205T, Y207S, N215S, I219T, R220Q, N224S, H255D, H225R, I232T, S238N, I242N, R252T, I253S, P259R, G261V, D264N, D264H, D264Y, M267T, L268S, V269M, V269A, I270M, S276G, T282I, M290L, M290I, A291T, L294S, F295R, S297C, L300P, R301G, R301Q, R301P, L310F, L311V, D313Y, V316I, I319T, N320I, Q321H, G325S, Q327E, G328A, S345P, W349R, R356W, R356Q, G360C, G361R, R363H, T385A, G395A, E398A, S405R, T410I, L415F, and E418G have been experimentally proven to respond to DGJ (Lukas et al., 2013; Shin et al., 2008). *N*-nonyl-deoxynojirimycin (NN-DNJ), a pharmacological chaperone, helps in increase of enzyme activity in the case of N370S variant. Another chaperone, *N*-octyl-2,5-anhydro-2,5-imino-D-glucitol enhances the activity of the G202R mutant. Others include naturally occurring calystegine B_2 which have been claimed to make a significant improvement with L444P mutant (Benito, García Fernández, & Ortiz Mellet, 2011; Yu et al., 2007). HEXA mutant G269S; HEXB mutants like R505Q, C137Y, G353R, T150P, and P417L (Boyd et al., 2013; Tropak et al., 2004) for Tay–Sachs disease and Sandhoff disease, respectively, have been experimentally proven to respond to two pharmacological chaperones, namely, PYR and *N*-acetyl glucosamine-thiazoline. Several studies have tried to increase the number of pharmacological chaperones known to man. Series of carbohydrate mimics as potential pharmacological chaperone for β-hexosaminidase, namely, acetamido dideoxy nojirimycin, a related acetamido analogue of castanospermine, and *N*-acetylglucosamine thiazoline (Boyd et al., 2013; Tropak et al., 2004). A more recent high-throughput screening approach narrowed down compounds from the original list of 50,000 compounds. They include pyrrol pyridazine, bisnapthalmide, and nitroindanone. All the mentioned compounds act as chaperones and showed a positive increase in enzyme activity (Clarke et al., 2011; Tropak et al., 2007). Pharmacological chaperones are not only used in LSDs but also used in amino acid metabolism disorders. Cystathionine β-synthase deficiency or more commonly known as homocystinuria and their variants like P422L, S466L, A114V, V180A, P78R, P49L, E302K, C165Y, R125Q, K102N, R369C, and R266K have been experimentally proven to respond to a pharmacological chaperone, namely, δ-aminolevulinic acid (δ-ALA) (Kopecká, Krijt, Raková, & Kožich, 2011). These pharmacological chaperones are favorable for designing genome-specific pharmacological chaperones (personalized pharmacological chaperones). This can be a future approach for treating LSDs. In most genetic diseases, mutational heterogeneity plays a significant role. A single amino acid substitution can have multiple effects on the protein. It can render the active site, lower the stability, decrease the activity, and change the protein conformation. Due to the reasons above, it is of utmost importance to know which mutation could respond to PCT. Few mutations only render the proteins that are unable to fold in the correct manner; they still retain the catalytic activity. Such proteins would respond to PCT. Defining potentially responsive mutations in individuals can be suggested to design personalized pharmacological chaperones.

7. COMPUTATIONAL STUDIES IN PCT

Computationally modeled structure of β-galactosidase (*GLB1*) responsible for G_{M1} gangliosidosis/MPS IVB-Morquio B docked with chaperone NOEV stipulated the binding fashion in different pH conditions, which can be theoretically helpful to design mutant-specific pharmacological chaperone (Suzuki et al., 2009). Action study of the chaperone *1-deoxynojirimycin* (DNJ) on mutant model of α-glucosidase (*GAA*) was responsive toward the mutated residues involving in catalytic activity and increased the lysosomal activity and trafficking (Flanagan et al., 2009). Homology model of *α-L-iduronidase* profiled with missense mutations for MPS I may contribute in designing a novel pharmacological chaperone (Chandar, 2012). In a recent study, the pharmacological chaperone, *PYR*, bounded with β-hexosaminidase responsible for Tay–Sachs disease and Sandhoff disease showed the need to adjust the enzyme structure to enhanced ligand binding (Bateman et al., 2011). Drug repositioning can also be carried out to design pharmacological chaperones (Hay Mele, Citro, Andreotti, & Cubellis, 2015). In this case, chaperones can be designed based on the catalytic domain of the mutant enzyme, so-called personalizing the pharmacological chaperones. Structural analysis of α-glucosidase (*GAA*) for Pompe disease docked with pharmacological chaperone DNJ revealed the folding behavior of the mutant protein based on the spatial orientation of the chaperone which can shed limelight to design effective chaperones enabled with proper folding of proteins for LSDs (Ling, 2011). Compounds like *2-O-caffeoyl tartaric acid* and *mumefural* from traditional Chinese medicine (TCM) database are found to be potential inhibitors of *phosphodiesterase 4D* (PDE4D) associated with stroke. TCM database (Chen & Chen, 2011) has an enormous collection of natural compounds which can be virtually screened to find novel pharmacological chaperones for LSDs. Molecular dynamics simulation study on two compounds from TCM, namely, *N-methylmescaline* and *shihunine* showed strong binding fashion enabling the mutant glucosylceramidase (*GBA*) to fold properly which handles Gaucher disease (Huang et al., 2014). From this all, personalizing pharmacological chaperones can be a future approach to treating LSDs.

8. ADVANTAGES AND DISADVANTAGES OF VARIOUS THERAPIES IN LSDs

ERT is efficient when it comes to treat patients with somatic symptoms. In case of mucopolysaccharidosis, there is an accumulation of GAGs in various tissues and leads to damage to the heart, lungs, bones, joints, and sometimes central nervous system. In such cases, the somatic symptoms are managed, but the CNS symptoms are not because the proteins have difficulty in crossing the BBB. Intrathecal administration of the deficient enzyme is the only way to get rid of this, but the invasive pressure caused by the frequent weekly administration is not pleasurable. In most cases of the mucopolysaccharidosis, affected people are children, and might not be compliant, and they might have further complications due to the stress involved in this procedure (Piotrowska et al., 2008). In such cases, treatment with pharmacological chaperones is very effective, as they can cross the BBB and manage the CNS symptoms. The administration of pharmacological chaperones does not require complex, invasive procedure. Pharmacological chaperones do not cause adverse immunological reactions (Khanna et al., 2010). Even though PCT has several advantages, there are also some inherent limitations for PCT. The misfolded/mutant enzyme should be active in the lysosome at least the residual activity should be seen. This is important in PCT to correctly fold the protein and evade from ERAD machinery and successfully transport into the lysosome. If the mutations affect the catalytic or a key region in the protein, the result will be a nonfunctional enzyme with no enzyme activity. At this stage, PCT is very likely to fail. In addition, this chaperoning activity at the ER should be achieved at subinhibitory concentrations under lysosomal conditions and, therefore, a balance between folding enhancement and enzyme inhibition should be reached for each particular case (Benjamin et al., 2009). Use of pharmacological chaperones has an impact on overcoming the limitations concerning biodistribution (crossing BBB), long-term administration, and high cost.

9. FUTURE OF PCTs

Several clinical trials are ongoing with pharmacological chaperones being the bulls eye of drug trial and future research in LSDs (Conn & Janovick, 2009; Ni & Lee, 2007; Reiss, Lesnik, Parvez, Parvez, & Ehrlich, 2000). But it has not created predominant impact due to long-term developmental process along with high cost. Generally, upon mutation, protein loses its catalytic activity or its ability to bind to substrate. In this case, a chaperone cannot rescue a protein from misfolding (Valenzano et al., 2011). Hence, it is important to document the mutations that are deleterious in nature. Clinical testing methods to identify these functional mutations and answering LSDs with PCT remains as a daunting task. Prediction of deleterious mutations occurring in different populations has the potential for prevention and treatment of any disease at an individual level. Recently, computational prediction methods based on structure and sequence information of a protein have helped in classifying the mutations into deleterious or neutral in an efficient and accurate manner. Drug-binding efficiency can also be determined computationally using docking and molecular dynamic simulation analysis. This computational platform may narrow down the mutations that would respond to PCT and tailoring the chaperones in favor of the altered binding site of the misfolded protein with responsive mutations. To achieve this, in such a demanding era, bioinformatics an emerging field that might have a fundamental role in exposing the molecular mechanism of the mutations and thereby help in personalizing a drug for an individual in fast and affordable manner. This enthralling application of computational power and the advent of other technologies have paved a promising way toward personalized medicine. From this, we highlight how a pharmacoperones can be personalized, and how it can be an approach for next-generation treatment in administering LSDs.

10. CONCLUSION

An increasing number of protein misfolding diseases are being managed better with pharmacological chaperones including *cystic fibrosis* (Hanrahan, Sampson, & Thomas, 2013). All these findings conclusively suggest that PCT could be applied to a wide variety of conformational protein-associated diseases. PCT is an alternative oral treatment that can be administered to patients who are unwilling or unable to undergo ERT or SRT. Combining different therapeutic approaches will be beneficial as each therapy has its own limitations. Such a path will help in managing, curing, or alleviating the symptoms of various protein-folding diseases. Significant research from recent clinical studies has led us to infer that a practical therapeutic approach is imminent. Pharmacological chaperones are the future of the drug industry, and they are here to stay. Such a path will help in managing, curing, or alleviating the symptoms of various LSDs. Pharmacological chaperones are the future of the drug industry, and this light on PCT can expel darkness in treatments for LSDs.

ACKNOWLEDGMENTS

The authors take this opportunity to thank the management of VIT for providing the facilities and encouragement to carry out this work. The authors have no conflict of interest to declare.

REFERENCES

Abian, O., Alfonso, P., Velazquez-Campoy, A., Giraldo, P., Pocovi, M., & Sancho, J. (2011). Therapeutic strategies for Gaucher disease: Miglustat (NB-DNJ) as a pharmacological chaperone for glucocerebrosidase and the different thermostability of velaglucerase alfa and imiglucerase. *Molecular Pharmaceutics*, *8*(6), 2390–2397.

Alfonso, P., Pampín, S., Estrada, J., Rodríguez-Rey, J. C., Giraldo, P., Sancho, J., et al. (2005). Miglustat (NB-DNJ) works as a chaperone for mutated acid beta-glucosidase in cells transfected with several Gaucher disease mutations. *Blood Cells, Molecules & Diseases*, *35*(2), 268–276.

Arvio, M., Sauna-aho, O., & Peippo, M. (2001). Bone marrow transplantation for aspartylglucosaminuria: Follow-up study of transplanted and non-transplanted patients. *The Journal of Pediatrics*, *138*(2), 288–290.

Balwani, M., Fuerstman, L., Desnick, R. J., Buckley, B., & McGovern, M. M. (2009). Use of complementary and alternative medicine by patients with lysosomal storage diseases. *Genetics in Medicine: Official Journal of the American College of Medical Genetics*, *11*(10), 722–727.

Bateman, K. S., Cherney, M. M., Mahuran, D. J., Tropak, M., & James, M. N. G. (2011). Crystal structure of β-hexosaminidase B in complex with pyrimethamine, a potential pharmacological chaperone. *Journal of Medicinal Chemistry*, *54*(5), 1421–1429.

Benito, J. M., García Fernández, J. M., & Ortiz Mellet, C. (2011). Pharmacological chaperone therapy for Gaucher disease: A patent review. *Expert Opinion on Therapeutic Patents, 21*(6), 885–903.

Benjamin, E. R., Flanagan, J. J., Schilling, A., Chang, H. H., Agarwal, L., Katz, E., et al. (2009). The pharmacological chaperone 1-deoxygalactonojirimycin increases alpha-galactosidase A levels in Fabry patient cell lines. *Journal of Inherited Metabolic Disease, 32*(3), 424–440.

Benjamin, E. R., Khanna, R., Schilling, A., Flanagan, J. J., Pellegrino, L. J., Brignol, N., et al. (2012). Co-administration with the pharmacological chaperone AT1001 increases recombinant human α-galactosidase A tissue uptake and improves substrate reduction in Fabry mice. *Molecular Therapy: The Journal of the American Society of Gene Therapy, 20*(4), 717–726.

Bolen, D. W., & Baskakov, I. V. (2001). The osmophobic effect: Natural selection of a thermodynamic force in protein folding. *Journal of Molecular Biology, 310*(5), 955–963.

Bonten, E. J., Yogalingam, G., Hu, H., Gomero, E., van de Vlekkert, D., & d'Azzo, A. (2013). Chaperone-mediated gene therapy with recombinant AAV-PPCA in a new mouse model of type I sialidosis. *Biochimica et Biophysica Acta, 1832*(10), 1784–1792.

Borgwardt, L., Dali, C. I., Fogh, J., Månsson, J. E., Olsen, K. J., Beck, H. C., et al. (2013). Enzyme replacement therapy for alpha-mannosidosis: 12 Months follow-up of a single centre, randomised, multiple dose study. *Journal of Inherited Metabolic Disease, 36*(6), 1015–1024.

Boyd, R. E., Lee, G., Rybczynski, P., Benjamin, E. R., Khanna, R., Wustman, B. A., et al. (2013). Pharmacological chaperones as therapeutics for lysosomal storage diseases. *Journal of Medicinal Chemistry, 56*(7), 2705–2725.

Braunlin, E., Rosenfeld, H., Kampmann, C., Johnson, J., Beck, M., Giugliani, R., et al. (2013). Enzyme replacement therapy for mucopolysaccharidosis VI: Long-term cardiac effects of galsulfase (Naglazyme®) therapy. *Journal of Inherited Metabolic Disease, 36*(2), 385–394.

Brown, C. R., Hong-Brown, L. Q., & Welch, W. J. (1997). Correcting temperature-sensitive protein folding defects. *The Journal of Clinical Investigation, 99*(6), 1432–1444.

Burrow, T. A., & Leslie, N. D. (2008). Review of the use of idursulfase in the treatment of mucopolysaccharidosis II. *Biologics: Targets & Therapy, 2*(2), 311–320.

Butters, T. D., Dwek, R. A., & Platt, F. M. (2000). Inhibition of glycosphingolipid biosynthesis: Application to lysosomal storage disorders. *Chemical Reviews, 100*(12), 4683–4696.

Byrne, B. J., Falk, D. J., Clément, N., & Mah, C. S. (2012). Gene therapy approaches for lysosomal storage disease: Next-generation treatment. *Human Gene Therapy, 23*(8), 808–815.

Canals, I., Benetó, N., Cozar, M., Vilageliu, L., & Grinberg, D. (2015). EXTL2 and EXTL3 inhibition with siRNAs as a promising substrate reduction therapy for Sanfilippo C syndrome. *Scientific Reports, 5*, 13654.

Cellitti, J., Zhang, Z., Wang, S., Wu, B., Guiney, D., & Pellecchia, M. (2010). *Inhibitors of protein folding: DnaK: In probe reports from the NIH Molecular Libraries Program.*

Chandar, S. S. (2012). Mucopolysaccharidosis type I: Homology modeling and docking analysis of the lysosomal enzyme, human α-L-iduronidase. *African Journal of Pharmacy and Pharmacology, 6*(27), 2027–2038.

Chen, K.-C., & Chen, C. Y.-C. (2011). Stroke prevention by traditional Chinese medicine? A genetic algorithm, support vector machine and molecular dynamics approach. *Soft Matter,* 7(8), 4001–4008.

Cheng, S. H., & Smith, A. E. (2003). Gene therapy progress and prospects: Gene therapy of lysosomal storage disorders. *Gene Therapy, 10*(16), 1275–1281.

Ciron, C., Desmaris, N., Colle, M.-A., Raoul, S., Joussemet, B., Vérot, L., et al. (2006). Gene therapy of the brain in the dog model of Hurler's syndrome. *Annals of Neurology*, *60*(2), 204–213.

Clarke, J. T. R., Mahuran, D. J., Sathe, S., Kolodny, E. H., Rigat, B. A., Raiman, J. A., et al. (2011). An open-label phase I/II clinical trial of pyrimethamine for the treatment of patients affected with chronic GM2 gangliosidosis (Tay-Sachs or Sandhoff variants). *Molecular Genetics and Metabolism*, *102*(1), 6–12.

Clarke, J. T. R., West, M. L., Bultas, J., & Schiffmann, R. (2007). The pharmacology of multiple regimens of agalsidase alfa enzyme replacement therapy for Fabry disease. *Genetics in Medicine*, *9*(8), 504–509.

Conn, P. M., & Janovick, J. A. (2009). Drug development and the cellular quality control system. *Trends in Pharmacological Sciences*, *30*(5), 228–233.

Cooper, G. M. (2000). *The cell: A molecular approach*. Washington, DC: ASM Press.

Cox, T. M. (2015). Innovative treatments for lysosomal diseases. *Best Practice & Research. Clinical Endocrinology & Metabolism*, *29*(2), 275–311.

Cox, T., Lachmann, R., Hollak, C., Aerts, J., van Weely, S., Hrebícek, M., et al. (2000). Novel oral treatment of Gaucher's disease with N-butyldeoxynojirimycin (OGT 918) to decrease substrate biosynthesis. *Lancet (London, England)*, *355*(9214), 1481–1485.

Dalmau Serra, J., Vitoria Miñana, I., CalderónFernández, R., Cortell Aznar, I., & Plaza Romero, S. (2015). Clinical response to long term enzyme replacement treatment in children, adolescent and adult patients with Hunter syndrome. *Medicina Clínica*, *145*(9), 392–398.

De Duve, C. (2005). The lysosome turns fifty. *Nature Cell Biology*, *7*(9), 847–849.

De la Mata, M., Cotán, D., Oropesa-Ávila, M., Garrido-Maraver, J., Cordero, M. D., Villanueva Paz, M., et al. (2015). Pharmacological chaperones and coenzyme Q10 treatment improves mutant β-glucocerebrosidase activity and mitochondrial function in neuronopathic forms of Gaucher disease. *Scientific Reports*, *5*, 10903.

Dornelles, A. D., de Camargo Pinto, L. L., de Paula, A. C., Steiner, C. E., Lourenço, C. M., Kim, C. A., et al. (2014). Enzyme replacement therapy for Mucopolysaccharidosis Type I among patients followed within the MPS Brazil Network. *Genetics and Molecular Biology*, *37*(1), 23–29.

Dwek, R. A., Butters, T. D., Platt, F. M., & Zitzmann, N. (2002). Targeting glycosylation as a therapeutic approach. *Nature Reviews. Drug Discovery*, *1*(1), 65–75.

Ellgaard, L., & Helenius, A. (2003). Quality control in the endoplasmic reticulum. *Nature Reviews. Molecular Cell Biology*, *4*(3), 181–191.

Elstein, D., Hollak, C., Aerts, J. M. F. G., van Weely, S., Maas, M., Cox, T. M., et al. (2004). Sustained therapeutic effects of oral miglustat (Zavesca, N-butyldeoxynojirimycin, OGT 918) in type I Gaucher disease. *Journal of Inherited Metabolic Disease*, *27*(6), 757–766.

Eng, C. M., Guffon, N., Wilcox, W. R., Germain, D. P., Lee, P., Waldek, S., et al. (2001). Safety and efficacy of recombinant human alpha-galactosidase A—Replacement therapy in Fabry's disease. *The New England Journal of Medicine*, *345*(1), 9–16.

Fan, J.-Q. (2003). A contradictory treatment for lysosomal storage disorders: Inhibitors enhance mutant enzyme activity. *Trends in Pharmacological Sciences*, *24*(7), 355–360.

Fan, J.-Q., & Ishii, S. (2007). Active-site-specific chaperone therapy for Fabry disease. Yin and Yang of enzyme inhibitors. *The FEBS Journal*, *274*(19), 4962–4971.

Fervenza, F. C., Torra, R., & Warnock, D. G. (2008). Safety and efficacy of enzyme replacement therapy in the nephropathy of Fabry disease. *Biologics: Targets & Therapy*, *2*(4), 823–843.

Flanagan, J. J., Rossi, B., Tang, K., Wu, X., Mascioli, K., Donaudy, F., et al. (2009). The pharmacological chaperone 1-deoxynojirimycin increases the activity and lysosomal trafficking of multiple mutant forms of acid alpha-glucosidase. *Human Mutation*, *30*(12), 1683–1692.

Franco, L. M., Sun, B., Yang, X., Bird, A., Zhang, H., Schneider, A., et al. (2005). Evasion of immune responses to introduced human acid alpha-glucosidase by liver-restricted expression in glycogen storage disease type II. *Molecular Therapy: The Journal of the American Society of Gene Therapy*, *12*(5), 876–884.

Gahl, W. A., Thoene, J. G., Schneider, J. A., O'Regan, S., Kaiser-Kupfer, M. I., & Kuwabara, T. (1988). NIH conference. Cystinosis: Progress in a prototypic disease. *Annals of Internal Medicine*, *109*(7), 557–569.

Garrido, C., Paul, C., Seigneuric, R., & Kampinga, H. H. (2012). The small heat shock proteins family: The long forgotten chaperones. *The International Journal of Biochemistry & Cell Biology*, *44*(10), 1588–1592.

Germain, D. P., & Fan, J.-Q. (2009). Pharmacological chaperone therapy by active-site-specific chaperones in Fabry disease: In vitro and preclinical studies. *International Journal of Clinical Pharmacology and Therapeutics*, *47*(Suppl. 1), S111–S117.

Giugliani, R., Waldek, S., Germain, D. P., Nicholls, K., Bichet, D. G., Simosky, J. K., et al. (2013). A phase 2 study of migalastat hydrochloride in females with Fabry disease: Selection of population, safety and pharmacodynamic effects. *Molecular Genetics and Metabolism*, *109*(1), 86–92.

Grabowski, G. A., Kacena, K., Cole, J. A., Hollak, C. E. M., Zhang, L., Yee, J., et al. (2009). Dose–response relationships for enzyme replacement therapy with imiglucerase/alglucerase in patients with Gaucher disease type 1. *Genetics in Medicine: Official Journal of the American College of Medical Genetics*, *11*(2), 92–100.

Greiner-Tollersrud, O., & Berg, T. (2005). Lysosomal storage disorders. In P. Saftig (Ed.), *Lysosomes* (pp. 60–73). New York: Springer Science, Landes Bioscience.

Grewal, S. S., Shapiro, E. G., Krivit, W., Charnas, L., Lockman, L. A., Delaney, K. A., et al. (2004). Effective treatment of α-mannosidosis by allogeneic hematopoietic stem cell transplantation. *The Journal of Pediatrics*, *144*(5), 569–573.

Gritti, A. (2011). Gene therapy for lysosomal storage disorders. *Expert Opinion on Biological Therapy*, *11*(9), 1153–1167.

Hahn, C. N., delPilar Martin, M., Zhou, X.-Y., Mann, L. W., & d'Azzo, A. (1998). Correction of murine galactosialidosis by bone marrow-derived macrophages overexpressing human protective protein/cathepsin A under control of the colony-stimulating factor-1 receptor promoter. *Proceedings of the National Academy of Sciences of the United States of America*, *95*(25), 14880–14885.

Hanrahan, J. W., Sampson, H. M., & Thomas, D. Y. (2013). Novel pharmacological strategies to treat cystic fibrosis. *Trends in Pharmacological Sciences*, *34*(2), 119–125.

Harada, H., Niiyama, H., Katoh, A., & Ikeda, H. (2014). Laronidase replacement therapy and left ventricular function in mucopolysaccharidosis I. *JIMD Reports*, *16*, 39–45.

Harmatz, P., Whitley, C. B., Waber, L., Pais, R., Steiner, R., Plecko, B., et al. (2004). Enzyme replacement therapy in mucopolysaccharidosis VI (Maroteaux-Lamy syndrome). *The Journal of Pediatrics*, *144*(5), 574–580.

Harmatz, P., Yu, Z.-F., Giugliani, R., Schwartz, I. V. D., Guffon, N., Teles, E. L., et al. (2010). Enzyme replacement therapy for mucopolysaccharidosis VI: Evaluation of long-term pulmonary function in patients treated with recombinant human N-acetylgalactosamine 4-sulfatase. *Journal of Inherited Metabolic Disease*, *33*(1), 51–60.

Hawkins-Salsbury, J. A., Reddy, A. S., & Sands, M. S. (2011). Combination therapies for lysosomal storage disease: Is the whole greater than the sum of its parts? *Human Molecular Genetics*, *20*(R1), R54–R60.

Hay Mele, B., Citro, V., Andreotti, G., & Cubellis, M. V. (2015). Drug repositioning can accelerate discovery of pharmacological chaperones. *Orphanet Journal of Rare Diseases*, *10*, 55.

Hegde, R. S., & Ploegh, H. L. (2010). Quality and quantity control at the endoplasmic reticulum. *Current Opinion in Cell Biology*, *22*(4), 437–446.

Hodges, B. L., & Cheng, S. H. (2006). Cell and gene-based therapies for the lysosomal storage diseases. *Current Gene Therapy*, *6*(2), 227–241.

Hollak, C. E. M., & Weinreb, N. J. (2015). The attenuated/late onset lysosomal storage disorders: Therapeutic goals and indications for enzyme replacement treatment in Gaucher and Fabry disease. *Best Practice & Research. Clinical Endocrinology & Metabolism*, *29*(2), 205–218.

Hoogerbrugge, P. M., Brouwer, O. F., Bordigoni, P., Ringden, O., Kapaun, P., Ortega, J. J., et al. (1995). Allogeneic bone marrow transplantation for lysosomal storage diseases. The European Group for Bone Marrow Transplantation. *Lancet*, *345*(8962), 1398–1402.

Hopwood, J., & Brooks, D. (1997). An introduction to the basic science and biology of the lysosome and storage diseases. In D. Applegarth, J. Dimmick, & J. Hall (Eds.), *Organelle diseases* (pp. 7–35). New York: Chapman and Hall Medical.

Hossain, M. A., Higaki, K., Shinpo, M., Nanba, E., Suzuki, Y., Ozono, K., et al. (2015). Chemical chaperone treatment for galactosialidosis: Effect of NOEV on β-galactosidase activities in fibroblasts. *Brain & Development*. http://dx.doi.org/10.1016/j.braindev.2015.07.006.

Huang, H.-J., Lee, C.-C., & Chen, C. Y.-C. (2014). Pharmacological chaperone design for reducing risk factor of Parkinson's disease from traditional Chinese medicine. *Evidence-Based Complementary and Alternative Medicine*, *2014*, 12 pages (Article ID 830490).

Ioannou, Y. A. (2000). Gene therapy for lysosomal storage disorders with neuropathology. *Journal of the American Society of Nephrology*, *11*(8), 1542–1547.

Ishii, S. (2012). Pharmacological chaperone therapy for Fabry disease. *Proceedings of the Japan Academy. Series B, Physical and Biological Sciences*, *88*(1), 18–30.

Jeyakumar, M., Norflus, F., Tifft, C. J., Cortina-Borja, M., Butters, T. D., Proia, R. L., et al. (2001). Enhanced survival in Sandhoff disease mice receiving a combination of substrate deprivation therapy and bone marrow transplantation. *Blood*, *97*(1), 327–329.

Jeyakumar, M., Smith, D. A., Williams, I. M., Borja, M. C., Neville, D. C. A., Butters, T. D., et al. (2004). NSAIDs increase survival in the Sandhoff disease mouse: Synergy with N-butyldeoxynojirimycin. *Annals of Neurology*, *56*(5), 642–649.

Johnson, F. K., Mudd, P. N., Bragat, A., Adera, M., & Boudes, P. (2013). Pharmacokinetics and safety of migalastat HCl and effects on agalsidase activity in healthy volunteers. *Clin Pharmacol Drug Dev*, *2*, 120–132.

Kakkis, E. D., Muenzer, J., Tiller, G. E., Waber, L., Belmont, J., Passage, M., et al. (2001). Enzyme-replacement therapy in mucopolysaccharidosis I. *The New England Journal of Medicine*, *344*(3), 182–188.

Khanna, R., Soska, R., Lun, Y., Feng, J., Frascella, M., Young, B., et al. (2010). The pharmacological chaperone 1-deoxygalactonojirimycin reduces tissue globotriaosylceramide levels in a mouse model of Fabry disease. *Molecular Therapy: The Journal of the American Society of Gene Therapy*, *18*(1), 23–33.

Kirkegaard, T. (2013). Emerging therapies and therapeutic concepts for lysosomal storage diseases. *Expert Opinion on Orphan Drugs*, *1*(5), 385–404.

Knight, E. M., Williams, H. N., Stevens, A. C., Kim, S. H., Kottwitz, J. C., Morant, A. D., et al. (2015). Evidence that small molecule enhancement of β-hexosaminidase activity corrects the behavioral phenotype in Dutch APP(E693Q) mice through reduction of ganglioside-bound Aβ. *Molecular Psychiatry*, *20*(1), 109–117.

Kopecká, J., Krijt, J., Raková, K., & Kožich, V. (2011). Restoring assembly and activity of cystathionine β-synthase mutants by ligands and chemical chaperones. *Journal of Inherited Metabolic Disease*, *34*(1), 39–48.

Krivit, W. (2004). Allogeneic stem cell transplantation for the treatment of lysosomal and peroxisomal metabolic diseases. *Springer Seminars in Immunopathology*, *26*(1–2), 119–132.

Krivit, W., Peters, C., & Shapiro, E. G. (1999). Bone marrow transplantation as effective treatment of central nervous system disease in globoid cell leukodystrophy, metachromatic leukodystrophy, adrenoleukodystrophy, mannosidosis, fucosidosis, aspartylglucosaminuria, Hurler, Maroteaux-Lamy, and Sly syndromes, and Gaucher disease type III. *Current Opinion in Neurology*, *12*(2), 167–176.

Laine, M., Ahtiainen, L., Rapola, J., Richter, J., & Jalanko, A. (2004). Bone marrow transplantation in young aspartylglucosaminuria mice: Improved clearance of lysosomal storage in brain by using wild type as compared to heterozygote donors. *Bone Marrow Transplantation*, *34*(11), 1001–1003.

Lee, J.-P., Jeyakumar, M., Gonzalez, R., Takahashi, H., Lee, P.-J., Baek, R. C., et al. (2007). Stem cells act through multiple mechanisms to benefit mice with neurodegenerative metabolic disease. *Nature Medicine*, *13*(4), 439–447.

Lee, W. C., Kang, D., Causevic, E., Herdt, A. R., Eckman, E. A., & Eckman, C. B. (2010). Molecular characterization of mutations that cause globoid cell leukodystrophy and pharmacological rescue using small molecule chemical chaperones. *The Journal of Neuroscience*, *30*(16), 5489–5497.

Leimig, T., Mann, L., Martin, M., del, P., Bonten, E., Persons, D., et al. (2002). Functional amelioration of murine galactosialidosis by genetically modified bone marrow hematopoietic progenitor cells. *Blood*, *99*(9), 3169–3178.

Ling, S. (2011). Structural bioinformatics analysis of acid alpha-glucosidase mutants with pharmacological chaperones. In M. A. Mahdavi (Ed.), *Bioinformatics - Trends and methodologies* (pp. 313–324). European Union: InTech.

Lodish, H., Berk, A., Zipursky, S. L., Matsudaira, P., Baltimore, D., & Darnell, J. (2000). *Molecular cell biology* (4th ed.). New York: W. H. Freeman and Company.

Lukas, J., Giese, A.-K., Markoff, A., Grittner, U., Kolodny, E., Mascher, H., et al. (2013). Functional characterisation of alpha-galactosidase a mutations as a basis for a new classification system in Fabry disease. *PLoS Genetics*, *9*(8), e1003632.

Maegawa, G. H. B., Tropak, M., Buttner, J., Stockley, T., Kok, F., Clarke, J. T. R., et al. (2007). Pyrimethamine as a potential pharmacological chaperone for late-onset forms of GM2 gangliosidosis. *The Journal of Biological Chemistry*, *282*(12), 9150–9161.

Mark, B. L., Mahuran, D. J., Cherney, M. M., Zhao, D., Knapp, S., & James, M. N. G. (2003). Crystal structure of human beta-hexosaminidase B: Understanding the molecular basis of Sandhoff and Tay-Sachs disease. *Journal of Molecular Biology*, *327*(5), 1093–1109.

Matsuda, J., Suzuki, O., Oshima, A., Yamamoto, Y., Noguchi, A., Takimoto, K., et al. (2003). Chemical chaperone therapy for brain pathology in GM1-gangliosidosis. *Proceedings of the National Academy of Sciences of the United States of America*, *100*(26), 15912–15917.

Meikle, P. J., Hopwood, J. J., Clague, A. E., & Carey, W. F. (1999). Prevalence of lysosomal storage disorders. *JAMA*, *281*(3), 249–254.

Morello, J. P., Petäjä-Repo, U. E., Bichet, D. G., & Bouvier, M. (2000). Pharmacological chaperones: A new twist on receptor folding. *Trends in Pharmacological Sciences*, *21*(12), 466–469.

Motabar, O., Sidransky, E., Goldin, E., & Zheng, W. (2010). Fabry disease—Current treatment and new drug development. *Current Chemical Genomics*, *4*, 50–56.

Moyses, C. (2003). Substrate reduction therapy: Clinical evaluation in type 1 Gaucher disease. *Philosophical Transactions of the Royal Society, B: Biological Sciences*, *358*(1433), 955–960.

Muenzer, J., Beck, M., Giugliani, R., Suzuki, Y., Tylki-Szymanska, A., Valayannopoulos, V., et al. (2011). Idursulfase treatment of Hunter syndrome in children younger than 6 years: Results from the Hunter Outcome Survey. *Genetics in Medicine*, *13*(2), 102–109.

Muenzer, J., Wraith, J. E., Beck, M., Giugliani, R., Harmatz, P., Eng, C. M., et al. (2006). A phase II/III clinical study of enzyme replacement therapy with idursulfase in mucopolysaccharidosis II (Hunter syndrome). *Genetics in Medicine*, *8*(8), 465–473.

Muntau, A. C., Leandro, J., Staudigl, M., Mayer, F., & Gersting, S. W. (2014). Innovative strategies to treat protein misfolding in inborn errors of metabolism: Pharmacological chaperones and proteostasis regulators. *Journal of Inherited Metabolic Disease*, *37*(4), 505–523.

Mynarek, M., Tolar, J., Albert, M. H., Escolar, M. L., Boelens, J. J., Cowan, M. J., et al. (2012). Allogeneic hematopoietic SCT for alpha-mannosidosis: An analysis of 17 patients. *Bone Marrow Transplantation*, *47*(3), 352–359.

Ni, M., & Lee, A. S. (2007). ER chaperones in mammalian development and human diseases. *FEBS Letters*, *581*(19), 3641–3651.

O'Leary, E. M., & Igdoura, S. A. (2012). The therapeutic potential of pharmacological chaperones and proteosomal inhibitors, Celastrol and MG132 in the treatment of sialidosis. *Molecular Genetics and Metabolism*, *107*(1–2), 173–185.

Parenti, G. (2009). Treating lysosomal storage diseases with pharmacological chaperones: From concept to clinics. *EMBO Molecular Medicine*, *1*(5), 268–279.

Parenti, G., Andria, G., & Valenzano, K. J. (2015). Pharmacological chaperone therapy: Preclinical development, clinical translation, and prospects for the treatment of lysosomal storage disorders. *Molecular Therapy: The Journal of the American Society of Gene Therapy*, *23*(7), 1138–1148.

Parenti, G., Fecarotta, S., la Marca, G., Rossi, B., Ascione, S., Donati, M. A., et al. (2014). A chaperone enhances blood α-glucosidase activity in Pompe disease patients treated with enzyme replacement therapy. *Molecular Therapy: The Journal of the American Society of Gene Therapy*, *22*(11), 2004–2012.

Parenti, G., Moracci, M., Fecarotta, S., & Andria, G. (2014). Pharmacological chaperone therapy for lysosomal storage diseases. *Future Medicinal Chemistry*, *6*(9), 1031–1045.

Passini, M. A., Dodge, J. C., Bu, J., Yang, W., Zhao, Q., Sondhi, D., et al. (2006). Intracranial delivery of CLN2 reduces brain pathology in a mouse model of classical late infantile neuronal ceroid lipofuscinosis. *The Journal of Neuroscience: The Official Journal of the Society for Neuroscience*, *26*(5), 1334–1342.

Pastores, G. M. (2007). Agalsidase alfa (Replagal™) in the treatment of Anderson-Fabry disease. *Biologics: Targets & Therapy*, *1*(3), 291–300.

Peltola, M., Kyttälä, A., Heinonen, O., Rapola, J., Paunio, T., Revah, F., et al. (1998). Adenovirus-mediated gene transfer results in decreased lysosomal storage in brain and total correction in liver of aspartylglucosaminuria (AGU) mouse. *Gene Therapy*, *5*(10), 1314–1321.

Perlmutter, D. H. (2002). Chemical chaperones: A pharmacological strategy for disorders of protein folding and trafficking. *Pediatric Research*, *52*(6), 832–836.

Piotrowska, E., Jakóbkiewicz-Banecka, J., Tylki-Szymanska, A., Liberek, A., Maryniak, A., Malinowska, M., et al. (2008). Genistin-rich soy isoflavone extract in substrate reduction therapy for Sanfilippo syndrome: An open-label, pilot study in 10 pediatric patients. *Current Therapeutic Research*, *69*(2), 166–179.

Pisani, A., Spinelli, L., Visciano, B., Capuano, I., Sabbatini, M., Riccio, E., et al. (2012). Effects of switching from agalsidase beta to agalsidase alfa in 10 patients with Anderson-Fabry disease. *JIMD Reports*, *9*, 41–48.

Platt, F. M., & Jeyakumar, M. (2008). Substrate reduction therapy. *Acta Paediatrica (Oslo, Norway: 1992). Supplement*, *97*(457), 88–93.

Platt, F. M., & Lachmann, R. H. (2009). Treating lysosomal storage disorders: Current practice and future prospects. *Biochimica et Biophysica Acta*, *1793*(4), 737–745.

Porto, C., Cardone, M., Fontana, F., Rossi, B., Tuzzi, M. R., Tarallo, A., et al. (2009). The pharmacological chaperone N-butyldeoxynojirimycin enhances enzyme replacement therapy in Pompe disease fibroblasts. *Molecular Therapy: The Journal of the American Society of Gene Therapy*, *17*(6), 964–971.

Porto, C., Ferrara, M. C., Meli, M., Acampora, E., Avolio, V., Rosa, M., et al. (2012). Pharmacological enhancement of α-glucosidase by the allosteric chaperone N-acetylcysteine. *Molecular Therapy*, *20*(12), 2201–2211.

Rastall, D. P., & Amalfitano, A. (2015). Recent advances in gene therapy for lysosomal storage disorders. *The Application of Clinical Genetics*, *8*, 157–169.

Reiss, C., Lesnik, T., Parvez, H., Parvez, S., & Ehrlich, R. (2000). Conformational toxicity and sporadic conformational diseases. *Toxicology*, *153*(1–3), 115–121.

Rigat, B. A., Tropak, M. B., Buttner, J., Crushell, E., Benedict, D., Callahan, J. W., et al. (2012). Evaluation of N-nonyl-deoxygalactonojirimycin as a pharmacological chaperone for human GM1 gangliosidosis leads to identification of a feline model suitable for testing enzyme enhancement therapy. *Molecular Genetics and Metabolism*, *107*(1–2), 203–212.

Ringdén, O., Remberger, M., Svahn, B.-M., Barkholt, L., Mattsson, J., Aschan, J., et al. (2006). Allogeneic hematopoietic stem cell transplantation for inherited disorders: Experience in a single center. *Transplantation*, *81*(5), 718–725.

Rohrbach, M., & Clarke, J. T. R. (2007). Treatment of lysosomal storage disorders: Progress with enzyme replacement therapy. *Drugs*, *67*(18), 2697–2716.

Rombach, S. M., Hollak, C. E., Linthorst, G. E., & Dijkgraaf, M. G. (2013). Cost-effectiveness of enzyme replacement therapy for Fabry disease. *Orphanet Journal of Rare Diseases*, *8*(1), 29.

Rovelli, A. M. (2008). The controversial and changing role of haematopoietic cell transplantation for lysosomal storage disorders: An update. *Bone Marrow Transplantation*, *41*(S2), S87–S89.

Sands, M. S., & Davidson, B. L. (2006). Gene therapy for lysosomal storage diseases. *Molecular Therapy: The Journal of the American Society of Gene Therapy*, *13*(5), 839–849.

Sands, M. S., & Haskins, M. E. (2008). CNS-directed gene therapy for lysosomal storage diseases. *Acta Paediatrica (Oslo, Norway: 1992). Supplement*, *97*(457), 22–27.

Sawkar, A. R., Adamski-Werner, S. L., Cheng, W.-C., Wong, C.-H., Beutler, E., Zimmer, K.-P., et al. (2005). Gaucher disease-associated glucocerebrosidases show mutation-dependent chemical chaperoning profiles. *Chemistry & Biology*, *12*(11), 1235–1244.

Sawkar, A. R., Cheng, W.-C., Beutler, E., Wong, C.-H., Balch, W. E., & Kelly, J. W. (2002). Chemical chaperones increase the cellular activity of N370S beta-glucosidase: A therapeutic strategy for Gaucher disease. *Proceedings of the National Academy of Sciences of the United States of America*, *99*(24), 15428–15433.

Sawkar, A. R., D'Haeze, W., & Kelly, J. W. (2006). Therapeutic strategies to ameliorate lysosomal storage disorders—A focus on Gaucher disease. *Cellular and Molecular Life Sciences: CMLS*, *63*(10), 1179–1192.

Schiffmann, R., Kopp, J. B., Austin, H. A., III, et al. (2001). Enzyme replacement therapy in Fabry disease: A randomized controlled trial. *JAMA*, *285*(21), 2743–2749.

Schneider, J. A., Clark, K. F., Greene, A. A., Reisch, J. S., Markello, T. C., Gahl, W. A., et al. (1995). Recent advances in the treatment of cystinosis. *Journal of Inherited Metabolic Disease*, *18*(4), 387–397.

Shin, S. H., Kluepfel-Stahl, S., Cooney, A. M., Kaneski, C. R., Quirk, J. M., Schiffmann, R., et al. (2008). Prediction of response of mutated alpha-galactosidase A to a pharmacological chaperone. *Pharmacogenetics and Genomics*, *18*(9), 773–780.

Staretz-Chacham, O., Lang, T. C., LaMarca, M. E., Krasnewich, D., & Sidransky, E. (2009). Lysosomal storage disorders in the newborn. *Pediatrics*, *123*(4), 1191–1207.

Suzuki, Y. (2013). Chaperone therapy update: Fabry disease, GM1-gangliosidosis and Gaucher disease. *Brain & Development*, *35*(6), 515–523.

Suzuki, Y., Ogawa, S., & Sakakibara, Y. (2009). Chaperone therapy for neuronopathic lysosomal diseases: Competitive inhibitors as chemical chaperones for enhancement of mutant enzyme activities. *Perspectives in Medicinal Chemistry*, *3*, 7–19.

Todde, V., Veenhuis, M., & van der Klei, I. J. (2009). Autophagy: Principles and significance in health and disease. *Biochimica et Biophysica Acta, 1792*(1), 3–13.
Tropak, M. B., Blanchard, J. E., Withers, S. G., Brown, E. D., & Mahuran, D. (2007). High-throughput screening for human lysosomal beta-N-Acetyl hexosaminidase inhibitors acting as pharmacological chaperones. *Chemistry & Biology, 14*(2), 153–164.
Tropak, M. B., Kornhaber, G. J., Rigat, B. A., Maegawa, G. H., Buttner, J. D., Blanchard, J. E., et al. (2008). Identification of pharmacological chaperones for Gaucher disease and characterization of their effects on beta-glucocerebrosidase by hydrogen/deuterium exchange mass spectrometry. *Chembiochem: A European Journal of Chemical Biology, 9*(16), 2650–2662.
Tropak, M. B., Reid, S. P., Guiral, M., Withers, S. G., & Mahuran, D. (2004). Pharmacological enhancement of beta-hexosaminidase activity in fibroblasts from adult Tay-Sachs and Sandhoff Patients. *The Journal of Biological Chemistry, 279*(14), 13478–13487.
Ulloa-Aguirre, A., Janovick, J. A., Brothers, S. P., & Conn, P. M. (2004). Pharmacologic rescue of conformationally-defective proteins: Implications for the treatment of human disease. *Traffic, 5*(11), 821–837.
Valenzano, K. J., Khanna, R., Powe, A. C., Boyd, R., Lee, G., Flanagan, J. J., et al. (2011). Identification and characterization of pharmacological chaperones to correct enzyme deficiencies in lysosomal storage disorders. *Assay and Drug Development Technologies, 9*(3), 213–235.
Van den Hout, J. M., Reuser, A. J., de Klerk, J. B., Arts, W. F., Smeitink, J. A., & Van der Ploeg, A. T. (2001). Enzyme therapy for Pompe disease with recombinant human α-glucosidase from rabbit milk. *Journal of Inherited Metabolic Disease, 24*(2), 266–274.
Van den Hout, H., Reuser, A. J., Vulto, A. G., Christa, B., Loonen, M., Cromme-Dijkhuis, A., et al. (2000). Recombinant human α-glucosidase from rabbit milk in Pompe patients. *Lancet, 356*(9227), 397–398.
van Karnebeek, C., Dunbar, M., Davis, J., Bowden, K., Chaplin, K., Chaplin, R., et al. (2014). Bone marrow transplant for a boy with alpha-mannosidosis illustrates a family's decision-making in rare diseases. *Rare Diseases and Orphan Drugs, 1*(2), 62–68.
Virta, S., Rapola, J., Jalanko, A., & Laine, M. (2006). Use of nonviral promoters in adenovirus-mediated gene therapy: Reduction of lysosomal storage in the aspartylglucosaminuria mouse. *The Journal of Gene Medicine, 8*(6), 699–706.
Vite, C. H., McGowan, J. C., Niogi, S. N., Passini, M. A., Drobatz, K. J., Haskins, M. E., et al. (2005). Effective gene therapy for an inherited CNS disease in a large animal model. *Annals of Neurology, 57*(3), 355–364.
Walkley, S. U., Thrall, M. A., Dobrenis, K., Huang, M., March, P. A., Siegel, D. A., et al. (1994). Bone marrow transplantation corrects the enzyme defect in neurons of the central nervous system in a lysosomal storage disease. *Proceedings of the National Academy of Sciences of the United States of America, 91*(8), 2970–2974.
Wang, R. Y., Bodamer, O. A., Watson, M. S., Wilcox, W. R., & ACMG Work Group on Diagnostic Confirmation of Lysosomal Storage Diseases. (2011). Lysosomal storage diseases: Diagnostic confirmation and management of presymptomatic individuals. *Genetics in Medicine: Official Journal of the American College of Medical Genetics, 13*(5), 457–484.
Wu, X., Katz, E., Valle, M. C. D., Mascioli, K., Flanagan, J. J., Castelli, J. P., et al. (2011). A pharmacogenetic approach to identify mutant forms of α-galactosidase A that respond to a pharmacological chaperone for Fabry disease. *Human Mutation, 32*(8), 965–977.
Xu, L., Haskins, M. E., Melniczek, J. R., Gao, C., Weil, M. A., O'Malley, T. M., et al. (2002). Transduction of hepatocytes after neonatal delivery of a Moloney murine leukemia virus based retroviral vector results in long-term expression of beta-glucuronidase in mucopolysaccharidosis VII dogs. *Molecular Therapy: The Journal of the American Society of Gene Therapy, 5*(2), 141–153.

Xu, S., Lun, Y., Brignol, N., Hamler, R., Schilling, A., Frascella, M., et al. (2015). Coformulation of a novel human α-galactosidase A with the pharmacological chaperone AT1001 leads to improved substrate reduction in Fabry mice. *Molecular Therapy: The Journal of the American Society of Gene Therapy*, *23*(7), 1169–1181.

Yam, G. H.-F., Bosshard, N., Zuber, C., Steinmann, B., & Roth, J. (2006). Pharmacological chaperone corrects lysosomal storage in Fabry disease caused by trafficking-incompetent variants. *American Journal of Physiology. Cell Physiology*, *290*(4), C1076–C1082.

Yu, Z., Sawkar, A. R., Whalen, L. J., Wong, C.-H., & Kelly, J. W. (2007). Isofagomine- and 2,5-anhydro-2,5-imino-D-glucitol-based glucocerebrosidase pharmacological chaperones for Gaucher disease intervention. *Journal of Medicinal Chemistry*, *50*(1), 94–100.

Zhang, P., Sun, B., Osada, T., Rodriguiz, R., Yang, X. Y., Luo, X., et al. (2012). Immunodominant liver-specific expression suppresses transgene-directed immune responses in murine Pompe disease. *Human Gene Therapy*, *23*(5), 460–472.

Zimran, A., & Elstein, D. (2003). Gaucher disease and the clinical experience with substrate reduction therapy. *Philosophical Transactions of the Royal Society of London. Series B, Biological Sciences*, *358*(1433), 961–966.

CHAPTER NINE

Investigating the Inhibitory Effect of Wortmannin in the Hotspot Mutation at Codon 1047 of PIK3CA Kinase Domain: A Molecular Docking and Molecular Dynamics Approach

D. Thirumal Kumar, C. George Priya Doss[1]

Medical Biotechnology Division, School of Biosciences and Technology, VIT University, Vellore, Tamil Nadu, India

[1]Corresponding author: e-mail addresses: georgecp77@yahoo.co.in; georgepriyadoss@vit.ac.in

Contents

Abstract

Oncogenic mutations in phosphatidylinositol-4,5-bisphosphate 3-kinase, catalytic subunit alpha (PIK3CA) are the most frequently reported in association with various forms of cancer. Several studies have reported the significance of hotspot mutations in a catalytic subunit of PIK3CA in association with breast cancer. Mutations are frequently observed in the highly conserved region of the kinase domain (797–1068 amino acids) of PIK3CA are activating or gain-of-function mutations. Mutation in codon 1047 occurs in the C-terminal region of the kinase domain with histidine (H) replaced by arginine (R),

Advances in Protein Chemistry and Structural Biology, Volume 102
ISSN 1876-1623
http://dx.doi.org/10.1016/bs.apcsb.2015.09.008

lysine (L), and tyrosine (Y). Pathogenicity and protein stability predictors PhD-SNP, Align GVGD, HANSA, iStable, and MUpro classified H1047R as highly deleterious when compared to H1047L and H1047Y. To explore the inhibitory activity of Wortmannin toward PIK3CA, the three-dimensional structure of the mutant protein was determined using homology modeling followed by molecular docking and molecular dynamics analysis. Docking studies were performed for the three mutants and native with Wortmannin to measure the differences in their binding pattern. Comparative docking study revealed that H1047R-Wortmannin complex has a higher number of hydrogen bonds as well as the best binding affinity next to the native protein. Furthermore, 100 ns molecular dynamics simulation was initiated with the docked complexes to understand the various changes induced by the mutation. Though Wortmannin was found to nullify the effect of H1047R over the protein, further studies are required for designing a better compound. As SNPs are major genetic variations observed in disease condition, personalized medicine would provide enhanced drug therapy.

1. INTRODUCTION

The new estimate released from World Cancer Research Fund International Breast Cancer in 2012 reports that nearly 1.67 million of new cases were diagnosed with breast cancer (Ferlay et al., 2012). Data also suggest that breast cancer is the first most prevalent cancer in women and second most common cancer in worldwide. In recent days, the mortality rate due to breast cancer is increasing rapidly with an increase in demand for a new drug (Breast Cancer Facts & Figures 2013–2014). Several genes such as *BRCA1* (Henneman et al., 2015; Yan et al., 2015), *BRCA2* (Aradottir et al., 2015; Parvez, Popovic, Elavathil, Okawara, & Hodgson, 2015), *PIK3CA* (Majewski et al., 2015; Micalizzi et al., 2015), *TP53* (Pal et al., 2015; Pfister et al., 2015), and *CDH1* (Mahmud, Ford, Longacre, Parent, & Norton, 2015) are associated with breast cancer. Among these, *PIK3CA* is found to be a second most predominant gene with mutations (Weng et al., 2015). This data suggest that the PIK3CA mutation is a more recurrent genetic alteration seen in breast cancer cases (Wu et al., 2005). PIK3CA, otherwise known as p110α, is an isoform of p110 subunit of PI3K protein (Firoozinia, Jahromi, Moghadamtousi, Nikzad, & Kadir, 2014; Zardavas, Phillips, & Loi, 2014). PI3K (phosphoinositol 3-kinase) are lipid kinases or enzymes that account for the significant cellular functions (cell growth, differentiation, motility, survival, proliferation, and intracellular trafficking; Ligresti et al., 2009). PI3K acts as a catalyst in producing second messengers such as PIP3 (lipid second messengers) at the cell membrane. This mechanism contributes to the activation of downstream targets that involve in helping the cell to do regular process such as serine-threonine protein kinase (Akt). This signaling process by which Akt activated through PI3K is called AKT/PI3K signaling pathway (Cantley, 2002; Luo, Manning, & Cantley, 2003). There are reports stating the deregulation of AKT/PI3K signaling pathway in breast cancer (Adams, Schachter, Liu, Zacksenhaus, & Egan, 2011; Campbell et al., 2004). PI3K consists of two major subunits, a catalytic subunit p110 and a regulatory subunit p85. A significant relationship between the catalytic subunit of PIK3CA and age association in women with breast cancer has been reported (Firoozinia et al., 2014). Frequent mutations were observed in the 9th and 20th exon of the helical (517–694 amino acids) and kinase domain (797–1068 amino acids) of PIK3CA (Bachman et al., 2004; Samuels et al., 2005). The mutations present in the helical domain are independent of binding to p85, but the interaction with RAS-GTP acts as a main criterion. Whereas, the mutations that occur in the kinase domain are dependent on p85 and gets activated in the absence of the RAS-GTP (Zhao & Vogt, 2008). Thus, this might be a reason for mutants present in the kinase region in elevating the kinase activity of the protein (Ikenoue, 2005). PIK3CA mutations are the most common aberrations with over one-third of the cases reported in breast cancer. Also, kinase domains are known to be conserved, and mutations associated with the conserved region are more relevant to the functionality of the protein. H1047R mutation is the most frequent hotspot mutation occurring in the C-terminal region of the kinase domain (Karakas, Bachman, & Park, 2006; Mangone, Bobrovnitchaia, Salaorni, Manuli, & Nagai, 2012; Vorkas et al., 2010). H1047R mutation has a higher stimulus of activating AKT than other mutations and likely to be representing the initiation and progression of cancer (Cizkova et al., 2012; Gymnopoulos, Elsliger, & Vogt, 2007; Ramirez-Ardila et al., 2013; Zardavas et al., 2014). In addition to H1047R, the incidence of H1047L and rare H1047Y mutations has also been reported (Levine et al., 2005).

A study by Gymnopoulos et al. (2007) investigated the oncogenic potential of the three variants occurring at codon 1047 with H1047R, H1074L, and H1047Y with severe, moderate, and weak oncogenic potency. Their group described that mutants located near the hinge of the activation loop might alter the confirmation in increasing the catalytic activity. In particular, PI3K catalytic subunit mutations upregulate PI3K/AKT/mTOR pathway and promote carcinogenesis providing a way for the development of PI3K inhibitors toward cancer therapy. Most of the inhibitors bind to the ATP-binding pocket of the catalytic domain. Based on the severity of the mutations, several drugs were designed to inhibit PI3K–AKT–mTOR pathway for cancer therapy. AZD6482 is known to be a potent, selective, and ATP-competitive PI3Kβ inhibitor, a first human target validated as antiplatelet therapy (Nylander et al., 2012). Bexarotene is a drug targeting the retinoid X receptor-selective retinoid that has shown to have antitumor activity in breast cancer in preclinical studies with limited efficacy (Esteva et al., 2003). Akt Inhibitor VIII is a lysozyme selective, cell permeable, and reversible quinoxaline lead compound that can inhibit the Akt1/Akt2 activity (Calleja, Laguerre, Parker, & Larijani, 2009). WO 2009093972 A1 selectively inhibits phosphoinositide (PI) 3-kinase β but fails neoplastic effect. Motesanib and lapatinib act as an antagonist of growth factor receptors such as VEGF and EGFR, respectively (Burris, 2004; Schilder et al., 2013). CI-1040 is found to be targeting the MEK1/2 pathways thereby blocking the ERK pathway (Allen, Sebolt-Leopold, & Meyer, 2003). PAC-1 targets the CASP3 receptor which initiates apoptosis (important anticarcinogenic property) and also shows antineoplastic activity (Al-Hujaily et al., 2011). VX-702 is a novel compound inhibiting p38 MAPK (Damjanov, Kauffman, & Spencer-Green, 2009). Methotrexate is an antifolate drug that can terminate a tumor cell but found to have an adverse effect such as liver damage and lowering blood cell count (Bannwarth, Labat, Moride, & Schaeverbeke, 1994). Furthermore, LY294002 is a chemical compound containing morpholine found to be a potent inhibitor of PI3Ks (Clark, West, Streicher, & Dennis, 2002; Gharbi et al., 2007). Among these compounds, Wortmannin was found to be the best selective inhibitory effect on PIK3 class of proteins with least side effects (Liu et al., 2005). This drug molecule is a covalently modified protein of the earlier stated compound that also has an inhibitory effect on PIK3 protein (Wymann et al., 1996). Wortmannin, a steroid metabolite, shows a better activity when compared with a broad range of medical assistance provided for breast cancer (Kuete et al., 2015). So far, several experimental studies have investigated the impact of hotspot mutation at codon 1047 and their inhibitory effect (Janku et al., 2014; Mankoo, Sukumar, & Karchin, 2009; Samuels et al., 2004; Sudhakar et al., 2015; Wu et al., 2005). Recently, Gkeka et al. (2014) mentioned the impact of H1047R mutation leading to enhanced activity of the protein at atomistic level. A previous study by our group also reported the deleterious nature of H1047R and H1047L over the protein in south Indian population (Sudhakar et al., 2015). In this chapter, authors investigated the impact of mutations H1047R and H1047L using molecular dynamics simulation of 30 ns. Till date, no study has been initiated to measure the inhibitory effect of Wortmannin in the hotspot mutation at codon 1047 of PIK3CA kinase domain. Therefore, in this report, for the first time, we have characterized the structural and functional consequence of the mutants H1047R, H1047L, and H1047Y occurring in the codon 1047 of PIK3CA by computational prediction methods. As a first step, pathogenic and protein stability predictors, namely, HANSA (Acharya & Nagarajaram, 2012), PhD-SNP (Capriotti, Calabrese, & Casadio, 2006), Align GVGD (Tavtigian et al., 2006), MUpro (Cheng, Randall, & Baldi, 2006), and iStable (Chen, Lin, & Chu, 2013) were used to assess the impact on protein structure and function. Next step, 3D model for the mutant proteins (H1047R, H1047L, and H1047Y) was constructed using SwissPDB Viewer (Johansson, Zoete, Michielin, & Guex, 2012). Furthermore, docking study by Autodock 4.2 (Morris et al., 2009) enabled us to predict the affinity of the drug Wortmannin toward the target native and mutant proteins. Lastly, molecular dynamics study was initiated using GROMACS (Hess, Bekker, Berendsen, & Fraaije, 1997) for a simulation period of 100 ns to study the atomic level conformational changes in protein–ligand interactions.

2. MATERIALS AND METHODS

2.1 *In Silico* Prediction of Pathogenicity and Stability

Many algorithms have been developed largely to predict the pathogenic nature of a nsSNP. The pathogenic characteristic of a mutation was analyzed using HANSA (Acharya & Nagarajaram, 2012), PhD-SNP (Capriotti et al., 2006), and Align GVGD (Tavtigian et al., 2006). HANSA has three broad features: (1) position specific, (2) sequence- and structure based, and (3) amino acid residue based. These features along with support vector machine help in predicting the nsSNPs as deleterious or benign (http://hansa.cdfd.org.in:8080) (Acharya & Nagarajaram, 2012). PhD-SNP (http://snps.biofold.org/phd-snp/phd-snp.html) predicts the pathogenicity either by using the protein sequence or the possible structure of PDB database. PhD-SNP predicts the deleterious nature of nsSNPs by considering the three datasets: (i) based on sequence information, (ii) based on profile information, and (iii) finally robustness was tested (Capriotti et al., 2006). Align GVGD, available at http://agvgd.iarc.fr/index.php, uses Gratham difference to serve as a plugin to execute multiple comparisons with the utility of multiple sequence alignments (Tavtigian et al., 2006). The scores of Align GVGD ranges from Class 0 to Class 65 with the spectrum of C0, C15, C25, C35, C45, C55, C65. C0 is classified as least likely to interfere with the function and C65 is classified as most likely to be interfering with function. The pathogenic nature of the SNPs greatly affects the stability of the protein structure. Furthermore, to analyze the consequence of protein stability upon mutation, we used MUpro (Cheng et al., 2006) and iStable servers (Chen et al., 2013). MUpro (http://www.igb.uci.edu/servers/servers.html) calculates the $\Delta\Delta G$ value to predict the change in protein stability for single-site mutations. The stability is known to be increasing if the value of $\Delta\Delta G$ is positive; on another hand, if $\Delta\Delta G$ is negative, then the mutation is known to destabilize the protein structure. iStable server predicts the stability from the experimental dataset found in ProTherm database (http://predictor.nchu.edu.tw/iStable). FASTA sequence along with mutant amino acids at the1047th position was provided as the input.

2.2 Molecule Preparation

The 3D structure of native PIK3CA protein molecule (PDB ID: 2RD0) with a resolution of 3.05 Å and 1068 amino acid length (Huang et al., 2007) was retrieved from Protein Data Bank (http://www.rcsb.org/). The protein molecule was further exposed to single individual mutations (H1047R, H1047L, and H1047Y) using SwissPDB Viewer (Johansson, Zoete, Michielin, & Guex, 2012). The SMILES string of Wortmannin was retrieved from the PubChem database and submitted to CORINA server to obtain the 3D structure. To avoid the distorted geometries, protein and inhibitor were subjected to energy minimization using GROMOS43a1 force field of GROMACS 4.5.6 (Hess et al., 1997). Energy-minimized structures were considered for further analysis.

2.3 Receptor–Ligand Docking Analysis

Autodock 4.2 (Morris et al., 2009) was used to study the binding affinity of Wortmannin toward the native and mutant (H1047R, H1047L, and H1047Y) protein molecules. More the binding affinity elucidates more the interaction of the drug toward the protein molecules. Semiflexible docking protocol was followed. The protein molecules were kept rigid, and ligand as flexible to attain a degree of freedom torsions bridged by the rotational parameter. The proteins and ligand PDBs were converted to .pdbqt after initial addition of hydrogen bonds and charges. These molecules were considered for docking analysis. The cubical grid box of $60 \times 60 \times 60$ size with 0.375 Å was fixed around the active sites of the protein [ILE-800, ILE-848, VAL-850, VAL-851, SER-919, MET-922, PHE-930, ILE-932, ASP-933, ILE-831, ILE-879, ILE-881, VAL-882, ASP-950, MET-953,

PHE-961, ILE-963, and ASP-964] (Mandelker, Schmidt-Kittler, Vogelstein, & Amzel, 2010). Autogrid4 parameter was used to attain a rigid grid box. Further to the auto grid, autodock4 with Lamarckian genetic algorithms was used to get best docking conformations. The complex with the best conformation was considered for molecular dynamics simulation.

2.4 Molecular Dynamics Simulation

Molecular dynamics simulation for the protein–ligand complexes were initiated using GROMOS96 43a1 force field of GROMACS 4.5.6 package (Hess et al., 1997). Pdb2gmx was used to achieve GROMACS file from protein. Additional hydrogen atoms were removed using –ignh command. The protein complexes were solvated in a cubic box with SPC water molecules maintaining the size at least 0.9 nm from all atoms. The prodrg server (Schüttelkopf & Alten, 2004) was used to create the ligand topology. The system was neutralized by adding $2Na^+$ to replace two solvent molecules. The neutralized system further subjected to energy minimization helps in removing the irregular torsions. The energy minimization was carried out using steepest descent minimization algorithm for 50000 steps using 1000 kJ/mol/nm as the maximum force. Further to reduce unrestrained dynamics, the system was well equilibrated before stimulation. The equilibration was carried out in two steps, NVT and NPT. The temperature was maintained constant at 300 K using Berendsen algorithm (Berendsen, Postma, van Gunsteren, DiNola, & Haak, 1984). Particle Mesh Ewald method (Cheatham, Miller, Fox, Darden, & Kollman, 1995) was used to treat long-range columbic interactions. SHAKE algorithm was utilized to measure the hydrogen bond length. The molecular dynamics simulation was carried out for 100 ns. The obtained trajectory files were analyzed using xmgrace and visualization using PyMOL (DeLano, 2002).

3. RESULTS

3.1 Prediction of nsSNP's Deleterious Nature

Examining the deleterious nature of nsSNPs has become easier with the aid of computational prediction methods. Especially, the algorithm such as HANSA, PhD-SNP, and Align GVGD predicts the pathogenic nature of a mutation. The results obtained from the tools mentioned above are listed in Table 1. HANSA is an automated tool that discriminates between the deleterious and neutral human nsSNPs using a novel feature, "nsSNP neutral

Table 1 Predicting the Pathogenicity and Stability Upon Mutation at 1047th Position

Prediction Tools	H1047R	H1047L	H1047Y
HANSA	Deleterious	Neutral	Neutral
PhD-SNP	Deleterious	Deleterious	Deleterious
Align GVGD	C45	C65	C65
MUpro	Decrease	Decrease	Decrease
Conf. Score	−1.752	−1.232	−1.593
iStable	Decrease	Decrease	Decrease
Conf. Score	−1.682	−1.123	−1.384

disease." Out of three nsSNPs, H1047R was predicted to be deleterious by HANSA. PhD-SNP predicted all the three mutants as deleterious. Align GVGD requires the result of multiple sequence analysis as an input to predict the pathogenicity. The interference is made based on the substitutions in the protein multiple sequence alignments. Align GVGD, classified H1047L, and H1047Y were predicted C65 and H1047R as C45. Both the classes (C65 and C45) are known to be possibly damaging scores. We used MUpro and iStable tools to predict the stability change upon mutation (Table 1). MUpro takes an advantage of comparing from large experimental data set deposited in ProTherm database (Bava, Gromiha, Uedaira, Kitajima, & Sarai, 2004). MUpro predicted that the stability of the protein as decreased in all the three mutants. By iStable, the stability of the protein was found to be decreasing for all the three mutants.

3.2 Molecular Docking Analysis

Several experimental techniques have shown the inhibitory activity of Wortmannin over the PIK3CA protein (Kuete et al., 2015; Liu et al., 2005). However, the efficiency of binding varies with the substitution of deleterious nsSNPs. The binding affinity has to be maxima to experience better drug effect over the protein. Lesser the binding energy higher will be the binding affinity of the drug toward the molecule. The energy changes based on the number of hydrogen bonds formed between the molecules. Thus, hydrogen bonds and the amino acids surrounding the drug in a protein play a major role in binding of the drug. The mutations may have negative effect on the binding pattern of the drug. A comparative docking analysis was carried out between the native and mutant proteins H1047R, H1047L, and H1047Y to assess the effect of the drug upon mutation. The binding energy of native was found to be −9.09 kcal/mol,

Table 2 Estimation of Binding Energy, Number of Polar Contacts, and Interacting Residues Between Wortmannin and PIK3CA (Native and Mutants) Using Autodock

Docked Complex	Binding Energy	No. of Polar Contacts	Interacting Residues
Native-Wortmannin	−9.09	6	LYS-802, SER-774, ILE-848, VAL-850, VAL-851
H1047R-Wortmannin	−8.84	5	LYS-802, VAL-850, VAL-851, TYR-836
H1047L-Wortmannin	−8.18	5	LYS-802, TYR-836, SER-774
H1047Y-Wortmannin	−7.85	4	ILE-800, THR-856, LYS-802

whereas the mutants H1047R, H1047L, and H1047Y showed −8.84, −8.18, and −7.85 kcal/mol, respectively (Table 2). From this analysis, we can clearly state that H1047R showed better binding energy when compared to the other mutants. Wortmannin-H1047R ranked at top had potential interaction with 15 residues, namely, MET-772, SER-774, TRP-780, ILE-800, LYS-802, TYR-836, ILE-848, GLU-849, VAL-850, VAL-851, ARG-852, SER-854, MET-922, ILE-932, and ASP-933 as same as that of native. Wortmannin-H1047L ranked second based on binding energy had potential interaction with 14 residues, namely, MET-772, SER-774, TRP-780, ILE-800, LYS-802, TYR-836, ILE-848, VAL-850, VAL-851, ARG-852, SER-854, MET-922, ILE-932, and ASP-933, respectively. Meanwhile, Wortmannin-H1047Y ranked last in binding energy score had potential interaction with 14 residues, namely, MET-772, TRP-780, ILE-800, LYS-802, TYR-836, VAL-850, VAL-851, ARG-852, SER-854, THR-856, SER-919, MET-922, ILE-932, and ASP-933 (Fig. 1A–D). We also observed the significant difference in number of hydrogen bonds formed in the native-Wortmannin and mutant-Wortmannin complexes. A highest number of six hydrogen bonds were formed in native-Wortmannin complex (two bonds with LYS-802, one bond with SER-774, one bond with ILE-848, one bond with VAL-851, and one bond with VAL-850). Followed by five hydrogen bonds in H1047L-Wortmannin complex (three hydrogen bonds with LYS-802, one bond with TYR-836, one bond with SER-774) and H1047R-Wortmannin complex (two bonds with LYS-802, VAL-850, VAL-851, TYR-836). Finally, four hydrogen bonds were formed between H1047Y and Wortmannin complex (two bonds with ILE-800, one bond with THR-856, and one bond with LYS-802).

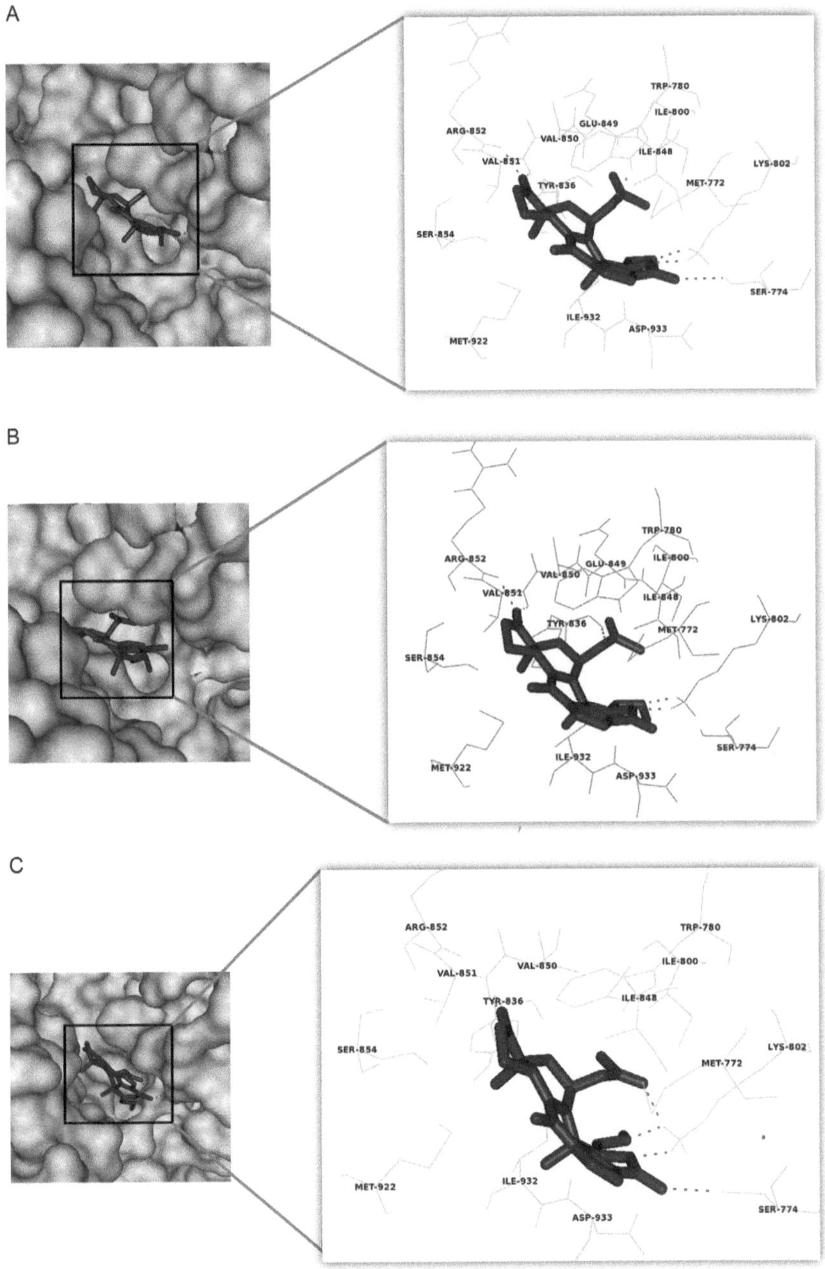

Figure 1 (A)–(D) PyMOL visualization showing surrounding residues and hydrogen bond interactions for native and mutants (native-Wortmannin, H1047R-Wortmannin, H1047L-Wortmannin, H1047Y-Wortmannin) docked complexes.

(Continued)

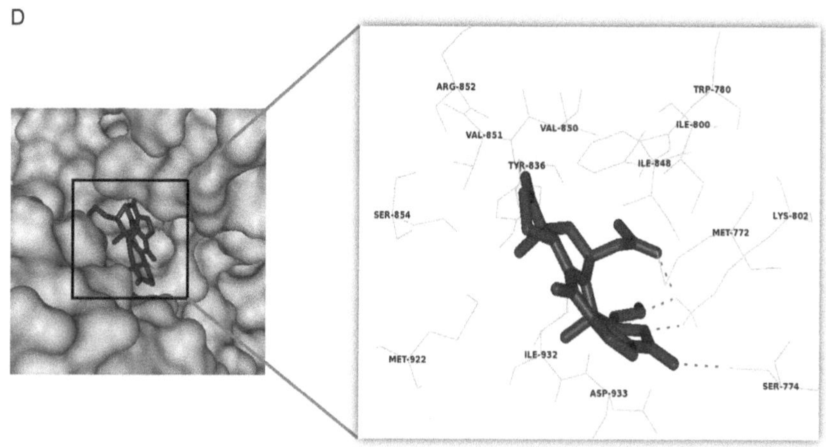

Figure 1—Cont'd

3.3 Molecular Dynamics Analysis

A simulation analysis was carried with the docked complexes (Wortmannin-native, Wortmannin-H1047R, Wortmannin-H1047Y, and Wortmannin-H1047L). g_rms was used to plot the root-mean-square deviation (RMSD) observed between the mutants and the native to measure the stability of the trajectories. RMSD graph predicted the overall stability of the docked molecule for 100 ns. Kinetic shock induced during the simulation reflects as fluctuation in the initial period. H1047R-Wortmannin complex showed a deviation between ~0.4 and ~0.45 nm, whereas the other mutants H1047L and H1047Y did not show visible deviation from the native complex (Fig. 2). Furthermore, to observe the residue fluctuation in the native and mutant complexes, g_rmsf was executed. In the overall time scale, H1047L showed maximum fluctuation by reaching ~0.5 nm. The fluctuation was found to be minimal in the kinase domain, where the mutation was induced. More specifically in the C-terminal region, the fluctuation was found to be least in H1047R when compared to the other mutants. This describes that the drug Wortmannin has nullified the mutation thereby decreasing the residue fluctuation and increasing the stability of the mutant H1047R (Fig. 3). Next, hydrogen bond analysis was carried out using g_hbond to compare the number of hydrogen bonds formed between drug and the protein. Hydrogen bond formation is responsible for steady conformation of the protein–ligand complex. Noticeably,

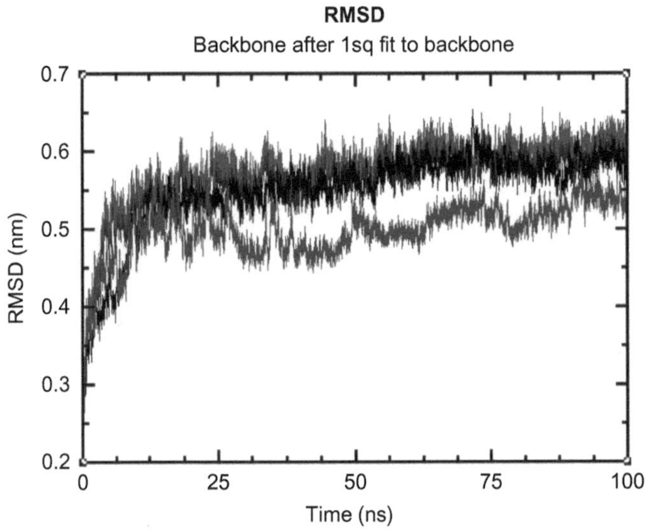

Figure 2 Comparative root-mean-square deviation for 100 ns of native and mutants docked complexes. Color scheme native-Wortmannin complex (red (light gray in the print version)), H1047R-Wortmannin complex (violet (gray in the print version)), H1047L-Wortmannin complex (black), H1047Y-Wortmannin complex (blue (dark gray in the print version)).

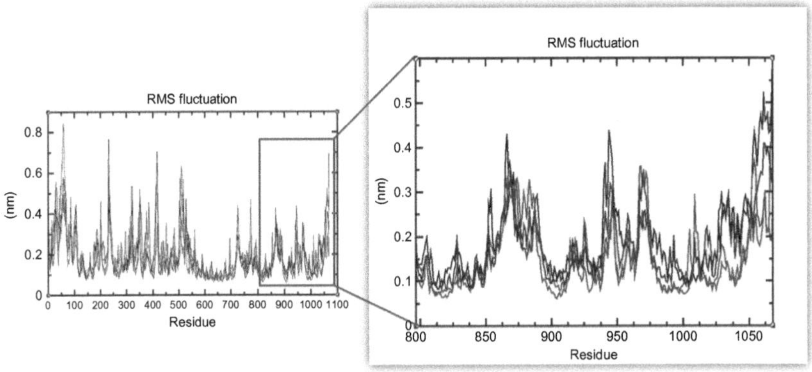

Figure 3 Comparative root-mean-square fluctuation for residues of native and mutants docked complexes. Enlarged image representing the fluctuation in kinase domain. Color scheme native-Wortmannin complex (red (light gray in the print version)), H1047R-Wortmannin complex (violet (gray in the print version)), H1047L-Wortmannin complex (black), H1047Y-Wortmannin complex (blue (dark gray in the print version)).

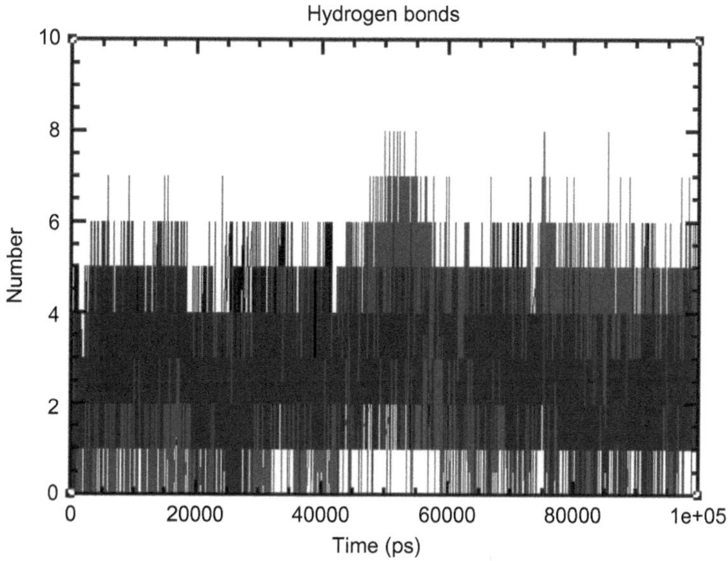

Figure 4 Analysis of intermolecular hydrogen bonds formed between native-Wortmannin and mutants-Wortmannin docked complexes for 100 ns. Color scheme native-Wortmannin complex (red (light gray in the print version)), H1047R-Wortmannin complex (violet (gray in the print version)), H1047L-Wortmannin complex (black), H1047Y-Wortmannin complex (blue (gray in the print version)).

number of hydrogen bonds was observed in the H1047R mutant complex (~eight hydrogen bonds) than the other mutants (Fig. 4). The compactness of the docked molecule can be understood by the radius of gyration that explains the arrangement of atoms around the central axis. g_gyrate was executed to plot radius of gyration. From the plot (Fig. 5), we can describe that the native showed a constant decrease in the value till 40 ns, and after that a consistency was observed between ~3.5 and ~3.8 nm. While comparing the mutants with the native, H1047Y showed a similar decrease as observed with native, but H1047L and H1047R showed a greater level of deviation. H1047L and H1047R were observed to be between ~3.14 to 3.16 nm and 3.11 to 3.13 nm at 100 ns, respectively (Fig. 5). g_mmpbsa was used to calculate the binding energy, polar, nonpolar, van der Waals forces, and electrostatic energy between the protein and drug for the last 10 ns (Table 3). These calculations have been greatly involved in the assessment of the binding energy pattern between the drug and the protein (native and mutants). Noncovalent bonds such as van der Waals and electrostatic energies were calculated to be −69.476 and −50.155 kJ/mol for the native complex, −292.494 and −64.993 kJ/mol for the mutant H1047R complex,

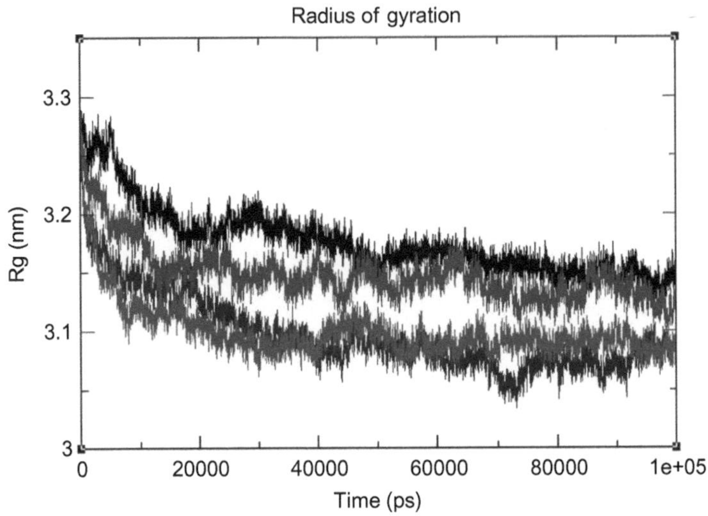

Figure 5 Analysis of Rg of native-Wortmannin and mutants-Wortmannin docked complexes for 100 ns. Color scheme native-Wortmannin complex (red (light gray in the print version)), H1047R-Wortmannin complex (violet (gray in the print version)), H1047L-Wortmannin complex (black), H1047Y-Wortmannin complex (blue (dark gray in the print version)).

−270.855 and −49.185 kJ/mol for the mutant H1047Y complex, and −287.706 and −41.763 kJ/mol for the mutant H1047L complex. The polar solvent was calculated to be 137.239, 173.558, 138.162, and 119.315 kJ/mol for the native, H1047R, H1047L, and H1047Y mutant complexes, respectively. The nonpolar energies were calculated to be −19.228, −17.789, −19.097, and −18.557 kJ/mol for the native, H1047R, H1047L, and H1047Y mutant complexes, respectively. The binding energies were calculated to be −198.62 kJ/mol for native complex, −201.719 kJ/mol for mutant H1047R complex, −200.975 kJ/mol for mutant H1047Y complex, and −208.712 kJ/mol for H1047L complex. g_covarandg_anaeig of GROMACS utilities command were executed to calculate the eigenvectors and PCA of the native and the mutant complexes. Principle component analysis of the native and the mutant complexes explains about the principle motions attained by the protein structure for the simulation period of 100 ns. Primarily, a covariance matrix is constructed from the trajectory of the MD simulation. The eigenvectors and eigenvalues were categorized by constructing a diagonal matrix. The *x*-axis explains the projections on eigenvector 1, and *y*-axis explains the projections on eigenvector 2. Projection variations observed between the native and the

Table 3 MM-PBSA Values (van der Waals Energy, Electrostatic Energy, Polar Solvation Energy, SASA Energy, and Binding Energy) of PIK3CA (Native and Mutants)-Wortmannin Complexes

Energy	Native (kJ/mol)	H1047R (kJ/mol)	H1047Y (kJ/mol)	H1047L (kJ/mol)
van der Waals energy	-269.476 ± 37.585	-292.494 ± 11.840	-270.855 ± 9.976	-287.706 ± 15.013
Electrostatic energy	-50.155 ± 9.032	-64.993 ± 8.097	-49.185 ± 12.560	-41.763 ± 9.164
Polar solvation energy	137.239 ± 31.144	173.558 ± 10.705	138.162 ± 15.359	119.315 ± 8.307
SASA energy	-19.228 ± 3.889	-17.789 ± 0.725	-19.097 ± 0.888	-18.557 ± 0.915
Binding energy	-198.621 ± 19.968	-201.719 ± 13.556	-200.975 ± 11.884	-208.712 ± 18.670

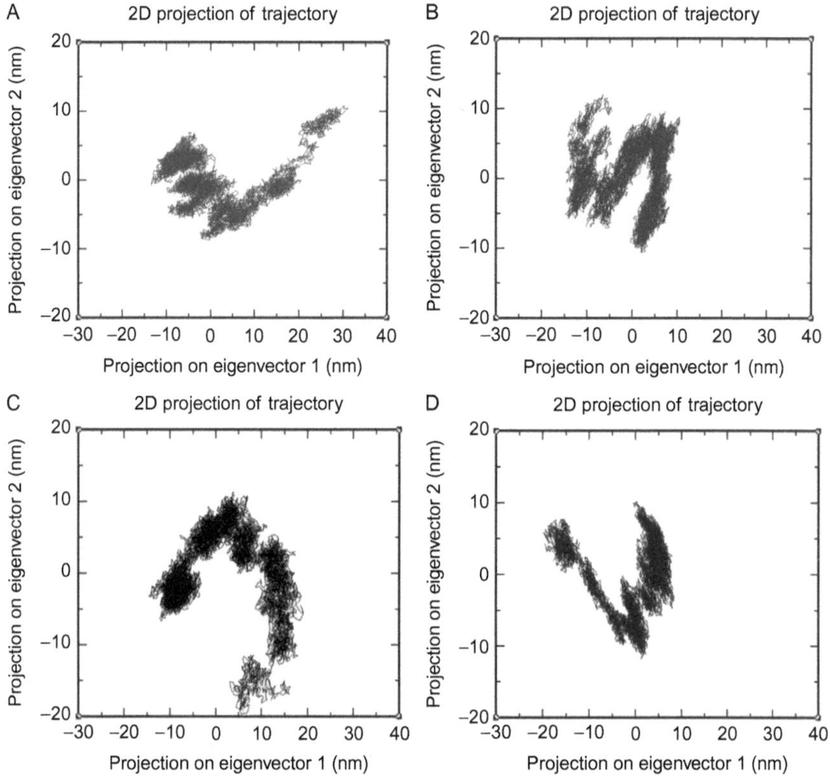

Figure 6 (A)–(D) Analysis of principal component analysis between native-Wortmannin and mutants-Wortmannin docked complexes for 100 ns. Color scheme native-Wortmannin complex (red (light gray in the print version)), H1047R-Wortmannin complex (violet (gray in the print version)), H1047L-Wortmannin complex (black), H1047Y-Wortmannin complex (blue (dark gray in the print version)).

mutant complex explain the effects of mutation. The native complex showed higher projection value of ~−15 to ~30 nm, whereas the mutant complexes showed the range of ~−15 to ~10 ns on the eigenvector 1. On other hand, in eigenvector 2, the projection of mutant complexes showed a projection range between ~−10 and ~10 nm, whereas the native showed the projection range between ~−5 and ~5 nm. Our results thus predicted that mutant H1047R covers the minimal area and thereby obtaining more thermodynamics stability (Fig. 6A–D).

4. DISCUSSION

Genetic variations are natural phenomena that occur rapidly to sustain the process of evolution. A genetic variation is caused by a change in the order of bases in the nucleotides of genes. Single-nucleotide polymorphisms are the most common type of genetic variations distributed throughout the human genome. Among the various kinds of SNPs, nsSNPs were found to be showing a deleterious impact on the structure and functionality of the protein by altering the hydrogen bond, disulfide bonds, salt bridge, and hydrophobic contacts. This kind of genomic instability has been identified in many human cancer types. Thus, identification and prioritization of the nsSNPs associated with cancer types may provide insight into better understanding the mechanism of disease-causing genes (George Priya Doss, Nagasundaram, Chakraborty, Chen, & Zhu, 2013; George Priya Doss, Rajasekaran, & Sethumadhavan, 2010; George Priya Doss & Rajith, 2012, 2013; George Priya Doss & Sethumadhavan, 2009). Combination of two major computational approaches, molecular docking and molecular docking dynamics, have made a positive impact in analyzing the impact of nsSNPs on the conformation of mutated proteins (Doss, Chakraborty, Chen, & Zhu, 2014; George Priya Doss, Rajith, Chiranjib, Balaji, et al., 2014; George Priya Doss, Rajith, Chiranjib, NagaSundaram, et al., 2014; Nagasundaram et al., 2015). These type of combinatorial have proved to be vital in drug designing scenario with reduction in manpower, cost, and time. Most of the *in silico* prediction methods consider evolutionary basis to decipher the impact of the mutation. Pathogenic and stability predictors are the two primary criteria used to investigate the deleterious nature of a nsSNP by computationally. In our present study, we employed PhD-SNP, Align GVGD, and HANSA to predict the pathogenic nature of the three mutations in PIK3CA protein, of which H1047R was reported to be deleterious (Table 1). MUpro and iStable showed the highest destabilization in H1047R with the score of −1.752 and −1.682, respectively. Both the tools designated all the three mutants as a decrease instability (Table 1). The native structure (PDB ID: 2RD0) retrieved from the protein data bank was taken for further analysis. The homology model was constructed using SwissPDB viewer to obtain the 3D structure of the mutants. The histidine residue in the 1047 position of native protein was mutated to arginine, leucine, and tyrosine. These modeled structures were well energy minimized to avoid the distorted torsion. Wortmannin drug widely used in the treatment of breast cancer that specifically inhibits the PIK3CA was retrieved from the PubChem database. Docking analysis was performed using AutoDock 4.2 for native, H1047R, H1047L, and H1047Y with Wortmannin. Least binding energy was observed with native and H1047R with −9.09 and −8.18 kcal/mol, respectively (Table 2). Lower binding energy defines better affinity of the drug to the target. Thus, Wortmannin shows better affinity with native and H1047R. The surrounding amino acids also play a crucial role in binding and hydrogen bond formation. Wortmannin forms two hydrogen bonds with LYS-802 and one bond with each VAL-851, VAL-850, and TYR-836 altogether forming five hydrogen bonds in the H1047R structure. Our results further correlate with experimental analysis stating that Wortmannin forms a better interaction with H1047R mutation (Mandelker et al., 2010). In other two mutants, four hydrogen bonds were formed. Lysine at 802 position was found to be a common amino acid in forming hydrogen bond in the native and mutant complexes. To understand the binding patterns and dynamic nature of the complexes, simulation analysis was performed to elucidate the difference at the atomic level.

The molecular dynamics studies were established in the late 1970s to overcome limitations in the atomic level studies. There are several studies citing the utilization of computational power molecular dynamics in providing molecular insight into mutational as well as protein–drug interaction (George Priya Doss et al., 2013; Miteva, Brugge, Rosing, Nicolaes, & Villoutreix, 2004; Nagasundaram & George Priya Doss, 2013; Witham, Takano, Schwartz, & Alexov, 2011; Zhang, Norris, Schwartz, & Alexov, 2011). In addition, results from experimental techniques are in correlation with molecular dynamics studies (John et al., 2013; Miteva et al., 2004; Steen, Miteva, Villoutreix, Yamazaki, & Dahlback, 2003; Sudhakar et al., 2015; Zhang, Teng, Wang, Schwartz, & Alexov 2010).

In molecular dynamics simulation, RMSD of the protein predicts the stability of the protein and the convergence pattern that helps us to proceed with the further analysis. In the RMSD plot of PIK3CA protein, we observed a convergence of native and mutants after ~50 ns providing a better data for the further analysis (Fig. 2). The RMSD of H1047R-Wortmannin complex showed lower deviation than the other mutants. Lower the deviation in RMSD explains the attainment of higher stability (Yun & Guy, 2011). Whereas RMSF showed similar fluctuation throughout the simulation period for 100 ns. It was observed that the fluctuation of mutant H1047R was significant lesser than the native and other mutants. Especially in the C-terminus region where the mutation was induced showed a minimal fluctuation (Fig. 3). This decrease in the fluctuation correlates with the RMSD graph. To explain the reason for the minimal deviation observed in RMSD, the number of hydrogen bonds formed during the simulation period was analyzed. Noticeably, we observed maximum number of hydrogen bonds in the mutant H1047R complex throughout the simulation (Fig. 4). The increase in number of hydrogen bonds formed between protein and ligand serves as a gain of stability which further elucidates the result obtained from the RMSD. The protein has to be in compact with a drug to exhibit a maximum efficiency. From the radius of gyration graph, the compactness of mutant H1047R was found to be comparatively superior to other mutant complexes (Fig. 5). g_mmpbsa calculates the binding energy between the protein and drug. This estimation helps in elucidating any difference that are observed in the binding pattern between the mutant complexes. The van der Waals forces did not show much difference between native and mutant complexes ranging from −292 to −270 kJ/mol. The polar energy of the H1047R complex was observed to be high when compared to the native and other mutant complexes (Table 3). This might be due to the change in the charge of the native and mutant amino acid from positive to neutral (Sudhakar et al., 2015). Solvent-accessible surface area (SASA) is the most popular method for calculating the nonpolar solute–solvent interactions; this is because a noncovalent energy of a structure is proportional to the SASA (Wagoner & Baker, 2006). Hence, we calculated SASA using mmpbsa which explains the noncovalent solvent interactions in the complex. The SASA for mutant H1047R, H1047L, and H1047Y ranged between −19 and −17 showing not much difference from that of native. Furthermore, we observed least binding energy of 201.719 and −208.712 kJ/mol in H1047R and H1047L when compared to native and H1047Y. The lower the binding energy more is the affinity of the drug toward the target. Hence, the binding affinity of the drug to the protein was higher for the mutant H1047R and H1047L (Table 3). The principle component analysis using molecular dynamics simulation explains the different motions of the complex during the simulation. Primarily, the translation and overall rotation from the trajectory files are removed to understand the internal motion of the protein system (Sittel, Jain, & Stock, 2014). Comparatively, huge degree of conformation change was observed with H1047L, whereas a least degree of conformation change was observed with H1047R, which illustrates the high thermodynamic stability and more compactness as supported by the results obtained from the radius of gyration. This further suggests that projection of H1047R complex shows better conformational change with the Wortmannin (Fig. 6A–D). Finally, dictionary of protein secondary structure (do_dssp) was used to predict the movement of secondary structure over a period (Fig. 7A–D). This tool uses "hydrogen bond estimation algorithm" for the prediction (Joosten et al., 2011). In DSSP analysis, we clearly observed a shift in secondary structure element from the coil to β-sheet at the C-terminus region (residues 1032–1068) of the H1047R mutant complex. This transition is visualized using PyMOL (Fig. 8). Perczel, Gáspári, and Csizmadia (2005) reported that presence of

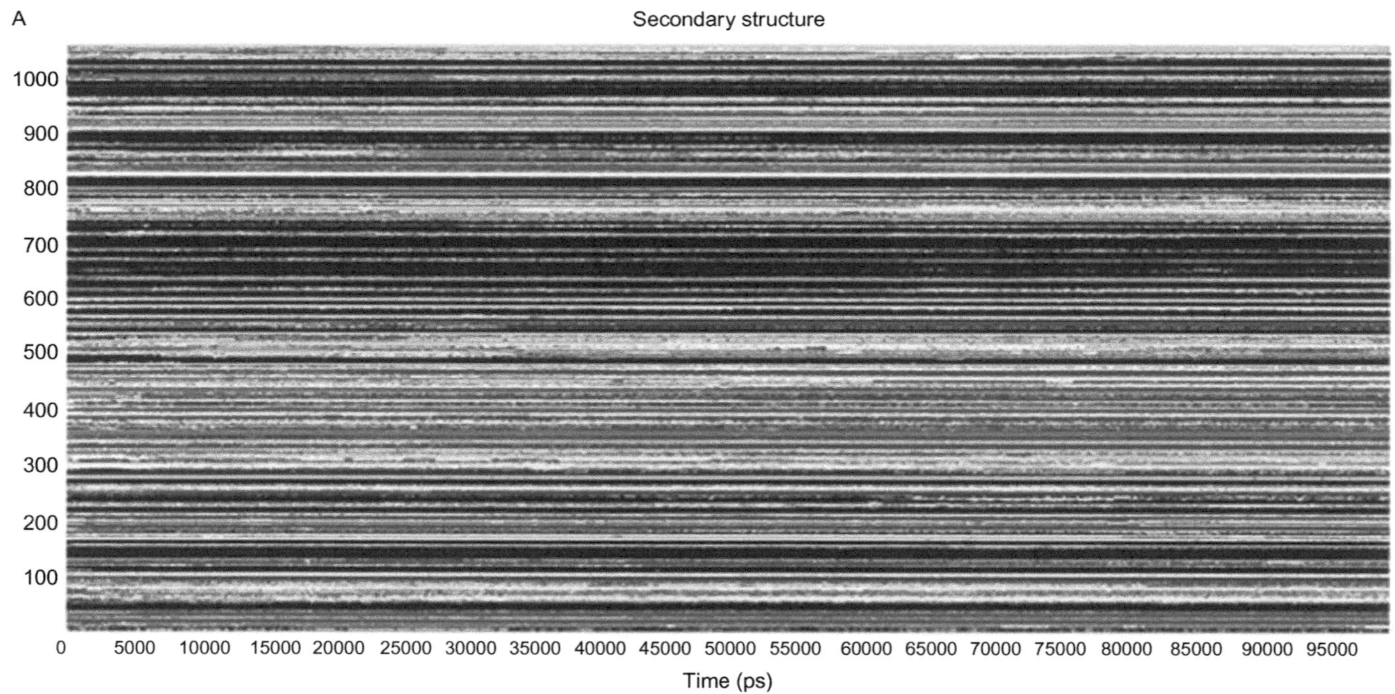

Figure 7 (A) DSSP plot analysis of native-Wortmannin complex.

(Continued)

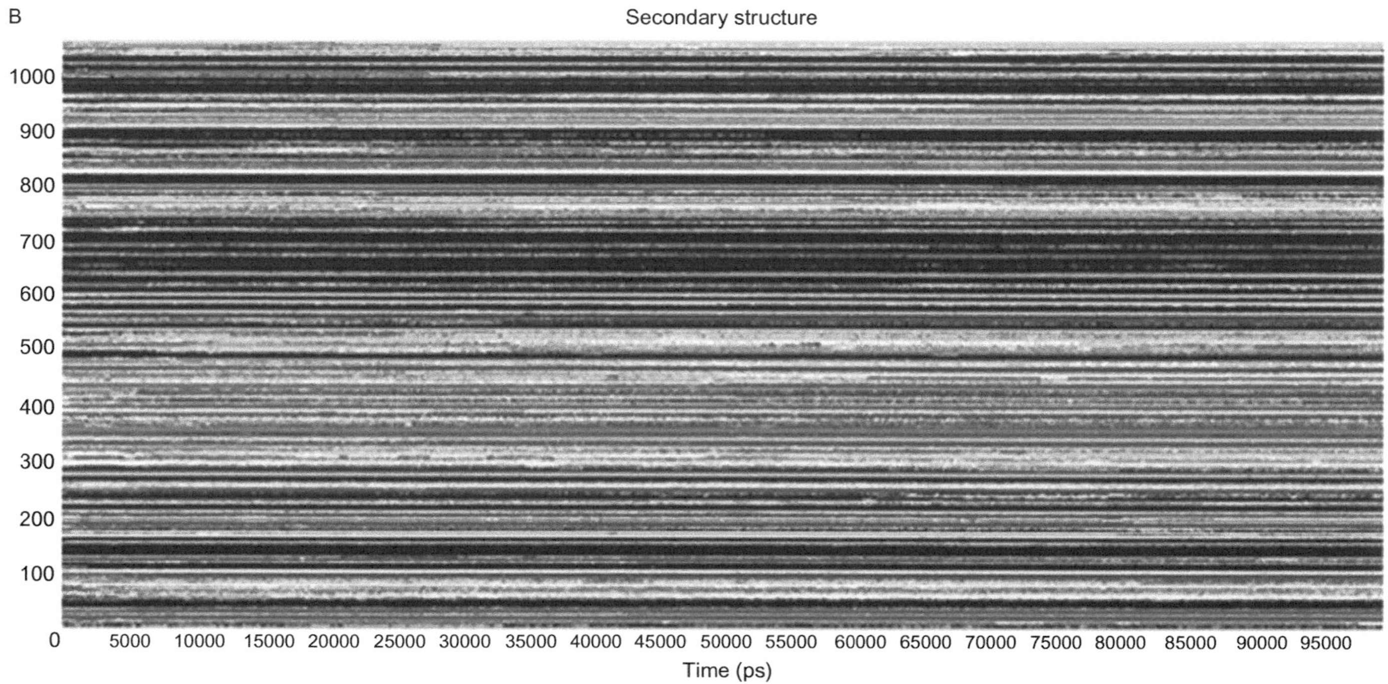

Figure 7—Cont'd (B) DSSP plot analysis of H1047R-Wortmannin complex.

(Continued)

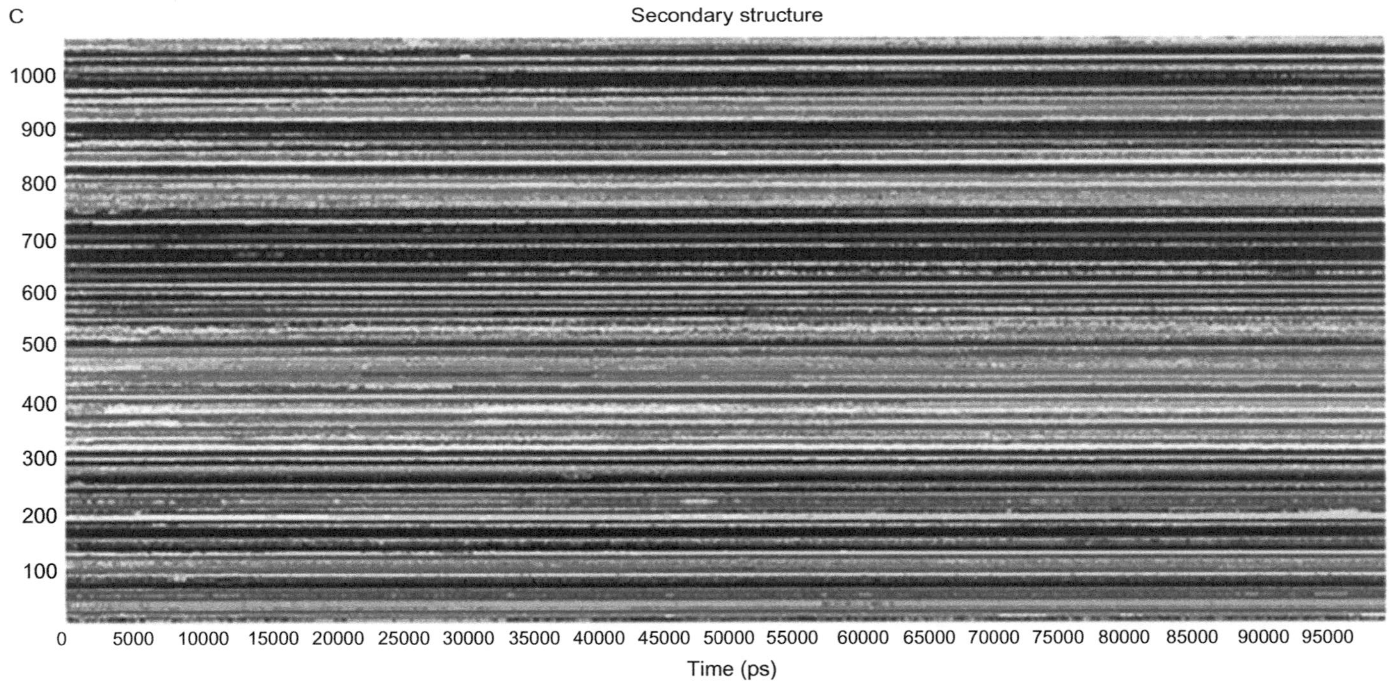

Figure 7—Cont'd (C) DSSP plot analysis of H1047L-Wortmannin complex.

(Continued)

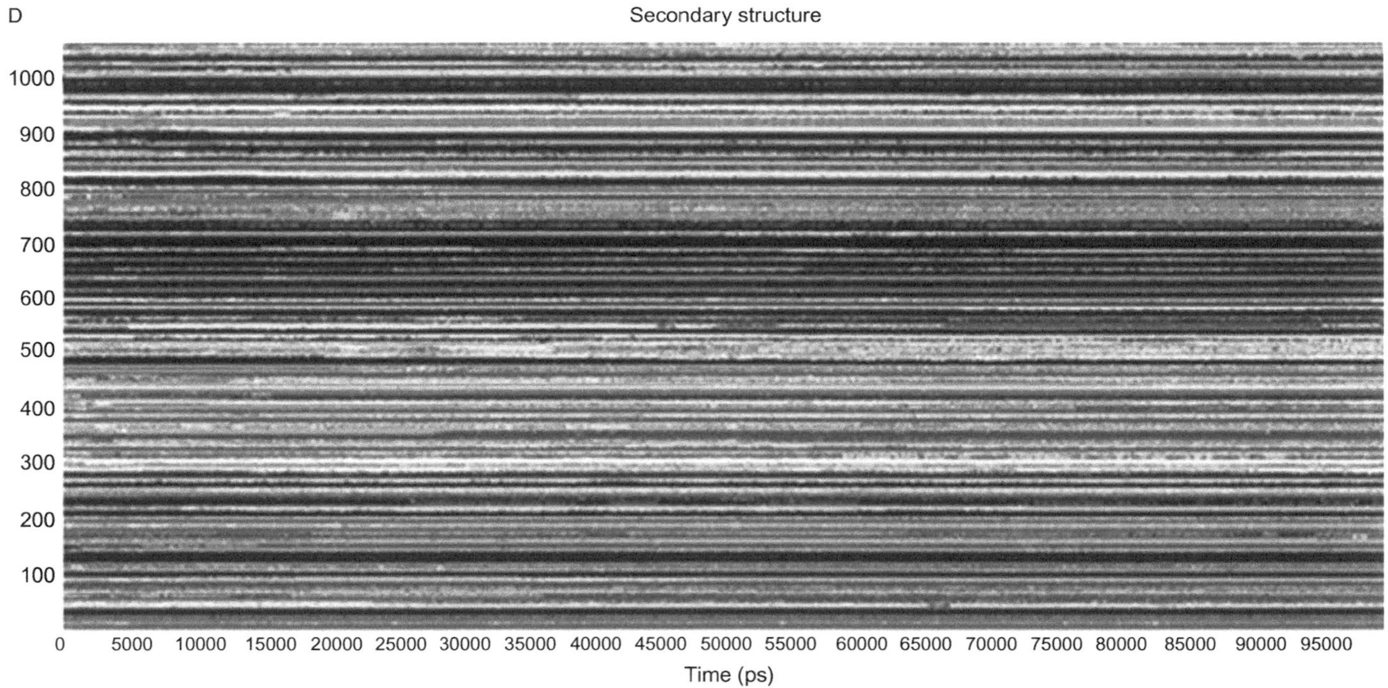

Figure 7—Cont'd (D) DSSP plot analysis of H1047Y-Wortmannin complex.

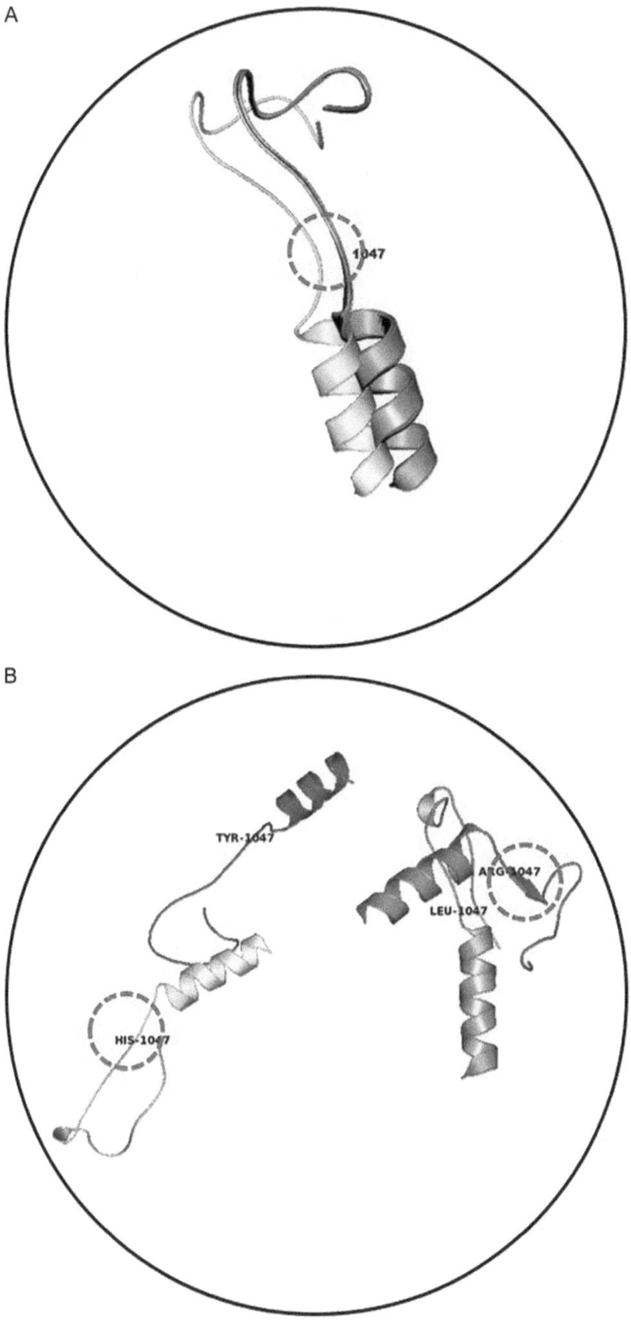

Figure 8 (A) and (B) PyMOL visualization of superimposed native-Wortmannin complex and mutants-Wortmannin complexes at 0 and 100 ns. Color scheme native-Wortmannin complex (yellow (light gray in the print version)), H1047R-Wortmannin complex (green (dark gray in the print version)), H1047L-Wortmannin complex (orange (gray in the print version)), H1047Y-Wortmannin complex (pink (dark gray in the print version)).

β-sheet in a protein provides an excellent opportunity to incorporate long-range stabilization (Perczel et al., 2005). This further confirms the increase in stability of the mutant H1047R complex when compared to other two mutants.

5. CONCLUSION

Mutations observed in PIK3CA result in gain of function with increased kinase activity thereby bypassing the natural cell death process leading to cancer. Several studies have pinpointed about the impact of hotspot mutations in the kinase domain of the PIK3CA. In this report, computational methods predicted H1047R to be highly deleterious when compared to the H1047L and H1047Y. Rigorous docking and dynamic molecular analysis was conducted to characterize the impact of mutation on drug-binding ability. H1047R exhibited better binding affinity and involvement of more hydrogen bonds leading to the stable complex formation. Obtained results from the extensive molecular dynamics simulation illustrate the involvement of mutation in protein–drug complex stability. This result can be further validated by experimental protocols to make the drug more efficient for better efficacy. We believe that this combinatorial computational approach might help in elucidating the action of the mutant over the drug and enhance the development of individualized medicine.

ACKNOWLEDGMENTS

The authors take this opportunity to thank the management of VIT University for providing the facilities and encouragement to carry out this work. The authors have no conflict of interest to declare.

REFERENCES

Acharya, V., & Nagarajaram, H. A. (2012). Hansa: An automated method for discriminating disease and neutral human nsSNPs. *Human Mutation, 33*(2), 332–337. http://doi.org/10.1002/humu.21642.

Adams, J. R., Schachter, N. F., Liu, J. C., Zacksenhaus, E., & Egan, S. E. (2011). Elevated PI3K signaling drives multiple breast cancer subtypes. *Oncotarget, 2*, 435–447.

Al-Hujaily, E. M., Mohamed, A. G., Al-Sharif, I., Youssef, K. M., Manogaran, P. S., Al-Otaibi, B., et al. (2011). PAC, a novel curcumin analogue, has anti-breast cancer properties with higher efficiency on ER-negative cells. *Breast Cancer Research and Treatment, 128*(1), 97–107. http://dx.doi.org/10.1007/s10549-010-1089-3.

Allen, L. F., Sebolt-Leopold, J., & Meyer, M. B. (2003). CI-1040 (PD184352), a targeted signal transduction inhibitor of MEK (MAPKK). *Seminars in Oncology, 30*(5 Suppl. 16), 105–116.

American Cancer Society, (2013). *Breast cancer facts & figures 2013–2014*. Atlanta: American Cancer Society, Inc.

Aradottir, M., Reynisdottir, S. T., Stefansson, O. A., Jonasson, J. G., Sverrisdottir, A., Tryggvadottir, L., et al. (2015). Aurora A is a prognostic marker for breast cancer arising in BRCA2 mutation carriers. *The Journal of Pathology. Clinical Research*, *1*(1), 33–40. http://doi.org/10.1002/cjp2.6.

Bachman, K. E., Argani, P., Samuels, Y., Silliman, N., Ptak, J., Szabo, S., et al. (2004). The PIK3CA gene is mutated with high frequency in human breast cancers. *Cancer Biology & Therapy*, *3*(8), 772–775. http://dx.doi.org/10.4161/cbt.3.8.994.

Bannwarth, B., Labat, L., Moride, Y., & Schaeverbeke, T. (1994). Methotrexate in rheumatoid arthritis. An update. *Drugs*, *47*(1), 25–50.

Bava, K. A., Gromiha, M. M., Uedaira, H., Kitajima, K., & Sarai, A. (2004). ProTherm, version 4.0: Thermodynamic database for proteins and mutants. *Nucleic Acids Research*, *32*(Database issue), D120–D121. http://doi.org/10.1093/nar/gkh082.

Berendsen, H. J. C., Postma, J. P. M., van Gunsteren, W. F., DiNola, A., & Haak, J. R. (1984). Molecular dynamics with coupling to an external bath. *The Journal of Chemical Physics*, *81*(8), 3684.

Burris, H. A. (2004). Dual kinase inhibition in the treatment of breast cancer: Initial experience with the EGFR/ErbB-2 inhibitor lapatinib. *The Oncologist*, *9*(3), 10–15. Retrieved from, http://www.ncbi.nlm.nih.gov/pubmed/15163842.

Calleja, V., Laguerre, M., Parker, P. J., & Larijani, B. (2009). Role of a novel PH-kinase domain interface in PKB/Akt regulation: Structural mechanism for allosteric inhibition. *PLoS Biology*, 7(1), e17. http://dx.doi.org/10.1371/journal.pbio.1000017.

Campbell, I. G., Russell, S. E., Choong, D. Y., Montgomery, K. G., Ciavarella, M. L., Hooi, C. S., et al. (2004). Mutation of the PIK3CA gene in ovarian and breast cancer. *Cancer Research*, *64*, 7678–7681.

Cantley, L. C. (2002). The phosphoinositide 3-kinase pathway. *Science (New York, N.Y.)*, *296*(5573), 1655–1657. http://dx.doi.org/10.1126/science.296.5573.1655.

Capriotti, E., Calabrese, R., & Casadio, R. (2006). Predicting the insurgence of human genetic diseases associated to single point protein mutations with support vector machines and evolutionary information. *Bioinformatics (Oxford, England)*, *22*(22), 2729–2734. http://doi.org/10.1093/bioinformatics/btl423.

Cheatham, T. E. I., Miller, J. L., Fox, T., Darden, T. A., & Kollman, P. A. (1995). Molecular dynamics simulations on solvated biomolecular systems: The particle mesh ewald method leads to stable trajectories of DNA, RNA, and proteins. *Journal of the American Chemical Society*, *117*(14), 4193–4194.

Chen, C.-W., Lin, J., & Chu, Y.-W. (2013). iStable: Off-the-shelf predictor integration for predicting protein stability changes. *BMC Bioinformatics*, *14*(2), S5. http://doi.org/10.1186/1471-2105-14-S2-S5.

Cheng, J., Randall, A., & Baldi, P. (2006). Prediction of protein stability changes for single-site mutations using support vector machines. *Proteins*, *62*(4), 1125–1132. http://doi.org/10.1002/prot.20810.

Cizkova, M., Susini, A., Vacher, S., Cizeron-Clairac, G., Andrieu, C., Driouch, K., et al. (2012). PIK3CA mutation impact on survival in breast cancer patients and in ERα, PR and ERBB2-based subgroups. *Breast Cancer Research*, *14*(1), R28. http://dx.doi.org/10.1186/bcr3113.

Clark, A. S., West, K., Streicher, S., & Dennis, P. A. (2002). Constitutive and inducible Akt activity promotes resistance to chemotherapy, trastuzumab, or tamoxifen in breast cancer cells. *Molecular Cancer Therapeutics*, *1*(9), 707–717.

Damjanov, N., Kauffman, R. S., & Spencer-Green, G. T. (2009). Efficacy, pharmacodynamics, and safety of VX-702, a novel p38 MAPK inhibitor, in rheumatoid arthritis: Results of two randomized, double-blind, placebo-controlled clinical studies. *Arthritis and Rheumatism*, *60*(5), 1232–1241. http://dx.doi.org/10.1002/art.24485.

DeLano, W. L. (2002). PyMOL: An open-source molecular graphics tool. *Ccp4 Newsletter on Protein Crystallography*, *40*.

Doss, C. G. P., Chakraborty, C., Chen, L., & Zhu, H. (2014). Integrating in silico prediction methods, molecular docking, and molecular dynamics simulation to predict the impact of ALK missense mutations in structural perspective. *BioMed Research International, 2014*, 895831. http://doi.org/10.1155/2014/895831.

Esteva, F. J., Glaspy, J., Baidas, S., Laufman, L., Hutchins, L., Dickler, M., et al. (2003). Multicenter phase II study of oral bexarotene for patients with metastatic breast cancer. *Journal of Clinical Oncology: Official Journal of the American Society of Clinical Oncology, 21*(6), 999–1006.

Ferlay, J., Soerjomataram, I., Ervik, M., Dikshit, R., Eser, S., Mathers, C., et al. (2012). Estimates of worldwide burden of cancer in 2008: GLOBOCAN 2008. *International Journal of Cancer, 127*, 2893–2917.

Firoozinia, M., Jahromi, M. Z., Moghadamtousi, S. Z., Nikzad, S., & Kadir, H. A. (2014). PIK3CA gene amplification and PI3K p110α protein expression in breast carcinoma. *International Journal of Medical Sciences, 11*(6), 620–625. http://dx.doi.org/10.7150/ijms.8251.

George Priya Doss, C., Nagasundaram, N., Chakraborty, C., Chen, L., & Zhu, H. (2013). Extrapolating the effect of deleterious nsSNPs in the binding adaptability of flavopiridol with CDK7 protein: A molecular dynamics approach. *Human Genomics*, 7, 10.

George Priya Doss, C., Rajasekaran, R., & Sethumadhavan, R. (2010). Computational identification and structural analysis of deleterious functional SNPs in MLL gene causing acute leukemia. *Interdisciplinary Sciences, Computational Life Sciences*, *2*(3), 247–255. http://doi.org/10.1007/s12539-010-0007-z.

George Priya Doss, C., & Rajith, B. (2012). Computational refinement of functional single nucleotide polymorphisms associated with ATM gene. *PloS One*, 7(4), e34573. http://doi.org/10.1371/journal.pone.0034573.

George Priya Doss, C., & Rajith, B. (2013). A new insight into structural and functional impact of single-nucleotide polymorphisms in PTEN gene. *Cell Biochemistry and Biophysics, 66*(2), 249–263. http://doi.org/10.1007/s12013-012-9472-9.

George Priya Doss, C., Rajith, B., Chiranjib, C., Balaji, V., Maitreya, D., Sneha, M., et al. (2014a). In silico profiling and structural insights of Missense mutations in RET protein kinase domain by molecular dynamics and docking approach. *Molecular BioSystems, 10*, 421–436.

George Priya Doss, C., Rajith, B., Chiranjib, C., NagaSundaram, N., Shabana, K. A., & Hailong, Z. (2014b). Structural signature of the G719S-T790M double mutation in the EGFR kinase domain and its response to inhibitors. *Scientific Reports, 4*, 5868.

George Priya Doss, C., & Sethumadhavan, R. (2009). Computational and structural analysis of deleterious functional SNPs in ARNT oncogene. *Interdisciplinary Sciences, Computational Life Sciences, 1*(3), 220–228. http://doi.org/10.1007/s12539-009-0042-9.

Gharbi, S. I., Zvelebil, M. J., Shuttleworth, S. J., Hancox, T., Saghir, N., Timms, J. F., et al. (2007). Exploring the specificity of the PI3K family inhibitor LY294002. *The Biochemical Journal, 404*(1), 15–21. http://dx.doi.org/10.1042/BJ20061489.

Gkeka, P., Evangelidis, T., Pavlaki, M., Lazani, V., Christoforidis, S., Agianian, B., et al. (2014). Investigating the structure and dynamics of the PIK3CA wild-type and H1047R oncogenic mutant. *PLoS Computational Biology, 10*(10), e1003895.

Gymnopoulos, M., Elsliger, M.-A., & Vogt, P. K. (2007). Rare cancer-specific mutations in PIK3CA show gain of function. *Proceedings of the National Academy of Sciences of the United States of America, 104*(13), 5569–5574. http://dx.doi.org/10.1073/pnas.0701005104.

Henneman, L., van Miltenburg, M. H., Michalak, E. M., Braumuller, T. M., Jaspers, J. E., Drenth, A. P., et al. (2015). Selective resistance to the PARP inhibitor olaparib in a mouse model for BRCA1-deficient metaplastic breast cancer. *Proceedings of the National Academy of Sciences of the United States of America, 112*(27), 8409–8414. http://doi.org/10.1073/pnas.1500223112.

Hess, B., Bekker, H., Berendsen, H. J. C., & Fraaije, J. G. E. M. (1997). LINCS: A linear constraint solver for molecular simulations. *Journal of Computational Chemistry*, *18*(12), 1463–1472. http://dx.doi.org/10.1002/(SICI)1096-987X(199709)18:12<1463::AID-JCC4>3.0.CO;2-H.

Huang, C. H., Mandelker, D., Schmidt-Kittler, O., Samuels, Y., Velculescu, V. E., Kinzler, K. W., et al. (2007). The structure of a human p110/p85 complex elucidates the effects of oncogenic PI3K mutations. *Science*, *318*, 1744–1748.

Ikenoue, T. (2005). Functional analysis of PIK3CA gene mutations in human colorectal cancer. *Cancer Research*, *65*(11), 4562–4567. http://dx.doi.org/10.1158/0008-5472.CAN-04-4114.

Janku, F., Wheler, J. J., Naing, A., Falchook, G. S., Hong, D. S., Stepanek, V. M., et al. (2014). PIK3CA mutation H1047R is associated with response to PI3K/AKT/mTOR signaling pathway inhibitors in early-phase clinical trials. *Cancer Research*, *73*(1), 276–284. http://dx.doi.org/10.1158/0008-5472.CAN-12-1726.

Johansson, M. U., Zoete, V., Michielin, O., & Guex, N. (2012). Defining and searching for structural motifs using DeepView/Swiss-PdbViewer. *BMC Bioinformatics*, *13*, 173.

John, A. M., George Priya Doss, C., Ebenazer, A., Seshadri, M. S., Nair, A., Rajaratnam, S., et al. (2013). p.Arg82Leu von Hippel-Lindau (VHL) gene mutation among three members of a family with familial bilateral pheochromocytoma in India: Molecular analysis and in silico characterization. *PloS One*, *8*, e61908.

Joosten, R. P., teBeek, T. A. H., Krieger, E., Hekkelman, M. L., Hooft, R. W. W., Schneider, R., et al. (2011). A series of PDB related databases for everyday needs. *Nucleic Acids Research*, *39*(Database issue). D411–D419, http://doi.org/10.1093/nar/gkq1105.

Karakas, B., Bachman, K. E., & Park, B. H. (2006). Mutation of the PIK3CA oncogene in human cancers. *British Journal of Cancer*, *94*(4), 455–459. http://dx.doi.org/10.1038/sj.bjc.6602970.

Kuete, V., Saeed, M. E. M., Kadioglu, O., Börtzler, J., Khalid, H., Greten, H. J., et al. (2015). Pharmacogenomic and molecular docking studies on the cytotoxicity of the natural steroid wortmannin against multidrug-resistant tumor cells. *Phytomedicine: International Journal of Phytotherapy and Phytopharmacology*, *22*(1), 120–127. http://doi.org/10.1016/j.phymed.2014.11.011.

Levine, D. A., Bogomolniy, F., Yee, C. J., Lash, A., Barakat, R. R., Borgen, P. I., et al. (2005). Frequent mutation of the PIK3CA gene in ovarian and breast cancers. *Clinical Cancer Research*, *11*(8), 2875–2878.

Ligresti, G., Militello, L., Steelman, L. S., Cavallaro, A., Basile, F., Nicoletti, F., et al. (2009). PIK3CA mutations in human solid tumors role in sensitivity to various therapeutic approaches. *Cell Cycle*, *8*(9), 1352–1358. http://dx.doi.org/10.4161/cc.8.9.8255.

Liu, Y., Shreder, K. R., Gai, W., Corral, S., Ferris, D. K., & Rosenblum, J. S. (2005). Wortmannin, a widely used phosphoinositide 3-kinase inhibitor, also potently inhibits mammalian polo-like kinase. *Chemistry & Biology*, *12*(1), 99–107. http://doi.org/10.1016/j.chembiol.2004.11.009.

Luo, J., Manning, B. D., & Cantley, L. C. (2003). Targeting the PI3K-Akt pathway in human cancer. *Cancer Cell*, *4*(4), 257–262. http://dx.doi.org/10.1016/S1535-6108(03)00248-4.

Mahmud, N., Ford, J. M., Longacre, T. A., Parent, R., & Norton, J. A. (2015). Metastatic lobular breast carcinoma mimicking primary signet ring adenocarcinoma in a patient with a suspected CDH1 mutation. *Journal of Clinical Oncology: Official Journal of the American Society of Clinical Oncology*, *33*(4), e19–e21. http://doi.org/10.1200/JCO.2013.49.1159.

Majewski, I. J., Nuciforo, P., Mittempergher, L., Bosma, A. J., Eidtmann, H., Holmes, E., et al. (2015). PIK3CA mutations are associated with decreased benefit to neoadjuvant human epidermal growth factor receptor 2-targeted therapies in breast cancer. *Journal of Clinical Oncology: Official Journal of the American Society of Clinical Oncology*, *33*(12), 1334–1339. http://doi.org/10.1200/JCO.2014.55.2158.

Mandelker, D., Schmidt-Kittler, O., Vogelstein, B., & Amzel, L. M. (2010). Somatic mutations in PI3Kalpha: Structural basis for enzyme activation and drug design. *Biochimica et Biophysica Acta*, *1804*(3), 533–540.

Mangone, F., Bobrovnitchaia, I., Salaorni, S., Manuli, E., & Nagai, M. (2012). PIK3CA exon 20 mutations are associated with poor prognosis in breast cancer patients. *Clinics*, *67*(11), 1285–1290. http://dx.doi.org/10.6061/clinics/2012(11)11.

Mankoo, P. K., Sukumar, S., & Karchin, R. (2009). PIK3CA somatic mutations in breast cancer: Mechanistic insights from Langevin dynamics simulations. *Proteins*, *75*(2), 499–508.

Micalizzi, D. S., Juric, D., Niemierko, A., Reynolds, K. L., Borger, D., Vora, S. R., et al. (2015). Abstract P1-13-03: Association of PIK3CA mutation with clinical response to specific endocrine therapies in metastatic hormone receptor positive (HR+) breast cancer. *Cancer Research*, *75*(9) http://doi.org/10.1158/1538-7445.SABCS14-P1-13-03.

Miteva, M. A., Brugge, J. M., Rosing, J., Nicolaes, G. A. F., & Villoutreix, B. O. (2004). Theoretical and experimental study of the D2194G mutation in the C2 domain of coagulation factor V. *Biophysical Journal*, *86*(1), 488–498.

Morris, G. M., Huey, R., Lindstrom, W., Sanner, M. F., Belew, R. K., Goodsell, D. S., et al. (2009). AutoDock4 and AutoDockTools4: Automated docking with selective receptor flexibility. *Journal of Computational Chemistry*, *30*(16), 2785–2791. http://dx.doi.org/10.1002/jcc.21256.

Nagasundaram, N., & George Priya Doss, C. (2013). Predicting the impact of single-nucleotide polymorphisms in CDK2-flavopiridol complex by molecular dynamics analysis. *Cell Biochemistry and Biophysics*, *66*(3), 681–695.

Nagasundaram, N., Hailong, Z., Jiming, L., Karthick, V., George Priya Doss, C., Chiranjib, C., et al. (2015). Analysing the effect of mutation on protein function and discovering potential inhibitors of CDK4: Molecular modelling and dynamics studies. *PloS One*, *10*(8), e0133969.

Nylander, S., Kull, B., Björkman, J. A., Ulvinge, J. C., Oakes, N., Emanuelsson, B. M., et al. (2012). Human target validation of phosphoinositide 3-kinase (PI3K)β: Effects on platelets and insulin sensitivity, using AZD6482 a novel PI3Kβ inhibitor. *Journal of Thrombosis and Haemostasis: JTH*, *10*(10), 2127–2136. http://dx.doi.org/10.1111/j.1538-7836.2012.04898.x.

Pal, A., Gonzalez-Malerva, L., Eaton, S., Guzman, M., Chow, D., Yin, H., et al. (2015). Abstract P4-05-07: Functional genomics of TP53 mutations and its impact in breast cancer progression. *Cancer Research*, *75*(9 Suppl.).

Parvez, E., Popovic, S., Elavathil, L., Okawara, G., & Hodgson, N. (2015). Early occurrence of angiosarcoma in a woman with a BRCA2 gene variation of unknown significance treated with breast-conserving therapy for bilateral ductal carcinoma: A case report. *Clinical Breast Cancer*. pii: S1526-8209(15)00145-7. http://doi.org/10.1016/j.clbc.2015.06.011.

Perczel, A., Gáspári, Z., & Csizmadia, I. G. (2005). Structure and stability of beta-pleated sheets. *Journal of Computational Chemistry*, *26*(11), 1155–1168.

Pfister, N. T., Fomin, V., Regunath, K., Zhou, J. Y., Zhou, W., Silwal-Pandit, L., et al. (2015). Mutant p53 cooperates with the SWI/SNF chromatin remodeling complex to regulate VEGFR2 in breast cancer cells. *Genes & Development*, *29*(12), 1298–1315. http://doi.org/10.1101/gad.263202.115.

Ramirez-Ardila, D. E., Helmijr, J. C., Look, M. P., Lurkin, I., Ruigrok-Ritstier, K., van Laere, S., et al. (2013). Hotspot mutations in PIK3CA associate with first-line treatment outcome for aromatase inhibitors but not for tamoxifen. *Breast Cancer Research and Treatment*, *139*(1), 39–49. http://doi.org/10.1007/s10549-013-2529-7.

Samuels, Y., Diaz, L. A., Schmidt-Kittler, O., Cummins, J. M., DeLong, L., Cheong, I., et al. (2005). Mutant PIK3CA promotes cell growth and invasion of human cancer cells. *Cancer Cell*, 7(6), 561–573. http://dx.doi.org/10.1016/j.ccr.2005.05.014.

Samuels, Y., Wang, Z., Bardelli, A., Silliman, N., Ptak, J., Szabo, S., et al. (2004). High frequency of mutations of the PIK3CA gene in human cancers. *Science*, *304*, 554.

Schüttelkopf, A. W., & van Aalten, D. M. F. (2004). PRODRG: A tool for high-throughput crystallography of protein-ligand complexes. *Acta Crystallographica*, *D60*, 1355–1363.

Schilder, R. J., Sill, M. W., Lankes, H. A., Gold, M. A., Mannel, R. S., Modesitt, S. C., et al. (2013). A phase II evaluation of motesanib (AMG 706) in the treatment of persistent or recurrent ovarian, fallopian tube and primary peritoneal carcinomas: A gynecologic oncology group study. *Gynecologic Oncology*, *129*(1), 86–91. http://dx.doi.org/10.1016/j.ygyno.2013.01.006.

Sittel, F., Jain, A., & Stock, G. (2014). Principal component analysis of molecular dynamics: On the use of Cartesian vs. Internal coordinates. *The Journal of Chemical Physics*, *141*(1), 014111. http://dx.doi.org/10.1063/1.4885338.

Steen, M., Miteva, M., Villoutreix, B. O., Yamazaki, T., & Dahlback, B. (2003). Factor V new brunswick: Ala221Val associated with FV deficiency reproduced in vitro and functionally characterized. *Blood*, *102*, 1316–1322.

Sudhakar, N., George Priya Doss, C., Thirumal Kumar, D., Chakraborty, C., Anand, K., & Suresh, M. (2015). Deciphering the impact of somatic mutations in exon 20 and exon 9 of PIK3CA gene in breast tumors among Indian women through molecular dynamics approach. *Journal of Biomolecular Structure and Dynamics*, 1–13. http://dx.doi.org/10.1080/07391102.2015.1007483.

Tavtigian, S. V., Deffenbaugh, A. M., Yin, L., Judkins, T., Scholl, T., Samollow, P. B., et al. (2006). Comprehensive statistical study of 452 BRCA1 missense substitutions with classification of eight recurrent substitutions as neutral. *Journal of Medical Genetics*, *43*(4), 295–305. http://doi.org/10.1136/jmg.2005.033878.

Vorkas, P. A., Poumpouridou, N., Agelaki, S., Kroupis, C., Georgoulias, V., & Lianidou, E. S. (2010). PIK3CA hotspot mutation scanning by a novel and highly sensitive high-resolution small amplicon melting analysis method. *The Journal of Molecular Diagnostics: JMD*, *12*(5), 697–704. http://dx.doi.org/10.2353/jmoldx.2010.100008.

Wagoner, J. A., & Baker, N. A. (2006). Assessing implicit models for nonpolar mean solvation forces: The importance of dispersion and volume terms. *Proceedings of the National Academy of Sciences of the United States of America*, *103*(22), 8331–8336. http://dx.doi.org/10.1073/pnas.0600118103.

Weng, Z., Spies, N., Zhu, S. X., Newburger, D. E., Kashef-Haghighi, D., Batzoglou, S., et al. (2015). Cell-lineage heterogeneity and driver mutation recurrence in pre-invasive breast neoplasia. *Genome Medicine*, 7(1), 1–10.

Witham, S., Takano, K., Schwartz, C., & Alexov, E. (2011). A missense mutation in CLIC2 associated with intellectual disability is predicted by in silico modeling to affect protein stability and dynamics. *Proteins: Structure, Function, and Bioinformatics*, *79*(8), 2444–2454.

Wu, G., Xing, M., Mambo, E., Huang, X., Liu, J., Guo, Z., et al. (2005). Somatic mutation and gain of copy number of PIK3CA in human breast cancer. *Breast Cancer Research: BCR*, 7(5), R609–R616. http://dx.doi.org/10.1186/bcr1262.

Wymann, M., Bulgarelli-Leva, G., Zvelebil, M., Pirola, L., Vanhaesebroeck, B., Waterfield, M., et al. (1996). Wortmannin inactivates phosphoinositide 3-kinase by covalent modification of Lys-802, a residue involved in the phosphate transfer reaction. *Molecular and Cellular Biology*, *16*(4), 1722–1733.

Yan, M., Shield-Artin, K., Byrne, D., Deb, S., Waddell, N., Haviv, I., et al. (2015). Comparative microRNA profiling of sporadic and BRCA1 associated basal-like breast cancers. *BMC Cancer*, *15*(1), 506. http://doi.org/10.1186/s12885-015-1522-4.

Yun, S., & Guy, H. R. (2011). Stability tests on known and misfolded structures with discrete and all atom molecular dynamics simulations. *Journal of Molecular Graphics & Modelling*, *29*(5), 663–675. http://doi.org/10.1016/j.jmgm.2010.12.002.

Zardavas, D., Phillips, W. A., & Loi, S. (2014). PIK3CA mutations in breast cancer: Reconciling findings from preclinical and clinical data. *Breast Cancer Research*, *16*(1), 201. http://dx.doi.org/10.1186/bcr3605.

Zhang, Z., Norris, J., Schwartz, C., & Alexov, E. (2011). In silico and in vitro investigations of the mutability of disease-causing missense mutation sites in spermine synthase. *PloS One*, *6*, 5.

Zhang, Z., Teng, S., Wang, L., Schwartz, C. E., & Alexov, E. (2010). Computational analysis of missense mutations causing Snyder-Robinson syndrome. *Human Mutation*, *31*(9), 1043–1049.

Zhao, L., & Vogt, P. K. (2008). Helical domain and kinase domain mutations in p110alpha of phosphatidylinositol 3-kinase induce gain of function by different mechanisms. *Proceedings of the National Academy of Sciences of the United States of America*, *105*(7), 2652–2657. http://dx.doi.org/10.1073/pnas.0712169105.

CHAPTER TEN

Role of von Willebrand Factor—A1 Domain Variants P1266L, H1268D, C1272R, and C1272F in VWD: A Molecular Modeling and Simulation Analysis Approach

C. George Priya Doss[1], Shabana Kouser Ali

Medical Biotechnology Division, School of Biosciences and Technology, VIT University, Vellore, Tamil Nadu, India

[1]Corresponding author: e-mail addresses: georgepriyadoss@vit.ac.in; georgecp77@yahoo.co.in

Contents

Abstract

von Willebrand disease (VWD) is an autosomal inherited disorder related to trauma-related bleeding in affected people. VWD results from deficiency of von Willebrand factor (VWF)—a glycoprotein involved in hemostasis and carrier for factor VIII (FVIII). Mutations in A1 domain of von Willebrand factor (*VWD*) gene are exceptionally polymorphic and associated with adhesion movement, clearance, and binding properties that could be interfaced with thrombosis. To address this issue, we implemented *in silico* prediction algorithms, namely, SIFT, PolyPhen 2.0, I-Mutant 3.0, SNAP, Align GVGD, PhD-SNP,

Advances in Protein Chemistry and Structural Biology, Volume 102
ISSN 1876-1623
http://dx.doi.org/10.1016/bs.apcsb.2015.09.006

SNPs&GO, and MutPred to classify the variants as pathogenic and affecting protein stability. Based on prediction scores, four variants, namely, P1266L, H1268D, C1272R, and C1272F, were predicted as highly deleterious from a pool of 72 nsSNPs/variants in A1 domain of VWD belonging to type 2A and 2B. Upon literature survey, amino acid substitution (P → L) at position 1266 is involved in improving the connection with platelets, substitution (C → F) at position 1272 results in extreme bleeding in patients, and substitution (H → D) at position 1268 disturbs the salt bridge scaffold were considered for further analysis. Through molecular dynamic simulation analysis over a period of 100 ns showed that four mutations near N-terminal region bring about a change in structure and function of the native VWD protein.

1. INTRODUCTION

von Willebrand factor (VWF) is a plasma glycoprotein synthesized in megakaryocytes and endothelial cells (Jaffe, Hoyer, & Nachman, 1974; Nachman, Levine, & Jaffe, 1977; Sporn, Chavin, Marder, & Wagner, 1985; Sussman & Rand, 1982). VWF has a fundamental role in primary hemostasis where it mediates platelet attachment to injured vascular subendothelium and leads to successive platelet aggregation. This is accomplished through binding of VWF to specific receptors on the platelet surface glycoproteins Ib and IIb/IIIa complex (Colombatti & Bonaldo, 1991; Fressinaud et al., 1994; Wise, Dorner, Krane, Pittman, & Kaufman, 1991). VWF also acts as a stabilizer of factor VIII (FVIII) in the circulation. Lower level of VWF is linked with reduced levels of circulating FVIII (Mazurier, 1992; Nishino, Girma, Rothschild, Fressinaud, & Meyer, 1989). The *VWF* gene, placed on the short arm of chromosome 12 (Ginsburg et al., 1985; Verweij, Diergaarde, Hart, & Pannekoek, 1986), resides within 178 kb of genomic DNA, with 52 exons. The *VWF* gene is transcribed into a 9-kb mRNA and translated into a 2813-amino acid product (Verweij et al., 1986), known to be a pre-pro-VWF (Bonthron et al., 1987). This precursor polypeptide consists of a 22-amino acid signal peptide, 741-amino acid propeptide, and a mature subunit of 2050 residues. The mature VWF consists of homologous sequence domains thought to be originated by gene segment duplication (Mancuso et al., 1991). The VWF protein comprises four types of repeating homologous areas, in the order D1-D2- D′-D3-A1-A2-A3-D4-B1-B2-B3-C1-C2 from amino (NH_2)- to carboxyl (COOH)-terminal end and each one having a trademarked binding properties (Tables 1 and 2). VWF multimers range in size from 0.5 to 20 MDa (Ruggeri & Zimmerman, 1981).

Table 1 Molecular Basis of von Willebrand Disease

VWD Type	Molecular Basis
Type 1	Several missense mutations reported; molecular basis unknown in most cases
Type 3	Throughout *VWF* gene nonsense, missense, frameshift mutations
Type 2A	Defect in intracellular transport and mutations increased proteolysis in plasma after secretion; mutations clustered within VWF A2
Type 2B	Missense mutations in VWF A1 result in increased binding to platelet GPIb
Type 2M	Missense mutations and small in-frame deletions in A1
Type 2N	Missense mutations interfere with FVIII binding

Table 2 The VWF Domain Structure, Representing the Complete 2812 Amino Acid Residues

Domains of VWF	Residues
Signal peptide	0–22
D1	34–240
D2	387–598
D′	764–859
D3	866–1074
A1	1277–1453
A2	1498–1665
A3	1691–1871
D4	1949–2153
B	2160–2243
C1	2255–2328
C2	2429–2495
C3	2580–2645
CTCK	2724–2812

The A1 domain of VWF contains binding site for the platelet membrane VWF receptor and GPIb-IX-V complex, and promotes the activation of GPIIb/IIIa resulting in GPIIb/IIIa dependent platelet aggregation (De Luca, et al., 2000). A1 domain also contains intramolecular disulfide bond (Cys-509–Cys-695) and salt-bridge network that wraps around the domain (Emsley, Cruz, Handin, & Liddington, 1998).

von Willebrand disease (VWD) is a congenital bleeding disorder occurs due to quantitative and qualitative defect in the VWF protein. It is studied to be most prevailed inherited bleeding disorder in humans as high as 1.3% of population (Bloom, 1991; Holmberg & Nilsson, 1985; Rodeghiero, Castaman, & Dini, 1987; Werner et al., 1993). The classification of VWD is reduced to six subgroups from previously distinct subtypes which retain three major types known to be type 1, type 2, and type 3 VWD (Sadler, 1994). Type 1 VWD attributes to partial quantitative deficiency of VWF. This is characterized by a reduction in (VWF: Ag) activity and levels, leading to bleeding diathesis. The bleeding tendency in type 2 variants is more prevalent than type 1 VWD. According to Sadler (1994), type 2A, 2B, and 2M VWD exhibit abnormal interaction with platelets, thereby affecting VWF platelet-dependent function, whereas 2N VWD is typified by normal platelet-dependent function, with substantially reduced binding capacity for FVIII. Type 2 VWD includes subtypes of 2A, 2B, 2M, and 2N characterized by a qualitative deficiency in VWF function. Of all VWDs, type 2 variants are thought to make up about 20–30% of diagnosis (Holmberg & Nilsson, 1985). Type 2A VWD is correlated with the lower levels of VWF ristocetin cofactor activity (VWF: Ricof) about VWF: Ag and the absence of largest VWF multimers. Type 2B VWD displays reduced high molecular weight multimers. Type 2M VWD is related with a defect in platelet/VWF interaction. Type 2N VWD is a recessive disorder of defective VWF binding to FVIII and resulting in a low level of circulating FVIII. Type 3 VWD is a clinically severe quantitative disorder marked by the absence of platelet and plasma VWF (normal range 50–150 U/dl), resulting in reduced FVIII activity. The *VWF* gene is highly polymorphic in nature (Ginsburg & Sadler, 1993; Sadler & Ginsburg, 1993) and most polymorphisms are closely related to VWD mutations. The fundamental basis for molecular genetic studies in VWD is limited. When necessitate, molecular diagnosis can possibly be accomplished in the qualitative VWD variants (types 2A, 2B, 2M, and 2N) by separating from common mutations. However, in the majority of cases of VWD, the causative mutation is unknown.

Several studies have reported the occurrence of mutations in the anterior region of exon 28 within A1 domain results in reducing GPIb binding affinity by the disruption of salt and disulfide bridges and decrease in hydrophobic contacts (Lars et al., 1993; Matsushita & Sadler, 1995; McKinnon et al., 2012; Meyer et al., 1997; Rabinowitz et al., 1993; Ribba et al., 1994; Woods et al., 2012).

Single-nucleotide polymorphisms (SNPs) occur in every 200–300 bp (Lee, Choi, & Lee, 2005). It is realized that 500,000 SNPs fall in the coding locale of the human genome (Collins, Brooks, & Chakravarti, 1998). Among them, nonsynonymous SNPs (nsSNPs) bring about the change in amino acid sequence. They are well on the way to be a huge variable that concurs to the practical uniqueness of the encoded proteins (Lander, 1996). Amino acids conserved across species are more prone to be functionally significant and deleterious (Zhu, Spitz, Amos, Lin, & Schabath, 2004). Last decade has witnessed a rapid deposit of SNP information especially nsSNPs in public databases. Henceforth, there is a need for a method to identify and classify the functionally significant nsSNPs/variants that may be deleterious to determine their molecular effects. Predicting the impact of SNPs (nsSNPs/variants) in coding region by computational methods provides an exceptional way to evaluate the relationship between functional nsSNPs/variants and their disease susceptibility (George Priya Doss, Rajasekaran, Arjun, & Sethumadhavan, 2010; George Priya Doss & Sethumadhavan, 2009; George Priya Doss et al., 2008; George & Rao, 2009a, 2009b; Rajasekaran, Sudandiradoss, George, & Rao, 2007). Research and computational science can assume a noteworthy part in giving valuable data in deciding the functional significance of each SNP (Frederic, Lalande, Boileau, Hamroun, & Claustres, 2009; Karchin, Diekhans, Kelly, Thomas, & Pieper, 2005; Ramensky, Bork, & Sunyaev, 2002; Wang et al., 2006; Yue, Melamud, & Moult, 2006). For molecular epidemiological studies, the application of molecular evolutionary methodology may be a compelling apparatus for prioritizing SNPs (Arbiza, Dopazo, & Dopazo, 2006; Burke et al., 2007; Dopazo, 2008). Therefore, selecting SNPs that are probable to have potential functional and disease susceptibility by computational analysis imparts useful information to biologists for further molecular studies (Bowang, Stefan, Marion, & Kari, 2009; George Priya Doss & Sethumadhavan, 2009; Mah & Chia, 2007).

VWD has been a major focus to molecular biologists and hematologists for a long duration. Based on this, we attempted to distinguish harmful SNPs from a pool by performing a computational examination of the mutant proteins to check their stability and functional activity. As a first step, we collected gene and SNP information from National Center for Biological Information (NCBI) and UniProt database to distinguish the deleterious nsSNPs with its structural and functional consequences on the protein structure. In this context, we connected a set of computational methods namely Sorting Intolerant from Tolerant (SIFT) (Kumar, Henikoff, & Ng, 2009), Polymorphism Phenotyping (PolyPhen 2.0) (Adzhubei, Schmidt, & Peshkin, 2010), I-Mutant 3.0 (Capriotti, Fariselli, Rossi, & Casadio, 2008), Screening for non-acceptable polymorphisms (SNAP) (Bromberg & Rost, 2007), Align GVGD (Tavtigian et al., 2005), PhD-SNP (Capriotti, Calabrese, & Casadio, 2006), SNPs&GO (Calabrese, Capriotti, Fariselli, Martelli, & Casadio, 2009), and MutPred (Li, Krishnan, & Mort, 2009) to predict the most disease associated nsSNP with structural consequence mutation in *VWF* gene. Predicted variants by applied *in silico* methods P1266L (P503L), H1268D (H505D), C1272F (C509F) and C1272R (C509R) (P1266L, H1268D, C1272R, and C1272F) belong to type 2 VWD, the result of a qualitative deficiency of VWF. Molecular analysis of VWD of qualitative (type 2) variants has given insight into structure and function of VWF protein. Among four predicted mutants (P1266L, H1268D, C1272R, and C1272F), two mutants C1272R (C509R) and C1272F belong to type 2A VWD described as the most common qualitative variant representing approximately 10% of all VWDs (Ginsburg & Bowie, 1992; Lavergne et al., 1992; Nichols & Ginsburg, 1997). Other two predicted mutants (P1266L and H1268D) fall in 2B type of VWD, related to increase in affinity of VWF for platelet GPIb (Cooney, Lyons, & Ginsburg, 1992; Ichinose et al., 1991). Till date, no computational analysis was attempted in A1 domain of VWF at the structural level. For the first time, dynamic molecular recreations with the detailed investigation were adapted for both the native and mutant proteins. This conformity helped us to comprehend the impact of alterations in the amino acid arrangement. The dynamics studies were conducted for 100 ns with the aid of GROMACS and Gromos43b1 force field (Hess, Kutzner, van der Spoel, & Lindahl, 2008). Amino acid substitution V1314F (V551F) (Ribba et al., 1994) and other mutants were recognized in the anterior region of exon 28 within A1 domain (Meyer et al., 1997). C1272G (C509G), R1308H (R545H), R1315C (R552C), and C1458Y (C695Y) were linked to platelet GPIb binding (Matsushita & Sadler, 1995). The overall strategy developed in this work is depicted in Fig. 1. Hence, our in-depth structural analysis might shed light on underlying molecular studies of VWD.

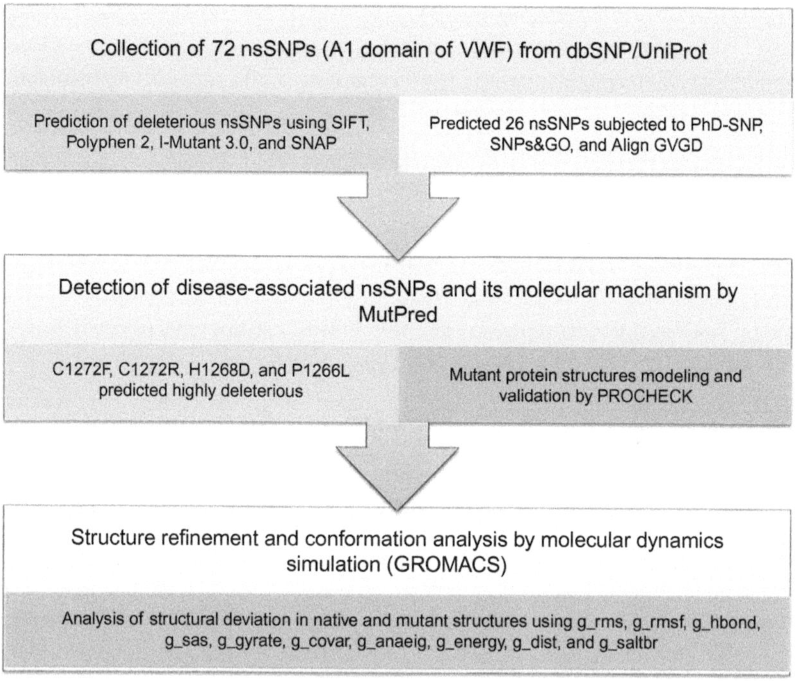

Figure 1 Mutational analysis strategies for A1 domain of VWF protein.

2. MATERIAL AND METHODS

2.1 Dataset

Human *VWF* gene and nsSNPs/variants related information was collected from OMIM (Amberger, Bocchini, Scott, & Hamosh, 2009), dbSNP (http://www.ncbi.nlm.nih.gov/snp/) (Sherry, Ward, & Kholodov, 2001), and UniProt database (Yip, Famiglietti, & Gos, 2008; Yip, Scheib, & Diemand, 2004). The amino acid sequence of VWF was retrieved in FASTA format from UniProt database (UniProt ID: P04275). The native protein crystal structure with PDB ID: 1AUQ (Emsley et al., 1998) was acquired from protein data bank (PDB) (Berman et al., 2002) for further modeling and dynamic studies.

2.2 Computational Methods for Finding Deleterious nsSNPs/Variants

The capacity to separate benign from pathogenic nsSNPs/variants from a pool is a difficult assignment by experimental methods. To address this issue, many computational algorithms were designed and developed, making it more feasible for accuracy and less time consuming. Among these methods, some classify deleterious variant as pathogenic, while other methods classify based on a change in protein stability. We utilized these methods for identifying deleterious variants in the A1 domain of *VWF* gene. The SNP occurring in the protein-coding region may have an effect on protein structure and function. To analyze the impact, we applied computational prediction methods, namely, PolyPhen 2.0 (Adzhubei et al., 2010), SIFT (Kumar et al., 2009), I-Mutant 3.0 (Capriotti et al., 2008), PhD-SNP (Capriotti et al., 2006), SNPs&GO (Calabrese et al., 2009), Align GVGD (Tavtigian et al., 2005), SNAP (Bromberg & Rost, 2007), and MutPred (Li et al., 2009). For predicting the functional context of a nsSNP/variant, we used sequence-based evolutionary method SIFT to predict a variant as neutral or deleterious using normal probability score. SIFT score of $\leq$0.05 and $\geq$0.05 is designated as intolerant or deleterious and tolerant. Measuring the mutational effect at sequence level by SIFT makes it applicable to any protein of human with known relatives. PolyPhen 2.0 utilizes a combination of sequence and structure information to classify a nsSNP/variant into three categories: benign (less than 0.14), possibly damaging (if score between 0.15 and 0.84), and probably damaging (affect protein function if score between 0.85 and 1). I-Mutant 3.0 is an SVM-based method to predict stability of a protein upon mutation. This method classifies a nsSNP/variant into three classes: neutral mutation ($-0.5 \leq$ DDG ≤ 0.5 kcal/mol), large decrease (≤ -0.5 kcal/mol), and large increase (>0.5 kcal/mol). SNAP based on neural network predicts whether a single amino acid alteration is neutral or nonneutral. We used PhD-SNP, SNPs&GO, Align GVGD, and MutPred to examine the disease-associated nsSNPs/variants. For each method, protein sequence information is given in the form of FASTA format along with native and mutant amino acid as input data with default parameters.

2.3 Modeling of the Mutant Protein Structure

To predict the impact of a nsSNP/variant in protein function, structure level analysis is proved to be vital. In this context, we used SAAPdb (Hurst et al., 2009) to map a nsSNP/variant with its corresponding three-dimensional (3D) structure for A1 domain of *VWF* gene. We used PDB ID coded by 1AUQ to construct mutant protein for MD analysis. Mutant structures were modeled at their corresponding positions P1266L, H1268D, C1272R, and C1272F of VWF protein using Swiss-PdbViewer (Kaplan & Littlejohn,

2001). Energy minimization was performed using 43a1 energy field of GROMACS (Hess et al., 2008) version 4.5.3. PROCHECK (Laskowski, Rullmann, MacArthur, Kaptein, & Thornton, 1996) was utilized to check the quality of the modeled mutant structures.

2.4 Molecular Dynamics Simulation

Molecular dynamics simulation of native and mutant protein structures was done with the aid of GROMACS 4.5.3 (Hess et al., 2008; Kutzner et al., 2007) that adopts GROMOS96 43a1 force field parameter for energy minimization (Van Gunsteren et al., 1996). The structures of native and four mutant models were solvated in a cubic box of 0.9 nm divider and stacked with simple point charge water molecules (Jorgensen, Chandrasekhar, Madura, Impey, & Klein, 1983). Periodic limit conditions were connected to the framework to maintain the quantity of particles, steady weight, and temperature. GENION method of GROMACS package was utilized to neutralize the structures by replacing water atom with Na^+ and Cl^- particles. The temperature was kept constant using Berendsen algorithm with a coupling time of 0.2 (Berendsen, Postma, Van Gunsteren, DiNola, & Haak, 1984). To soak the macromolecules into the water molecules, both the native and mutant-minimized structure were equilibrated for 50,000 steps at 300 K. At that point, the equilibrated structures were subjected to 100 ns molecular dynamics simulation. For analysis of structural deviation, root-mean-square deviation (RMSD), root-mean-square fluctuation (RMSF), radius of gyration (Rg), and solvent accessible region in native and mutant structures were performed by utilizing g_rms, g_rmsf, g_sas, and g_gyrate individually. Number of hydrogen bonds formed during the simulation by particular residue to other amino acids was calculated by utilizing g_hbond. g_energy, g_dist, and g_saltbr were used for the analysis of total energy, disulfide bond, and salt bridges involved in native and mutant proteins. All the graphs were plotted by utilizing XMGRACE program (Turner, 2005).

2.5 Principal Component Analysis

Computational advances have proved to be useful in exploring the changes that occur at various levels of structural details (Karplus, 2002). It is now well established that these motions can be efficiently expressed in terms of small number of collective coordinates describing and predicting the motion in a low-dimensional active phase space (Kitao & Go, 1999; Kitao, Hayward, &

Go, 1998). A transparent way of identifying these generalized directions of motion is through principal component analysis of the fluctuation covariance matrix obtained from the dynamic molecular simulations (Amadei, Linssen, & Berendsen, 1993). Using the GROMACS software, the eigenvectors and eigenvalues calculation was carried out. The matrix was constructed from the trajectory files. Using the g_covar backbone of C-alpha bond trajectories, the matrix was constructed.

3. RESULTS

3.1 Screening of Deleterious nsSNPs/Variants

A total of 490 SNPs were retrieved from the public database under the annotation nsSNP/variant. Among them, 72 nsSNPs/variants were found to be located in A1 domain of VWF. By SIFT, 49 nsSNPs/variants, namely, E1447Q, V1360A, N1435S, I1329M, V1318L, G1324A, L1279I, M1304V, P1337L, I1343V, A1377V, P1266L, H1268D, C1272S, C1272R, C1272F, L1276P, L1282R, S1285F, S1285P, L1288R, F1293L, D1302G, R1306W, L1307P, R1308C, S1310F, W1313C, V1314D, R1315C, R1341W, R1342C, E1359K, L1361S, K1362T, I1368T, F1369I, R1374H, R1374S, R1374C, R1374H, R1379C, L1383P, G1415D, I1416N, I1425F, C1458Y, V1360A, and I1308V, were predicted to be deleterious with tolerance index of $\leq$0.05 (Table 3). Among them, 36 nsSNPs/variants namely P1266L, H1268D, C1272S, C1272R, C1272F, L1276P, L1282R, S1285F, S1285P, L1288R, F1293L, D1302G, R1306W, L1307P, R1308C, S1310F, W1313C, V1314D, R1315C, R1341W, R1342C, E1359K, L1361S, K1362T, I1368T, F1369I, R1374H, R1374S, R1374C, R1374H, R1379C, L1383P, G1415D, I1416N, I1425F, and C1458Y were reported to be highly deleterious with SIFT score of 0.00 as shown in Table 3. Of 72 SNPs as input dataset, 60 nsSNPs/variants were found to be "damaging (0.5–1.000) to protein structure and function" and the remaining 12 nsSNPs/variants were characterized as benign by PolyPhen 2.0. Among these 60 deleterious nsSNPs/variants, 35 nsSNPs/variants namely P1266L, H1268N, C1272R, C1272F, L1276P, L1282R, S1285F, S1285P, L1288R, F1293L, L1296P, R1306W, L1307P, S1310F, R1315C, V1316M, R1341W, G1324A, G1324S, R1342C, E1359K, L1361S, K1362T, F1369I, R1374H, A1372V, R1379C, L1383P, R1399C, G1415D, I1416N, I1425F, C1458Y, R1374S, and R1374C were reported to be highly deleterious with PolyPhen score of 1.000 (Table 3). To validate

Table 3 nsSNPs/Variants Analyzed by Four Different *In Silico* Prediction Methods SIFT, PolyPhen 2.0, I-Mutant 3.0, and SNAP in VWF

nsSNP/Variant ID	Positions	SIFT	PolyPhen	I-Mutant 3.0	SNAP
rs61749370	P1266Q	0.01	0.998	−1.68	Nonneutral
VAR_005791	**P1266L**	0.00	1.000	−0.78	Nonneutral
rs61749371	H1268N	0.48	0.987	−0.67	Nonneutral
VAR_005792	**H1268D**	0.00	0.687	−0.33	Nonneutral
rs63524161	C1272S	0.00	0.997	−0.92	Nonneutral
rs61749372	**C1272R**	0.00	1.000	−0.41	Nonneutral
VAR_067340	**C1272F**	0.00	1.000	−0.20	Nonneutral
rs61749374	**L1276P**	0.00	1.000	−1.42	Nonneutral
rs61749376	V1279I	0.09	0.880	−0.84	Neutral
rs61749378	**L1282R**	0.00	1.000	−1.45	Nonneutral
rs61749380	**S1285F**	0.00	1.000	0.15	Nonneutral
rs61749379	**S1285P**	0.00	1.000	−0.21	Nonneutral
rs267607334	**L1288R**	0.00	1.000	−1.70	Nonneutral
rs138900040	E1290K	0.01	0.997	−0.99	Nonneutral
rs267607335	**F1293L**	0.00	1.000	−1.25	Nonneutral
rs61749381	L1296P	0.01	1.000	−1.82	Nonneutral
rs61749382	D1302G	1.00	0.001	−1.26	Nonneutral
rs267607336	M1304V	0.26	0.998	−0.98	Neutral
rs61749385	R1306Q	0.01	0.970	−1.06	Nonneutral
VAR_005794	**R1306W**	0.00	1.000	−0.49	Nonneutral
rs61749386	**L1307P**	0.00	1.000	−1.78	Nonneutral
rs61749388	R1308H	1.00	0.001	−1.59	Nonneutral
rs61749387	R1308C	0.00	0.992	−1.10	Nonneutral
rs61749389	I1309V	0.01	0.981	−0.95	Neutral
rs61749390	**S1310F**	0.00	1.000	0.41	Nonneutral
rs61749392	W1313C	0.00	0.972	−1.46	Nonneutral
rs61749392	V1314D	0.00	0.938	−1.37	Nonneutral

Continued

Table 3 nsSNPs/Variants Analyzed by Four Different *In Silico* Prediction Methods SIFT, PolyPhen 2.0, I-Mutant 3.0, and SNAP in VWF—cont'd

nsSNP/Variant ID	Positions	SIFT	PolyPhen	I-Mutant 3.0	SNAP
rs61749393	V1314L	0.03	0.002	−0.75	Neutral
rs61749396	R1315H	0.05	0.998	−1.55	Nonneutral
rs61749395	**R1315C**	0.00	1.000	−1.14	Nonneutral
rs61749397	V1316M	0.01	1.000	−1.09	Nonneutral
VAR_005799	V1318L	0.09	0.622	−0.61	Nonneutral
rs61749399	G1324A	0.07	1.000	−1.04	Nonneutral
rs61749398	G1324S	0.01	1.000	−1.40	Neutral
rs138413641	I1329M	0.05	0.618	−1.72	Neutral
rs61749400	P1337L	0.15	0.997	−0.53	Neutral
VAR_005801	R1341Q	0.03	0.996	−1.12	Neutral
rs61749402	**R1341W**	0.00	1.000	−0.49	Neutral
rs61749404	R1342C	0.00	1.000	−1.12	Neutral
rs150923481	I1343V	0.05	0.071	−1.15	Neutral
rs61749407	**E1359K**	0.00	1.000	−0.82	Nonneutral
rs267607338	V1360A	0.54	0.976	−1.79	Neutral
rs61749408	L1361S	0.00	1.000	−2.08	Neutral
rs61750067	K1362T	0.00	1.000	−0.39	Neutral
rs199703097	I1368V	0.33	0.000	−1.07	Neutral
rs199675425	I1368T	0.00	0.035	−2.38	Neutral
rs61750069	**F1369I**	0.00	1.000	−0.81	Nonneutral
rs61750071	I1372S	0.19	0.901	−2.49	Nonneutral
VAR_005803	**R1374H**	0.00	1.000	−1.38	Nonneutral
rs61750071	**R1374S**	0.00	1.000	−1.22	Nonneutral
VAR_005802	**R1374C**	0.00	1.000	−0.99	Nonneutral
rs141211612	A1377V	0.05	1.000	−0.01	Nonneutral
rs61750073	S1378F	0.01	0.995	0.17	Nonneutral
rs61750074	**R1379C**	0.00	1.000	−0.86	Nonneutral

Table 3 nsSNPs/Variants Analyzed by Four Different *In Silico* Prediction Methods SIFT, PolyPhen 2.0, I-Mutant 3.0, and SNAP in VWF—cont'd

nsSNP/Variant ID	Positions	SIFT	PolyPhen	I-Mutant 3.0	SNAP
rs11063988	I1380V	0.59	0.004	−0.85	Neutral
rs216311	T1381A	1.00	0.000	−0.84	Nonneutral
rs61750075	**L1383P**	0.00	1.000	−1.45	Nonneutral
rs149950364	S1394F	0.01	0.989	0.43	Nonneutral
rs61750077	R1399C	0.01	1.000	−1.12	Nonneutral
rs1800382	R1399H	0.03	0.922	−1.41	Neutral
rs141990425	K1408Q	0.03	0.993	−0.43	Nonneutral
rs61750079	P1413L	0.01	0.994	−0.69	Neutral
rs61750080	**G1415D**	0.00	1.000	−0.96	Nonneutral
rs61750081	**I1416N**	0.00	1.000	−1.76	Nonneutral
rs61750082	N1421K	0.01	0.998	0.00	Nonneutral
rs61750083	**I1425F**	0.00	1.000	−1.19	Nonneutral
rs11063987	N1435S	0.17	0.997	−0.37	Neutral
rs61750084	A1437T	0.01	0.998	−0.78	Nonneutral
rs150077670	V1439M	0.01	0.685	−0.69	Nonneutral
rs61750085	E1447Q	0.16	0.482	−0.86	Nonneutral
rs61750087	**C1458Y**	0.00	1.000	−0.33	Nonneutral
rs61750088	L1460V	0.01	0.996	−1.57	Nonneutral

nsSNP/variant highlighted in bold found to be highly deleterious.

the obtained results, all the nsSNPs/variants submitted to PolyPhen 2.0 and SIFT were submitted as input to I-Mutant 3.0 server. By protein stability predictor, 42 nsSNPs/variants were predicted to affect the stability of a protein, 28 nsSNPs/variants showed to be neutral, and remaining 2 nsSNPs/variants predicted to increase the stability of a protein. Hence, out of 72 SNPs, 52 and 20 nsSNPs/variants were predicted to be nonneutral and neutral by improved neural network-based method SNAP, respectively.

Based on the obtained scores from the above prediction methods, 26 nsSNPs/variants, namely, P1266L, H1268D, C1272R, C1272F, L1276P, L1282R, S1285F, S1285P, L1288R, F1293L, R1306W,

L1307P, S1310F, R1315C, R1341W, E1359K, F1369I, R1374H, R1374S, R1374C, R1379C, L1383P, C1458Y, I1425F, I1416N, and G1415D, were designated as highly deleterious and damaging (Table 3).

To further classify the predicted deleterious nsSNPs/variants as a disease associated, we utilized two more support vector machine-based methods. Twenty-five and 26 nsSNPs/variants were predicted to be a disease associated by PhD-SNP and SNPs&GO (Table 4).

3.2 Biophysical Validation of nsSNPs/Variants

Align GVGD prediction is based on grade classifiers (C0–C65) with range of spectrum of deleterious (C65) to neutral (C0) combining biophysical characteristics of protein such as chain composition, polarity, and volume of amino acids and protein numerous sequence alignment. Total of 26 nsSNPs/variants predicted to be disease associated with all the above *in silico* prediction methods were submitted as input to Align GVGD. Among them, 12 nsSNPs/variants namely P1266L, H1268D, C1272R, C1272F, S1285F, S1285P, L1288R, R1315C, R1374S, R1374C, L1383P, and C1458Y were classified as class C65 (Table 4). Twelve nsSNPs/variants classified as pathogenic were further analyzed by MutPred to differentiate the substitution as disease related or neutral with molecular cause of disease (Table 5). Predictions based on general scores with $g > 0.5$ and $p < 0.05$ are referred to as actionable hypothesis and scores with $g > 0.75$ and $p < 0.05$ are referred to as a confident hypothesis. It is found mutant P1266l to be highly deleterious with g score of 0.934 and predicted to loss of the disorder with p score of 0.0431 having a confident hypothesis. Mutant H1268D predicted to have g score of 0.915 and mutation involved in the loss of stability with p score of 0.0101 showing a confident hypothesis. The nsSNPs/variants C1272R and C1272F were predicted to have g score of 0.826 and 0.850 with p score of 0.0213 and 0.0257, respectively, showing a confident hypothesis. Based on the scores obtained from the above all *in silico* prediction methods, we filtered four nsSNPs/variants P1266L, H1268D, C1272R, and C1272F as most deleterious and disease-associated mutation in *VWF* gene for further structure analysis (Table 5).

3.3 Modeling of nsSNPs/Variants

The disease-associated nsSNPs/variants P1266L, H1268D, C1272R, and C1272F can alter the native conformation of VWF protein. To analyze the conformational change, mutant structures were modeled using native

Table 4 Disease-Associated nsSNPs/Variants and Biophysical Properties Prediction

nsSNP/Variant ID	Position	PhD-SNP	SNPs&GO	Align GVGD
VAR_005791	**P1266L**	Disease	Disease	**Class C65**
VAR_005792	**H1268D**	Disease	Disease	**Class C65**
rs61749372	**C1272R**	Disease	Disease	**Class C65**
VAR_067340	**C1272F**	Disease	Disease	**Class C65**
rs61749374	**L1276P**	Disease	Disease	Class C45
rs61749378	**L1282R**	Disease	Disease	Class C35
rs61749380	**S1285F**	Disease	Disease	**Class C65**
rs61749379	**S1285P**	Disease	Disease	**Class C65**
rs267607334	**L1288R**	Disease	Disease	**Class C65**
rs267607335	**F1293L**	Disease	Disease	Class C15
VAR_005794	**R1306W**	Disease	Disease	Class C0
rs61749386	**L1307P**	Disease	Disease	Class C35
rs61749390	**S1310F**	Neutral	Disease	Class C25
rs61749395	**R1315C**	Disease	Disease	**Class C65**
rs61749402	**R1341W**	Disease	Disease	Class C0
rs61749407	**E1359K**	Disease	Disease	Class C0
rs61750069	**F1369I**	Disease	Disease	Class C0
VAR_005803	**R1374H**	Disease	Disease	Class C25
rs61750071	**R1374S**	Disease	Disease	**Class C65**
VAR_005802	**R1374C**	Disease	Disease	**Class C65**
rs61750074	R1379C	Disease	Disease	Class C35
rs61750075	**L1383P**	Disease	Disease	**Class C65**
rs61750080	**G1415D**	Disease	Disease	Class C15
rs61750081	**I1416N**	Disease	Disease	Class C35
rs61750083	**I1425F**	Disease	Disease	Class C0
rs61750087	**C1458Y**	Disease	Disease	**Class C65**

Highlighted in bold are disease associated.

Table 5 Prediction Reliability by MutPred Server

nsSNP/Variant ID	Positions	G Score	P Score	Prediction Reliability
VAR_005791	**P1266L**	0.934	0.0431	Confident hypothesis
VAR_005792	**H1268D**	0.915	0.0101	Confident hypothesis
rs61749372	**C1272R**	0.81	0.0213	Confident hypothesis
VAR_067340	**C1272F**	0.85	0.0841	Confident hypothesis
rs61749380	S1285F	0.583	0.0823	No reliable inference
rs61749379	S1285P	0.673	0.0719	No reliable inference
rs267607334	L1288R	0.611	0.06	No reliable inference
rs61749395	R1315C	0.553	0.0831	No reliable inference
rs61750071	R1374S	0.524	0.0976	No reliable inference
VAR_005802	R1374C	0.733	0.703	No reliable inference
rs61750075	L1383P	0.74	0.0622	No reliable inference
rs61750087	C1458Y	0.754	0.0811	No reliable inference

Table 6 Quality of the Structure of Native and Modeled Mutant VWF Proteins Analyzed Independently by PROCHECK Server

Proteins	Favored Region (%)	Allowed Region (%)	Outlier Region (%)
Native	94.2	5.8	0.0
C1272F	89.8	9.7	0.5
C1272R	92.7	6.8	0.5
P1266L	89.8	9.7	0.5
H1268D	92.6	6.9	0.5

VWF protein at their corresponding positions 1266 (Proline 503 Leucine), 1268 (Histidine 505 Aspartic acid), 1272 (Cysteine 509 Arginine), and 1272 (Cysteine 509 Phenylalanine). The quality of the modeled structure of native and mutant VWF proteins was assessed independently by PROCHECK (Table 6). The obtained results showed good stereochemical properties of the modeled protein were considered for further MD analysis.

3.4 Molecular Dynamics Simulation Studies

Conformational change in VWF protein residues upon mutation were analyzed using various factors such as tolerances index, DDG value, biophysical properties, pathogenicity index, general score (*g*), and property score (*p*). The results obtained from *in silico* prediction analysis motivated us to study the atomic-level dynamic behavior of the native and mutant VWF proteins. Simulation analysis by GROMACS was initiated to understand the structural and functional behavior of the disease-related mutants (P1266L, H1268D, C1272F, and C1272R). We analyzed RMSD, RMSF, Rg, SASA, and number of hydrogen bonds between the native and mutant structures from the obtained trajectory files of 100 ns. To study the convergence of the protein system from the initial structure, RMSD backbone atoms of native and four mutant structures were plotted (Fig. 2). In A1 domain of VWF, native structure retained the maximum deviation till the end of the simulation resulting in the RMSD of ~0.10 to 0.22 nm. Meanwhile, H1268D, P1266L, and C1272R mutant showed different pattern of deviation till the end of the simulation resulting in the RMSD of ~0.10 to 0.25 nm, ~0.10 to 0.25 nm, and ~0.10 to 0.23 nm, respectively (Fig. 2). Among four mutants, H1268D, P1266L, and C1272R mutants

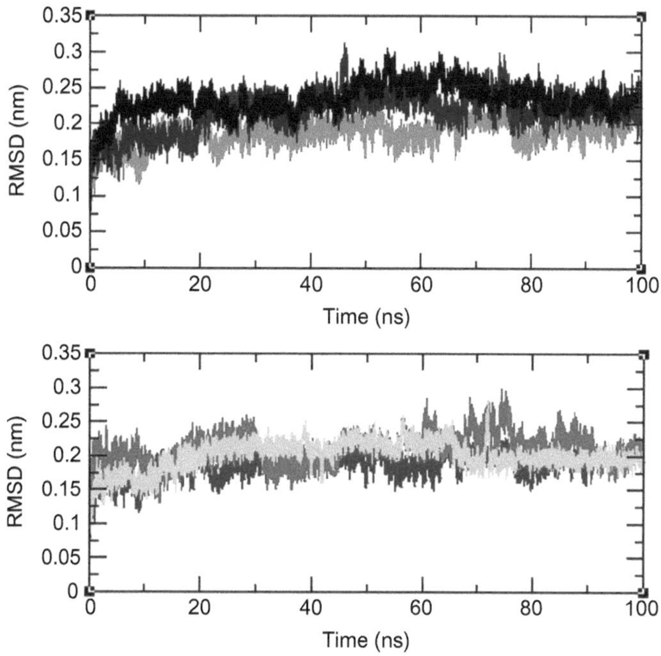

Figure 2 Time evolution of backbone RMSDs shown as a function of time for native and mutant models at 100 ns. Color-coding scheme is as follows: native (green color, gray in the print version), mutant C1272F (indigo color, dark gray in the print version), C1272R (magenta color, light gray in the print version), P1266L (blue color, dark gray in the print version), and H1268D (maroon color, dark gray in the print version).

showed higher deviation and C1272F showed similar deviation pattern as of native protein (~0.10 to 0.22 nm). Four mutants showed convergence till 100 ns indicated that the mutant structures produced stable trajectories and slightly deviated from the native. This provided an appropriate basis for further analysis. RMSF (C-α atom) graph was plotted to measure the residue-wise mobility of the protein upon mutation. The marked regions in graph displays the residues involved in the catalytic regions, mutations sites, and disulfide bond forming regions of VWF protein (Fig. 3). We observed fluctuation of ~0.25 nm for the native protein. Mutant P1266L, H1268D, C1272F, and C1272R showed fluctuation of ~0.40, ~0.35, ~0.38, and ~0.15 nm throughout the simulation time, respectively. However, at particular mutational residue position of the protein in Fig. 4, the mutants showed an abrupt change. Out of four mutants, P1266L showed maximum fluctuation of ~0.25 nm as proline is replaced by leucine (Fig. 4). Whereas mutant C1272F showed least fluctuation of ~0.10 nm as cysteine is replaced by phenylalanine (Fig. 4). Fluctuation score analysis showed a higher degree of flexibility in three mutant structures (P1266L, H1268D,

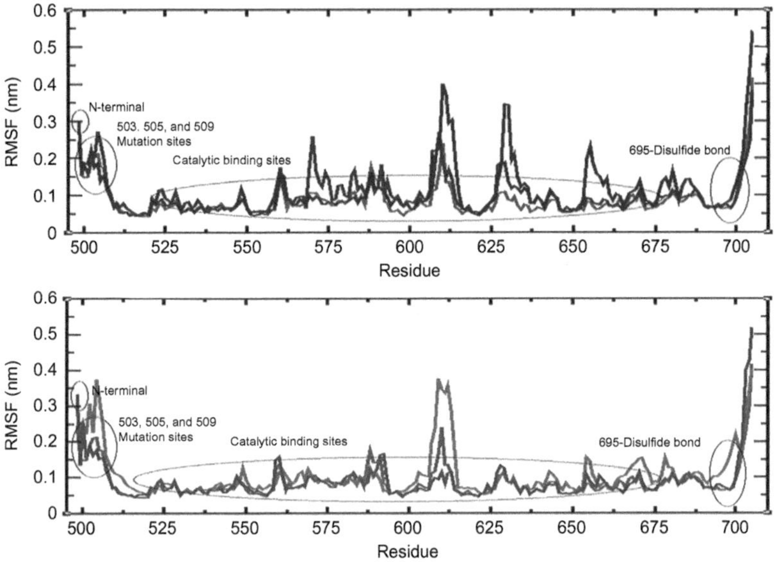

Figure 3 RMSF of the backbone carbon alpha of A1 domain of VWF over the entire simulation. The ordinate is RMSF (nm), and the abscissa is residue. Color-coding scheme is as follows: native (green color, gray in the print version), mutant C1272F (indigo color, dark gray in the print version), C1272R (magenta color, light gray in the print version), P1266L (blue color, dark gray in the print version), and H1268D (maroon color, dark gray in the print version).

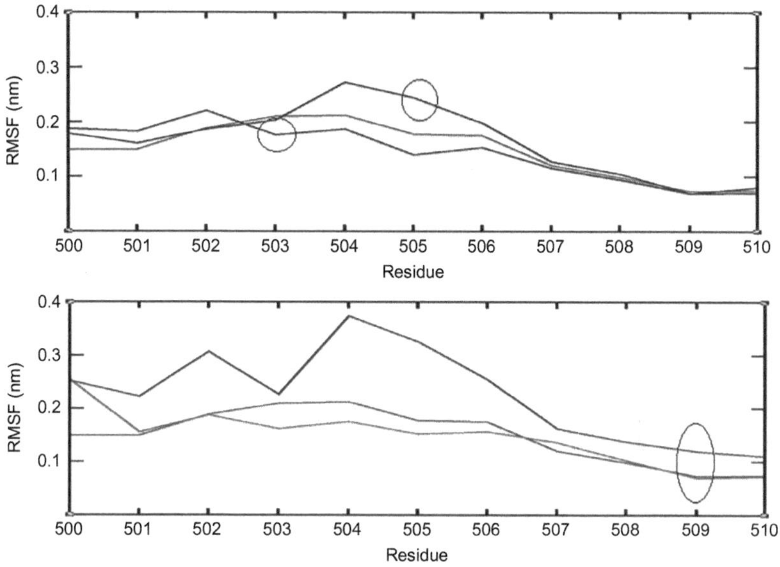

Figure 4 Comparison of the RMS fluctuation at their corresponding positions. Color-coding scheme is as follows: native (green color, gray in the print version), mutant C1272F (indigo color, dark gray in the print version), C1272R (magenta color, light gray in the print version), P1266L (blue color, dark gray in the print version), and H1268D (maroon color, dark gray in the print version).

and C1272F) than of native VWF structure. Hydrogen bonds are the main driving force to stabilize the protein structure and also play a vital role in biological interaction with respect to protein stability (Dill, 1990; Honig & Yang, 1995; Myers & Pace, 1996). Mainly intrahydrogen bond interaction (NH) is more significant for protein stability than the degree of bond bending; bond may be negligible, removed, or reversed (Van Zandt & Schroll, 2012). We observed the difference in NH bond formation pattern during simulation. Mutant proteins showed less participation in hydrogen bond formation than the native. With an increase in few hydrogen bonds throughout the simulation, the native in Fig. 5 was found to be adequately stabilized with ~180 intrahydrogen bonds. Mutants P1266L, H1268D, C1272F, and C1272R were observed to be stable with intrahydrogen bond formation of ~170, ~173, ~178, and ~165, respectively (Fig. 5). This indicates mutation resulted in lowering the number of hydrogen bonds formation in mutants as compared to that of native protein leading to loss of stability and conformational change. The radius of gyration measure provides a valuable insight into an overall dimension of the protein. C-α atom of protein versus time at

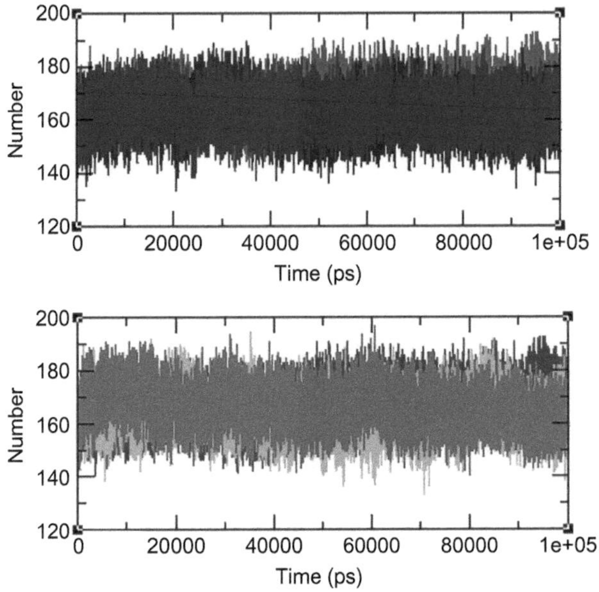

Figure 5 Number of hydrogen bonds in native and mutant model versus time. Color-coding scheme is as follows: native (green color, gray in the print version), mutant C1272F (indigo color, dark gray in the print version), C1272R (magenta color, light gray in the print version), P1266L (blue color, dark gray in the print version), and H1268D (maroon color, dark gray in the print version).

300 K for Rg plots are shown in Fig. 6. At the end of the simulation, native showed higher Rg value than mutant proteins. The native protein showed Rg value of ~1.64 nm at end of 100 ns simulation. Mutants C1272F and C1272R showed rg values of ~1.63 and ~1.64 nm indicating the shape and compactness of the protein as similar to the native. In the case of other two mutants P1266L and H1268D exhibited Rg value of ~1.62 and ~1.61 nm indicated the change in shape and compactness of the protein as compared to that of native. Solvent accessibility surface area (SASA) accounts for the biomolecular surface area that is accessible to solvent molecules. Native protein and C1272F obtained SASA of ~58.5 nm in the entire simulation period of 100 ns. This indicates mutant C1272F retained functional property of protein as that of native. Notably increase and decrease in SASA value were observed in the mutants P1266L (~60 nm) and C1272R (~58.9 nm), and H1268D (~55.1 nm), respectively (Fig. 7). Mutant C1272F retained functional property of protein as the native. To monitor the overall change in motion of the protein, the principal component analysis (PCA) was performed on C-α atom of the protein (Fig. 8). The PCA spectrum results

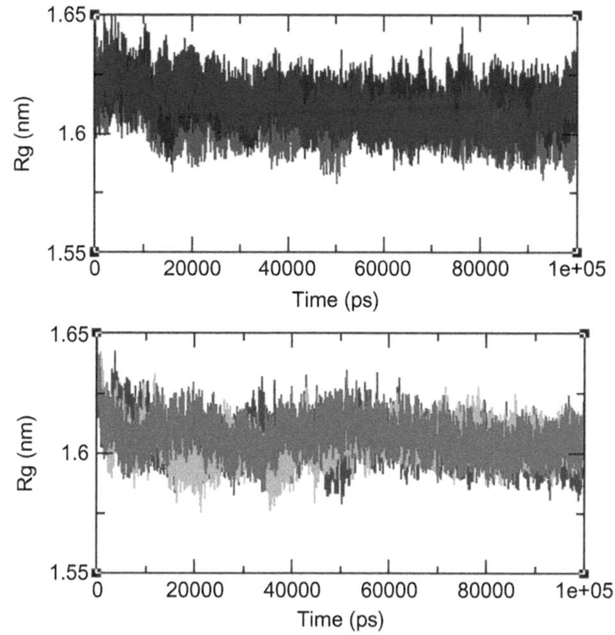

Figure 6 Radius of gyration of A1 domain for VWF protein. Color-coding scheme is as follows: native (green color, gray in the print version), mutant C1272F (indigo color, dark gray in the print version), C1272R (magenta color, light gray in the print version), P51266L (blue color, dark gray in the print version), and H1268D (maroon color, dark gray in the print version).

indicated the level of fluctuation and dynamic behavior of protein by eigenvalues derived from eigenvector 1 versus eigenvector 2. C1272F showed a slight change in motion of protein with well-defined clusters as in native (Fig. 8). In mutant C1272R change in motion of the protein with extended clusters was observed due to mutation. Whereas, P1266L and H1268D mutants showed change in overall motion with extended clusters as compared to that of native protein. Experimental studies have described the formation of disulfide loop between C1272 and C1458 (Lavergne et al., 1992). To analyze the change in distance due to mutation, 100 ns trajectory files of the both mutants C1272R and C1272F and native were analyzed using g_dist. It was observed that the C1272F showed maximum deviation of ~0.53 nm from that of native ~0.50 nm when compared to other mutant C1272R with ~1.0 nm as shown in Fig. 9. Similarly, to measure the change in distance in the formation of salt bridge upon mutation, trajectory files of mutant H1268D and native analyzed using g_saltbr. Native showed distance

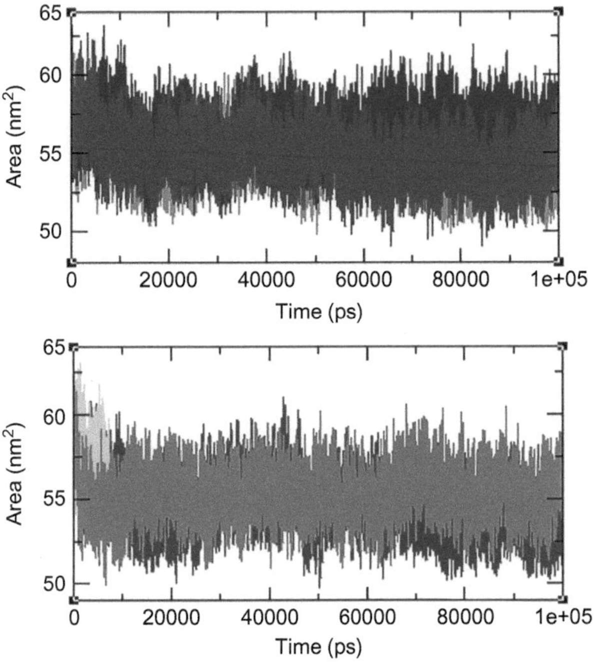

Figure 7 SASA of native and the mutants over 100 ns. Color-coding scheme is as follows: native (green color, gray in the print version), mutant C1272F (indigo color, dark gray in the print version), C1272R (magenta color, light gray in the print version), P1266L (blue color, dark gray in the print version), and H1268D (maroon color, dark gray in the print version).

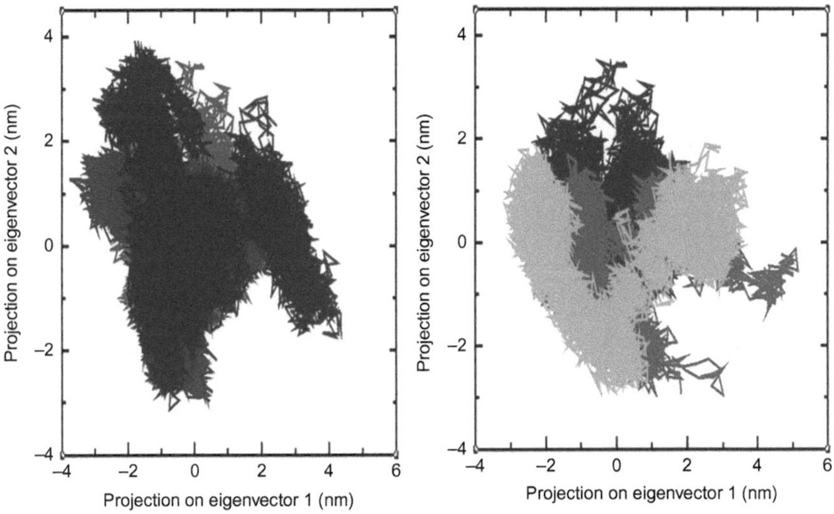

Figure 8 Projection of the motion of protein. Color-coding scheme is as follows: native (green color, gray in the print version), mutant C1272F (indigo color, dark gray in the print version), C1272R (magenta color, light gray in the print version), P1266L (blue color, dark gray in the print version), and H1268D (maroon color, dark gray in the print version).

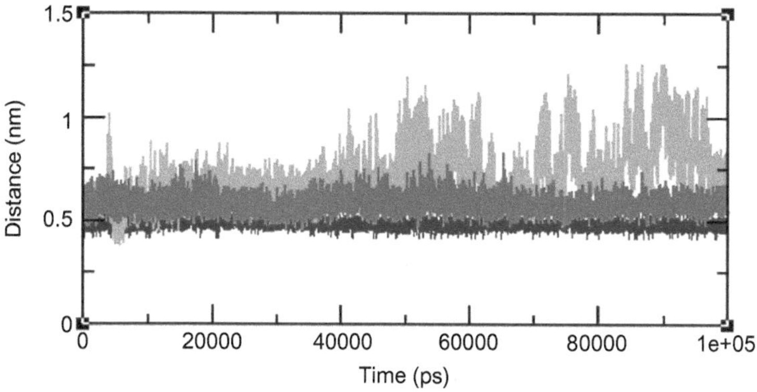

Figure 9 Change in distance due to mutation between native, C1272R, and C1272F.

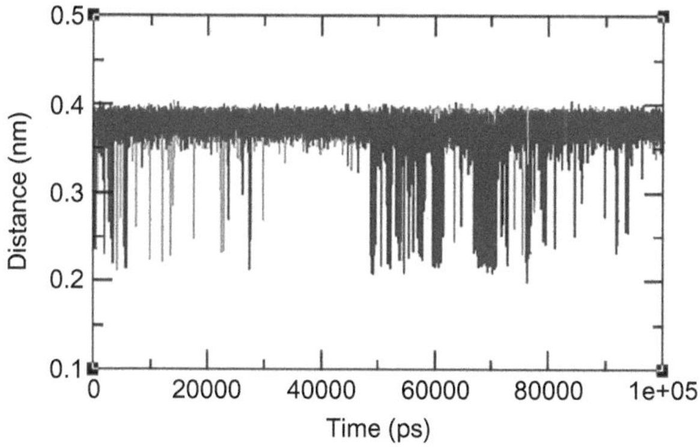

Figure 10 Change in stability of the native protein due to mutation H1268D.

of ~0.38 nm whereas mutant H1268D showed decrease in distances of ~0.35 nm. Indicating upon mutation at H1268 (505) position, there is loss of stability of the protein (Fig. 10). Furthermore, total energy was calculated to elucidate if a mutation cause any damage to the structural stability of the protein using g_energy. The results in Fig. 11 indicated slight fall in total energy of the mutant structures C1272R (−1.078) and P1266L (−1.078), and H1268D (−1.075) and C1272F (−1.075) as compared to that of native protein with value of −1.070 stating the loss of stability induced by the mutations. Results obtained from RMSD, RMSF, Rg, and SASA of native and mutant VWF proteins depicted that mutants became rigid upon mutation resulting in the disturbance of the functional behavior of the protein. This point was further supported by PCA analysis of mutants.

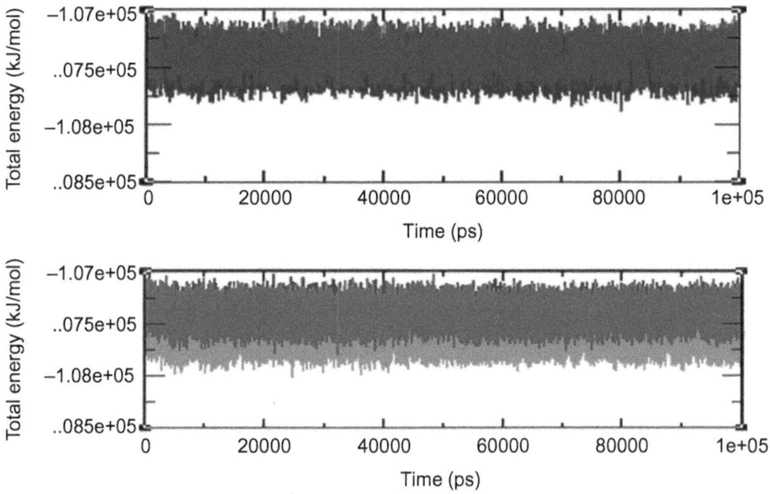

Figure 11 The total energy fluctuation in native and mutants of VWF A1 domain versus time at 300 K. Color-coding scheme is as follows: native (green color, gray in the print version), mutant C1272F (indigo color, dark gray in the print version), C1272R (magenta color, light gray in the print version), P1266L (blue color, dark gray in the print version), and H1268D (maroon color, dark gray in the print version).

4. DISCUSSION

Genetic data on various protein genotype, haplotype, gene expression, and protein expression are made available in public databases. A significant use of measured genetic data might be useful in identifying finding the polymorphisms that underline human diseases (Shastry, 2003). One of the most known types of genetic data comes in the form of SNPs that occur approximately to every 600–1000 bp in the human genome (Kimura, 1983; Shastry, 2003). Understanding the genetic basis of disease requires detailed knowledge about genetic polymorphisms that contribute such as SNPs (Stenson et al., 2009). Databases such as National Center for Biotechnology Information (NCBI), Online OMIM, and UniProt provide explicit information about the variations observed in disease conditions along with experimental data. Elucidating each SNP contributing to disease using experimental analysis on molecular basis may be time consuming and laborious. To address this, more sophisticated computational methods with different algorithms were made available on web to make accurate predictions. In recent years many researchers have utilized the computational methods to predict the impact of SNP on genes associated with disease (George and

Rao, 2010; George Priya Doss & Sethumadhavan, 2009; George & Rao, 2009a, 2009b; George Priya Doss et al., 2008; Gnad, Baucom, Mukhyala, Manning, & Zhang, 2013; Gray, Kukurba, & Kumar, 2012; Peterson, Doughty, & Kann, 2013). In addition, combination of MD and *in silico* prediction methods has proved to be vital assessing the impact of nsSNPs/variants in context to structural and functional level of the proteins (John et al., 2013; Miteva, Brugge, Rosing, Nicolaes, & Villoutreix, 2004; Steen, Miteva, Villoutreix, Yamazaki, & Dahlback, 2003; Sudhakar et al., 2015). The molecular dynamic analysis simulates the motion of given particle to result in a fluctuation in the relative position of atoms in macromolecules or DNA. This computational analysis plays a vital role in modern biology in analyzing protein structures from NMR to protein X-ray crystal structures and also calculates free energy change, fluctuation, and stability change upon mutation. VWD is the most common inherited genetically and clinically heterogeneous bleeding disorder. People with VWD have a problem with a protein present in the blood called VWF that enables to control bleeding. During injury, VWF helps platelets to mesh together and form a clot to stop bleeding. People with VWD do not have enough VWF and takes longer time to clot. A1 domain forms the binding site for platelet glycoprotein Ib (GPIb). Mutations in this region might play a significant role in causing a disease. Existence of nsSNP in VWF protein has already known to be affecting blood clotting process leading to VWD (Ribba et al., 1994; Matsushita & Sadler, 1995; Meyer et al., 1997). Studying nsSNPs based on amino acid properties is considered to be a fundamental characteristic in determining the protein folding, function, and stability (Teng, Madej, Panchenko, & Alexov, 2009). Keeping this in note, a comprehensive computational approach was applied to prioritize and predict the functional nsSNPs in A1 domain of the VWF protein. SIFT predicted 86.1% of nsSNPs to be deleterious, PolyPhen 2.0 predicted 62.5% to be damaging, I-Mutant 3.0 predicted 58.3% nsSNPs in decreasing the stability of the protein, and SNAP predicted 72.2% as nonneutral from the total of 72 nsSNPs. Combing the prediction of all four methods, 26 nsSNPs were designated as pathogenic. Furthermore, 26 nsSNPs were submitted to PhD-SNP, SNPs&GO, and Align GVGD for further validation. Using computational methods, we filtered four nsSNPs namely P1266L, H1268D, C1272R, and C1272F to be highly deleterious from a pool of 72 nsSNPs/variants. Position 1266, 1268, 1272 present in the C-terminal part of A1 domain shown to have mutations that may hinder the function of the protein were selected for further simulation analysis. Results obtained from RMSD, RMSF, Rg, and NH bond analysis showed higher deviation and fluctuation in P1266L and H1268D indicating toward loss of stability and function of native VWF. Hydrogen bond analysis showed less number of intrahydrogen bond formation in mutants P1266L, H1268D, and C1272R resulting in loss of native function. Decrease in radius of gyration value and hydrogen bonds formation in H1268D resulted in increase in rigidity with change in shape and compactness of native VWF protein. Drift in hydrophobic to hydrophilic or vice verse upon mutation indicates the deleteriousness upon mutation. Solvent accessible surface area (SASA) analysis showed decrease in value for mutant H1268D resulting in change with decrease in hydrophilic and increase in hydrophobic nature of protein. Whereas, mutant P1266L showed increase in SASA values comparing to native making the mutant protein more hydrophilic toward the solvent molecules. Furthermore, in principle component analysis, mutant P1266L and H1268D showed extended cluster formations increasing the overall motion of the protein resulting in instability. Decrease in total energy value and less participation of intrahydrogen bond formation indicates the increase in rigidity of the mutant proteins H1268D and P1266L. In native VWF protein position C1272 is involved in disulfide bonds formation. Due to mutation native cysteine is replaced by phenylalanine and arginine results with increase in distance making the mutant proteins unstable and disturbs the bonding pattern (Fig. 9). Native residue H1268 involved in salt bridge formation, with replacement of native histidine with aspartic acid decreases the distance making protein unstable and loss of function (Fig. 10). These results illustrate how various effects of amino acid substitutions can change the affinity of A1 domain to GPIb receptor and local destabilization of the protein scaffold. The obtained results with this computational pipeline can aid in prioritizing the functional nsSNPs/variants in any disease.

5. CONCLUSION

In human genome, SNPs are the most common single amino acid substitutions. Identifying the significant SNPs that result in clinically relevant phenotype is important for prompt diagnosis and treatment. With various SNP predictions methods existing, quality of prediction can be improved by combinatorial approaches. These computational methods not only allow scientists to cut down testing time but also make predictions in accurate manner. In present study, we have studied impact of mutations in A1 domain of VWF for the first time with simulation approach. These results illustrate how various effect of amino acid substitution can change the affinity of A1 domain to GPIb receptor. We suggest that these studies provide the relationship between the genetic and clinical measurements to increase information about clinical characteristics of the reported cases. It is likely that this study might shed light on the underlying molecular pathology of this disease and also provide a way for structure-based drug design.

ACKNOWLEDGMENT

The authors take this opportunity to thank the management of VIT University for providing the facilities and encouragement to carry out this work.

REFERENCES

Adzhubei, I. A., Schmidt, S., & Peshkin, L. (2010). A method and server for predicting damaging missense mutations. *Nature Methods*, *7*, 248–249.

Amadei, A., Linssen, A. B. M., & Berendsen, H. J. C. (1993). Essential of proteins. *Proteins: Structure, Function, and Genetics*, *17*, 412–425.

Amberger, J. S., Bocchini, C. A., Scott, A. F., & Hamosh, A. (2009). McKusick's Online Mendelian Inheritance in Man (OMIM). *Nucleic Acids Research*, *37*, 793–796.

Arbiza, L., Dopazo, J., & Dopazo, H. (2006). Selective pressures at a codon-level predict deleterious mutations in human disease genes. *Journal of Molecular Biology*, *358*, 1390–1404.

Berendsen, H. J. C., Postma, J. P. M., Van Gunsteren, W. F., DiNola, A., & Haak, J. R. (1984). Molecular dynamics with coupling to an external bath. *Journal of Chemical Physics*, *81*, 3684–3690.

Berman, H. M., Westbrook, J., Feng, Z., Gilliland, G., Bhat, T. N., Weissig, H., et al. (2002). The protein data bank. *Nucleic Acids Research*, *28*, 235–242.

Bloom, A. L. (1991). von Willebrand factor: Clinical features of inherited and acquired disorders. *Mayo Clinic Proceedings*, *66*, 743–751.

Bonthron, D., Orr, E. C., Mitsock, L. M., Ginsburg, D., Handin, R. I., & Orkin, S. H. (1987). Nucleotide sequence of pre-pro-von Willebrand factor cDNA. *Nucleic Acids Research*, *14*, 7125–7127.

Bowang, C., Stefan, W., Marion, D., & Kari, H. (2009). SNP_tools: A compact tool package for analysis and conversion of genotype data for MS-Excel. *BMC Research*, *2*, 214.

Bromberg, Y., & Rost, B. (2007). SNAP: Predict effect of non-synonymous polymorphisms on function. *Nucleic Acids Research*, *35*, 3823–3835.

Burke, D. F., Worth, C. L., Priego, E. M., Cheng, T., Smink, L. J., Todd, J. A., et al. (2007). Genome bioinformatic analysis of nonsynonymous SNPs. *BMC Bioinformatics*, *8*, 301–315.

Calabrese, R., Capriotti, E., Fariselli, P., Martelli, P. L., & Casadio, R. (2009). Functional annotations improve the predictive score of human disease-related mutations in proteins. *Journal of Human Mutation*, *30*, 1237–1244.

Capriotti, E., Calabrese, R., & Casadio, R. (2006). Predicting the insurgence of human genetic diseases associated to single point protein mutations with support vector machines and evolutionary information. *Bioinformatics*, *22*, 2729–2734.

Capriotti, E., Fariselli, P., Rossi, I., & Casadio, R. (2008). A three state prediction of single point mutations on protein stability changes. *BMC Bioinformatics*, *9*, S6.

Collins, F. S., Brooks, L. D., & Chakravarti, A. (1998). A DNA polymorphism discovery resource for research on human genetic variation. *Genome Research*, *8*, 1229–1231.

Colombatti, A., & Bonaldo, P. (1991). The superfamily of proteins with von Willebrand factor type A-like domains: One theme common to components of extracellular matrix, haemostasis, cellular adhesion and defence mechanisms. *Blood*, *77*, 2305–2315.

Cooney, K. A., Lyons, S. E., & Ginsburg, D. (1992). Functional analysis of a type IIB von Willebrand disease missense mutation: Increased binding of large von Willebrand factor multimers to platelets. *Proceedings of the National Academy of Sciences of the United States of America*, *89*, 2869–2872.

De Luca, M., Facey, D. A., Favaloro, E. J., Hertzberg, M. S., Whisstock, J. C., McNally, T., et al. (2000). Structure and function of the vonWillebrand factorA1 domain: Analysis with monoclonal antibodies reveals distinct binding sites involved in recognition of the platelet membrane glycoprotein Ib-IX-V complex and ristocetin-dependent activation. *Blood*, *95*, 164–172.

Dill, K. A. (1990). Dominant forces in protein folding. *Biochemistry*, *29*, 7133.

Dopazo, H. (2008). Selective constraints and human disease genes: Evolutionary and bioinformatics approaches. In *Encyclopedia of life sciences (ELS)*.Chichester: John Wiley & Sons, Ltd. http://dx.doi.org/10.1002/9780470015902.a0020762.

Emsley, J., Cruz, M., Handin, R., & Liddington, R. (1998). Crystal structure of the von Willebrand Factor A1 domain and implications for the binding of platelet glycoprotein Ib. *Journal of Biological Chemistry*, *273*, 10396–10401.

Frederic, M. Y., Lalande, M., Boileau, C., Hamroun, D., & Claustres, M. (2009). UMD-predictor, a new prediction tool for nucleotide substitution pathogenicity—Application to four genes: FBN1, FBN2, TGFBR1, and TGFBR2. *Human Mutation*, *30*, 952–959.

Fressinaud, E., Federici, A. B., Castaman, G., Rothschild, C., Rodeghiero, F., Baumgartner, H. R., et al. (1994). The role of platelet von Willebrand factor in platelet adhesion and thrombus formation: A study of 34 patients with various subtypes of type I von Willebrand disease. *British Journal of Haematology*, *86*, 327–332.

George Priya Doss, C., Rajasekaran, R., Arjun, P., & Sethumadhavan, R. (2010). Prioritization of candidate SNPs in colon cancer using bioinformatics tools: An alternative approach for a cancer biologist. *Interdisciplinary Sciences*, *2*, 320–346.

George Priya Doss, C., & Sethumadhavan, R. (2009). Investigation on the role of nsSNPs in HNPCC genes—A bioinformatics approach. *Journal of Biomedical Science*, *24*(16), 42.

George Priya Doss, C., Sudandiradoss, C., Rajasekaran, R., Choudhury, P., Sinha, P., Hota, P., et al. (2008). Applications of computational algorithm tools to identify functional SNPs. *Functional & Integrative Genomics*, *8*, 309–316.

George, P. D., & Rao, S. (2009a). Impact of single nucleotide polymorphisms in HBB gene causing haemoglobinopathies: In silico analysis. *New Biotechnology*, *25*, 214–219.

George, P. D., & Rao, S. (2009b). Functional and structural characterization of polymorphisms in MSH2 gene using computational tools. *Journal of Bionanoscience*, *3*, 7–15.

George, P. D., & Rao, S. (2010). Computational identification and structural analysis of deleterious functional SNPs in MLL gene causing acute Leukemia. *Interdisciplinary Sciences: Computational Life Sciences*, *2*, 247–255.

Ginsburg, D., & Bowie, E. J. W. (1992). Molecular genetics of von Willebrand disease. *Blood*, *79*, 2507–2519.

Ginsburg, D., Handin, R. I., Bonthron, D. T., Donlon, T. A., Bruns, G. A. P., Latt, S. A., et al. (1985). Human von Willebrand factor (vWF): Isolation of complementary DNA (cDNA) clones and chromosomal localization. *Science*, *228*, 1401–1406.

Ginsburg, D., & Sadler, J. E. (1993). von Willebrand disease: A database of point mutations, insertions and deletions. *Thrombosis and Haemostasis*, *69*, 177–184.

Gnad, F., Baucom, A., Mukhyala, K., Manning, G., & Zhang, Z. (2013). Assessment of computational methods for predicting the effects of missense mutations in human cancers. *BMC Genomics*, *14*(3), S7.

Gray, V. E., Kukurba, K. R., & Kumar, S. (2012). Performance of computational tools in evaluating the functional impact of laboratory-induced amino acid mutations. *Bioinformatics*, *28*, 2093–2096.

Hess, B., Kutzner, C., van der Spoel, D., & Lindahl, E. (2008). GROMACS 4: Algorithms for highly efficient, load-balanced, and scalable molecular simulation. *Journal of Chemical Theory and Computation*, *4*, 435–447.

Holmberg, L., & Nilsson, I. M. (1985). von Willebrand disease. *Clinical Haematology*, *14*, 461–488.

Honig, B., & Yang, A. S. (1995). Free energy balance in protein folding. *Advances in Protein Chemistry*, *46*, 27.

Hurst, J. M., McMillan, L. E., Porter, C. T., Allen, J., Fakorede, A., & Martin, A. C. (2009). The SAAPdb web resource: A large-scale structural analysis of mutant proteins. *Human Mutation*, *30*, 616–624.

Ichinose, A., Espling, E. S., Takamatsu, J., Saito, H., Shinmyozu, K., Maruyama, I., et al. (1991). Two types of abnormal genes for plasminogen in families with a predisposition for thrombosis. *Proceedings of the National Academy of Sciences of the United States of America*, *88*, 115–119.

Jaffe, E. A., Hoyer, L. W., & Nachman, R. L. (1974). Synthesis of von Willebrand factor by cultured human endothelial cells. *Proceedings of the National Academy of Sciences of the United States of America*, *71*, 1906–1909.

John, A. M., George Priya Doss, C., Ebenazer, A., Seshadri, M. S., Nair, A., Rajaratnam, S., et al. (2013). p.Arg82Leu von Hippel-Lindau (VHL) gene mutation among three members of a family with familial bilateral pheochromocytoma in India: Molecular analysis and in silico characterization. *PLoS One*, *8*, e61908.

Jorgensen, W. L., Chandrasekhar, J., Madura, J. D., Impey, R. W., & Klein, M. L. (1983). Comparison of simple potential functions for simulating liquid water. *Journal of Chemical Physics*, *79*, 926.

Kaplan, W., & Littlejohn, T. G. (2001). Swiss-PDB viewer (deep view). *Briefings in Bioinformatics*, *2*, 195–197.

Karchin, R., Diekhans, M., Kelly, L., Thomas, D. J., & Pieper, U. (2005). LS-SNP: Large-scale annotation of coding non-synonymous SNPs based on multiple information sources. *Bioinformatics*, *21*, 2814–2820.

Karplus, M. (2002). Molecular dynamics simulations of biomolecules. *Accounts of Chemical Research*, *35*, 321–323.

Kimura, M. (1983). *The neutral theory of molecular evolution*. Cambridge: Cambridge University Press.

Kitao, A., & Go, N. (1999). Investigating protein dynamics in collective coordinate space. *Current Opinion in Structural Biology*, *9*, 164–169.

Kitao, A., Hayward, S., & Go, N. (1998). Energy landscape of a native protein: Jumping-among-minima model. *Proteins*, *33*, 496–517.

Kumar, P., Henikoff, S., & Ng, P. C. (2009). Predicting the effects of coding non-synonymous variants on protein function using the SIFT algorithm. *Nature Protocols*, *4*, 1073–1082.

Kutzner, C., Van Der Spoel, D., Fechner, M., Lindahl, E., Schmittn, U. W., De Groot, B. L., et al. (2007). Speeding up parallel GROMACS on high-latency networks. *Journal of Computational Chemistry*, *28*, 2075–2084.

Lander, E. S. (1996). The new genomics: Global views of biology. *Science*, *274*, 536–539.

Lars, H., Judith, A. D., Reinhard, S., Ulrich, B., Jerry, W., & von Zaverio, M. R. (1993). von Willebrand factor mutation enhancing interaction with platelets in patients with normal multimeric structure. *Journal of Clinical Investigation*, *91*, 2169–2177.

Laskowski, R. A., Rullmann, J. A. C., MacArthur, M. W., Kaptein, R., & Thornton, J. M. (1996). AQUA and PROCHECK-NMR: Programs for checking the quality of protein structures solved by NMR. *Journal of Biomolecular NMR*, *8*, 477–486.

Lavergne, J. M., De Pailette, L., Bahnak, B. R., Ribba, A. S., Fressinaud, E., Pietu, G., et al. (1992). Defects in type IIA von Willebrand's disease: A cysteine 509 to arginine substitution in the mature von Willebrand factor disrupts a disulphide loop involved in the interaction with platelet glycoprotein Ib-IX. *British Journal of Haematology*, *82*, 66–72.

Lee, J. E., Choi, J. H., & Lee, M. G. (2005). Gene SNPs and mutations in clinical genetic testing: Haplotype-based testing and analysis. *Mutation Research*, *573*, 195–204.

Li, B., Krishnan, V. G., & Mort, M. E. (2009). Automated inference of molecular mechanisms of disease from amino acid substitutions. *Bioinformatics*, *25*, 2744–2750.

Mah, J. T., & Chia, K. S. (2007). A gentle introduction to SNP analysis: Resources and tools. *Journal of Bioinformatics and Computational Biology*, *5*, 1123–1138.

Mancuso, D. J., Tuley, E. A., Westfield, L. A., Lester-Mancuso, T. L., Lebeau, M. M., Sorace, J. M., et al. (1991). Human von Willebrand factor gene and pseudogene: Structural analysis and differentiation by polymerase chain reaction. *Biochemistry*, *30*, 253–269.

Matsushita, & Sadler, J. E. (1995). Identification of amino acid residues essential for von Willebrand factor binding to platelet glycoprotein Ib. Charged-to-alanine scanning mutagenesis of the A1 domain of human von Willebrand factor. *Journal of Biological Chemistry*, *270*, 13406–13414.

Mazurier, C. (1992). von Willebrand disease masquerading as haemophilia A. *Thrombosis and Haemostasis*, *67*, 391–396.

McKinnon, T. A., Nowak, A. A., Cutler, J., Riddell, A. F., Laffan, M. A., & Millar, C. M. (2012). Characterisation of von Willebrand factor A1 domain mutants I1416N and I1416T: Correlation of clinical phenotype with flow-based platelet adhesion. *Journal of Thrombosis Haemostasis*, 7, 1409–1416.

Meyer, D., Fressinaud, E., Gaucher, C., Lavergne, J. M., Hilbert, L., Ribba, A. S., et al. (1997). Gene defects in 150 unrelated french cases with type 2 von Willebrand disease: From the patient to the gene. INSERM network on molecular abnormalities in von Willebrand disease. *Thrombosis and Haemostasis*, *78*, 451–456.

Miteva, M. A., Brugge, J. M., Rosing, J., Nicolaes, G. A. F., & Villoutreix, B. O. (2004). Theoretical and experimental study of the D2194G mutation in the C2 domain of coagulation factor V. *Biophysical Journal*, *86*, 488–498.

Myers, J. K., & Pace, C. N. (1996). Hydrogen bonding stabilizes globular proteins. *Biophysical Journal*, *71*, 2033–2039.

Nachman, R. L., Levine, R., & Jaffe, E. A. (1977). Synthesis of factor VIII antigen by cultured guinea pig megakaryocytes. *Journal of Clinical Investigation*, *60*, 914–921.

Nichols, W. C., & Ginsburg, D. (1997). von Willebrand disease. *Medicine (Baltimore)*, *76*, 1–20.

Nishino, M., Girma, J. P., Rothschild, C., Fressinaud, E., & Meyer, D. (1989). New variant of von Willebrand disease with defective binding to factor VIII. *Blood*, *74*, 1591–1599.

Peterson, T. A., Doughty, E., & Kann, M. G. (2013). Towards precision medicine: Advances in computational approaches for the analysis of human variants. *Journal of Molecular Biology*, *425*, 4047–4063.

Rabinowitz, I., Randi, A. M., Shindler, K. S., Tuley, E. A., Rustagi, P. K., & Evan Sadler, J. (1993). Type IIB mutation His-505 Asp implicates a new segment in the control of von Willebrand factor binding to platelet glycoprotein Ib. *The Journal of Biological Chemistry*, *27*, 20497–20501.

Rajasekaran, R., Sudandiradoss, C., George, P. D., & Rao, S. (2007). Identification and in silico analysis of functional SNPs of the BRCA1 gene. *Genomics*, *4*, 447–452.

Ramensky, V., Bork, P., & Sunyaev, S. (2002). Human non-synonymous SNPs: Server and survey. *Nucleic Acids Research*, *30*, 3894–3900.

Ribba, A. S., Christophe, O., Derlon, A., Cherel, G., Siguret, V., Lavergne, J. M., et al. (1994). Discrepancy between IIA-phenotype and IIB-genotype in a patient with a variant of von Willebrand disease. *Blood*, *83*, 833–841.

Rodeghiero, F., Castaman, G., & Dini, E. (1987). Epidemiological investigation of the prevalence of von Willebrand's disease. *Blood*, *69*, 454–459.

Ruggeri, Z. M., & Zimmerman, T. S. (1981). The complex multimeric composition of factor VIII/von Willebrand factor. *Blood*, *57*, 1140–1143.

Sadler, J. E. (1994). A revised classification of von Willebrand disease. *Thrombosis and Haemostasis*, *71*, 520–525.

Sadler, J. E., & Ginsburg, D. (1993). A database of polymorphisms in the von Willebrand factor gene and pseudogene. *Thrombosis and Haemostasis*, *69*, 185–191.

Shastry, B. S. (2003). SNPs and haplotypes: Genetic markers for disease and drug response. *International Journal of Molecular Medicine*, *11*, 379–382.

Sherry, S. T., Ward, M. H., & Kholodov, M. (2001). DbSNP: The NCBI database of genetic variation. *Nucleic Acids Research*, *29*, 308–311.

Sporn, L. A., Chavin, S. I., Marder, V. J., & Wagner, D. D. (1985). Biosynthesis of von Willebrand protein by human megakaryocytes. *Journal of Clinical Investigation*, *76*, 1102–1106.

Stenson, P. D., Mort, M., Ball, E. V., Howells, K., Phillips, A. D., Thomas, N. S., et al. (2009). The human gene mutation database: 2008 update. *Genome Medicine*, *1*, 13.

Steen, M., Miteva, M., Villoutreix, B. O., Yamazaki, T., & Dahlback, B. (2003). Factor V new brunswick: Ala221Val associated with FV deficiency reproduced in vitro and functionally characterized. *Blood*, *102*, 1316–1322.

Sudhakar, N., George Priya Doss, C., Thirumal Kumar, D., Chakraborty, C., Anand, K., & Suresh, M. (2015). Deciphering the impact of somatic mutations in exon 20 and exon 9 of PIK3CA gene in breast tumors among Indian women through molecular dynamics approach. *Journal of Biomolecular Structure and Dynamics*, 1–13.

Sussman, I. I., & Rand, J. R. (1982). Subendothelial deposition of von Willebrand factor requires the presence of endothelial cells. *Journal of Laboratory and Clinical Medicine*, *100*, 526–532.

Tavtigian, S. V., Deffenbaugh, A. M., Yin, L., Judkins, T., Scholl, T., Samollow, P. B., et al. (2005). Comprehensive statistical study of 452 BRCA1 missense substitutions with classification of eight recurrent substitutions as neutral. *Journal of Medical Genetics*, *43*, 295–305.

Teng, S., Madej, T., Panchenko, A., & Alexov, E. (2009). Modeling effects of human single nucleotide polymorphisms on protein-protein interactions. *Biophysical Journal*, *96*, 2178–2188.

Turner, P. J. (2005). *XMGRACE, version 5.1.19*. Beaverton, OR: Center For Coastal and Land-Margin Research, Oregon Graduate Institute of Science and Technology.

Van Gunsteren, W. F., Billeter, S. R., Eising, A. A., Hünenberger, P. H., Kruger, P., Mark, A. E., et al. (1996). *Biomolecular simulation: The GROMOS96 manual and user guide* (pp. 1–1042). Zürich, Switzerland: Vdf Hochschulverlag AG an der ETH Zürich.

Van Zandt, L. L., & Schroll, W. K. (2012). Modeling hydrogen bonds in three dimensions. *Journal of Biomolecular Structure and Dynamics*, *8*, 431–438.

Verweij, C. L., Diergaarde, P. J., Hart, M., & Pannekoek, H. (1986). Full-length von Willebrand factor (vWF) cDNA encodes a highly repetitive protein considerably larger than the mature vWF subunit. *EMBO Journal*, *5*, 1839–1847.

Wang, P., Dai, M., Xuan, W., McEachin, R. C., Jackson, A. U., Scott, L. J., et al. (2006). SNP Function Portal: A web database for exploring the function implication of SNP alleles. *Bioinformatics*, *22*, 523–529.

Werner, E. J., Broxson, E. H., Tucker, E. L., Giroux, D. S., Shults, J., & Abshire, T. C. (1993). Prevalence of von Willebrand disease in children: A multiethnic study. *Journal of Pediatrics*, *123*, 893–898.

Woods, A. I., Sanchez-Luceros, A., Kempfer, A. C., Powazniak, Y., Calderazzo Pereyra, J. C., Blanco, A. N., et al. (2012). C1272F: A novel type 2A von Willebrand's disease mutation in A1 domain; its clinical significance. *Haemophilia*, *18*, 112–116.

Wise, R. J., Dorner, A. J., Krane, M., Pittman, D. D., & Kaufman, R. J. (1991). The role of von Willebrand factor multimers and propeptide cleavage in binding and stabilization of factor VIII. *Journal of Biological Chemistry*, *266*, 21948–21955.

Yip, Y. L., Famiglietti, M., & Gos, A. (2008). Annotating single amino acid polymorphisms in the UniProt/Swiss-Prot knowledgebase. *Human Mutation*, *29*, 361–366.

Yip, Y. L., Scheib, H., & Diemand, A. V. (2004). The Swiss-Prot variant page and the ModSNP database: A resource for sequence and structure information on human protein variants. *Journal of Human Mutation*, *23*, 464–470.

Yue, P., Melamud, E., & Moult, J. (2006). SNPs3D: Candidate gene and SNP selection for association studies. *BMC Bioinformatics*, *7*, 166.

Zhu, Y., Spitz, M. R., Amos, C. I., Lin, J., & Schabath, M. B. (2004). An evolutionary perspective on single-nucleotide polymorphism screening in molecular cancer epidemiology. *Cancer Research*, *64*, 2251–2257.

AUTHOR INDEX

Note: Page numbers followed by "*f*" indicate figures, and "*t*" indicate tables.

B

C

D

E

F

G

H

K

L

S

U

V

W

Z

SUBJECT INDEX

Note: Page numbers followed by "*f*" indicate figures and "*t*" indicate tables.

M

N

R

FSC
www.fsc.org
MIX
Papier | Fördert
gute Waldnutzung
FSC® C083411

Kolibri 360 GmbH
Ferdinand-Jühlke-Straße 7
99095 Erfurt, Deutschland
produktsicherheit@kolibri360.de